Circum Mare: Themes in Ancient Warfare

Mnemosyne
Supplements

HISTORY AND ARCHAEOLOGY
OF CLASSICAL ANTIQUITY

Series Editor

Hans van Wees (*University College London*)

Associate Editors

Jan Paul Crielaard (*Vrije Universiteit Amsterdam*)
Benet Salway (*University College London*)

VOLUME 388

The titles published in this series are listed at *brill.com/mns-haca*

Circum Mare:
Themes in Ancient Warfare

Edited by

Jeremy Armstrong

With a Foreword by

Lee L. Brice

BRILL

LEIDEN | BOSTON

Cover illustration: Grande Ludovisi 'Battle Sarcophagus', Rome, Palazzo Altemps (Museo Nazionale Romano) Inv. 8574. Image reproduced with the permission of the Ministero dei beni e delle attività culturali e del turismo—Soprintendenza Speciale per il Colosseo, il Museo Nazionale Romano e l'Area archeologica di Roma. Photo by Jeremy Armstrong.

Library of Congress Cataloging-in-Publication Data

Names: Armstrong, Jeremy, editor of compilation.
Title: Circum mare : themes in ancient warfare / edited by Jeremy Armstrong ; with a foreword by Lee L. Brice.
Other titles: Themes in ancient warfare
Description: Boston : Brill, [2016] | Series: Mnemosyne / supplements ; volume 388 | Includes bibliographical references and index.
Identifiers: LCCN 2016011219 (print) | LCCN 2016011917 (ebook) | ISBN 9789004284845 (hbk. : alk. paper) | ISBN 9789004284852 (e-book)
Subjects: LCSH: Mediterranean Region–History, Military–To 1500. | Military art and science–Mediterranean Region–History–To 1500 .
Classification: LCC DE84 .C57 2016 (print) | LCC DE84 (ebook) | DDC 355.020937–dc23
LC record available at https://lccn.loc.gov/2016011219

Want or need Open Access? Brill Open offers you the choice to make your research freely accessible online in exchange for a publication charge. Review your various options on brill.com/brill-open.

Typeface for the Latin, Greek, and Cyrillic scripts: "Brill". See and download: brill.com/brill-typeface.

ISSN 2352-8656
ISBN 978-90-04-28484-5 (hardback)
ISBN 978-90-04-28485-2 (e-book)

Copyright 2016 by Koninklijke Brill NV, Leiden, The Netherlands.
Koninklijke Brill NV incorporates the imprints Brill, Brill Hes & De Graaf, Brill Nijhoff, Brill Rodopi and Hotei Publishing.
All rights reserved. No part of this publication may be reproduced, translated, stored in a retrieval system, or transmitted in any form or by any means, electronic, mechanical, photocopying, recording or otherwise, without prior written permission from the publisher.
Authorization to photocopy items for internal or personal use is granted by Koninklijke Brill NV provided that the appropriate fees are paid directly to The Copyright Clearance Center, 222 Rosewood Drive, Suite 910, Danvers, MA 01923, USA. Fees are subject to change.

This book is printed on acid-free paper.

Contents

Foreword   VII
  Lee L. Brice
Acknowledgments   IX
Abbreviations   X

1   War and Society in the Ancient World: An Introduction   1
    Jeremy Armstrong

Military Narratives

2   Simple Words, Simple Pictures: The Link between the Snapshots of
    Battle and the War Diary Entries in Ancient Egypt   13
    Anthony Spalinger

3   Caesar's *Exempla* and the Role of Centurions in Battle   34
    David Nolan

The Economics of Warfare

4   Coinage and the Economics of the Athenian Empire   65
    Matthew Trundle

5   *Tributum* in the Middle Republic   80
    Nathan Rosenstein

Military Cohesion

6   The Ties that Bind: Military Cohesion in Archaic Rome   101
    Jeremy Armstrong

7   *Sacramentum Militiae*: Empty Words in an Age of Chaos   120
    Mark Hebblewhite

Military Authority

8 Circumscribing *Imperium*: Power and Regulation in the Republican Province 145
   Ralph Covino

9 The Delian and Second Athenian Leagues: The Perspective of Collective Action 164
   James Kierstead

Irregular Warfare

10 'Warlordism' and the Disintegration of the Western Roman Army 185
   Jeroen W.P. Wijnendaele

11 The Significance of Insignificant Engagements: Irregular Warfare during the Punic Wars 204
   Louis Rawlings

Fortifications and Sieges

12 'Siege Warfare' in Ancient Egypt, as Derived from Select Royal and Private Battle Scenes 237
   Brett H. Heagren

13 Tissaphernes and the Achaemenid Defense of Western Anatolia, 412–395 BC 262
   John W.I. Lee

Bibliography 283
Index 317

Foreword

My own involvement with this volume is as the series editor for Brill's new *Warfare in the Ancient Mediterranean World—Brill Companions to the Classical World*. Although this present volume preceded that series, it is indicative of a new commitment Brill has made to publish more works on the military history of the ancient world. These volumes will appear in several series, including Ancient Egypt, the Ancient Near East and Persia, the aforementioned series, and *Mnemosyne* supplements—such as this volume. Taken as a whole these volumes, companions, and monographs will continue to present the rich production of military historical work being done and stimulate new work as the field continues to evolve. *Circum Mare: Themes in Ancient Warfare* demonstrates some of the ways in which military history remains a fascinating and lively topic. It is my immense honor to introduce this new contribution to the field.

Circum mare, "Around the Sea," may seem like an unusual name for a volume that addresses military history on land, but it captures the thrust of this volume in a couple useful ways. The sea around which these chapters have focused is the Mediterranean. Readers and practitioners of military history tend to think of the field in geographically narrow terms. We can pick up books on Greece, Rome, Egypt, or Persia, among other topics, and each one tends to focus on warfare and militaries from one state's or power's perspective, much as do our sources. Even when we find books that bring together warfare from multiple regions, cultures, or periods, like Greece and Rome, they still tend to do so in ways that focus on the differences between them rather than the connections between how people in different contexts solved similar problems. Certainly there are obviously consistent themes across time and culture—victory and loss, hand-to-hand combat, spoils, missed opportunities, death, etc.—but there remain many topics we do not usually consider. These essays provide readers a chance to compare how people in different ancient contexts approached related aspects of warfare and the military. The Mediterranean basin connects them and provides a unifying geographic context.

A second way in which *Circum mare* fits this volume is in the array of methodologies it brings together. In that sense, the metaphorical "sea" around which these chapters are arranged is traditional military history. Despite forty-five years of the so-called New Military History, a ginormous amount of written work in military history remains focused on the traditional areas of biography and 'drums and trumpets'. Academic military history has moved well beyond these approaches. This volume brings together, in pairs, chapters that demonstrate the continuing evolution of military history methodology. Attempting to shed light on various aspects of the military past, the authors develop new twists on older topics, like economics and literary/visual culture, as well as more recent work on collective action, rational actors, irregular warfare, military cohesion, and warlordism. The resulting combination provides examples of the ways in which new approaches, and asking new questions of older approaches, continue to shed light on aspects of ancient warfare and militaries.

Lee L. Brice
Western Illinois University

Acknowledgments

This edited volume, like so many others, was born out of a conference—in this particular instance, one focused on the topic of ancient warfare and held at the University of Auckland in July 2012. The 'Call for Papers' which was issued for the conference was extremely (and consciously) vague, hoping to pull in the widest possible range of speakers—and indeed, scholars representing a wide range of research interests (including Ancient Egypt, Greece, Rome, and late Antiquity) presented papers on various aspects of warfare in the ancient world. However, despite the variety of the offerings and the myriad approaches and methodologies which were employed, during the discussions which followed each paper it became clear that all of the scholars were working on similar themes and issues and that the papers, although varied in geographic region and temporal context, could be brought together to present a surprisingly cohesive 'snapshot' of current directions in ancient military history. The decision was then made to publish the papers from the conference which best represented this 'snapshot.' These papers were supplemented with a few invited chapters to fill out any areas or themes which were initially underrepresented, although preference was given to thematic unity over geographic or temporal coverage. The collected papers were then paired based on thematic choices to form, what the editor hopes is, a provocative examination of current directions in this exciting field from around the Mediterranean.

As always with these projects, there are far too many people to thank, as every stage of this project represented a collaboration of one sort or another. However, I would first of all like to thank all of the contributors to the volume for their initial contributions, for their diligent reworking of those contributions following comments from myself and the reviewers, and for their patience and understanding as this project slowly trundled forward. I would also like to thank the Faculty of Arts of the University of Auckland for their financial support of the conference in 2012, Prof. Lee L. Brice and Tessel Jonquière for their editorial support, input, and acumen, and all of the peer reviewers, whose work was more often than not paid in mere 'thanks'. Finally, the present volume would also not have been possible without the editorial support of Ms. Sheira Cohen and Ms. Julia Hamilton, whose tireless work on the minutiae (and particularly the referencing and bibliographic style) I am immensely grateful for.

Jeremy Armstrong
University of Auckland

Abbreviations

Abbreviations, both ancient and modern, generally follow those in the *Oxford Classical Dictionary*. A few additional abbreviations are:

URK   *Urkunden der 18. Dynastie*
KRI   K.A. Kitchen, *Ramesside Inscriptions, Historical and Biographical. I–VII* (Oxford, 1969–1990)
Wresz.   Atlas Wreszinski, *Atlas zur altaegyptischen Kulturgeschichte I–V*.

CHAPTER 1

War and Society in the Ancient World: An Introduction[1]

Jeremy Armstrong

Warfare is an incredibly complex phenomenon. Indeed, it is arguably the *most* complex of human activities.

War can be argued to be the most social activity in which human beings engage, as the creation of 'in-group' and 'out-group' biases through conflict (and particularly armed and violent conflict) has been put forward as one of the fundamental bases of human society—in fact, it has been suggested that one cannot have a 'society' without some form of 'war' (although, in this context, the term can be broadly construed).[2] However, war is also one of the most anti-social activities which humans engage in, as it forces individuals to break even the most fundamental social rules—like the prohibition on killing/murder. As a result, warfare seems to inhabit, or perhaps define, that important liminal zone between two groups or societies—between 'us' and 'them'—delineating social and cultural boundaries in the most absolute terms possible. By killing an enemy or protecting a comrade, a combatant demonstrates—in the most primitive fashion—who is in his 'group' or society and who is not. War, therefore, touches on some of the most basic human instincts and emotions as it breaks down and re-establishes social norms and positions. It indicates what is most sought after and what is worth defending in a society. It also indicates what is honoured and respected, as well as what is disdained or taboo. Warfare is, effectively, the most basic and ancient means of identity creation and manipulation—a social dialogue which human beings engage in every day, although thankfully not always in its most violent form. This is why a 'state of

---

1 For anyone familiar with ancient warfare, this title would likely bring to mind the two seminal works by John Rich and Graham Shipley, *War and Society in the Greek World* (Routledge 1993) and *War and Society in the Roman World* (Routledge 1993), and indeed this is intentional. Although by no means a book of the same type or focus, the present volume can hopefully be seen as a descendent of these great books (along with Brice and Roberts 2011a, etc.)— touching on similar themes and questions, and demonstrating how the field has continued to evolve over the past 20 plus years.

2 Keeley (1997) 3–25.

war' was famously offered by Hobbes as humanity's natural state. And yet, as argued by Margret Mead, violent war may also be a cultural construct, which (at least it was hoped in the 1940s and 1950s) might belie its necessity.[3] But, despite its (possibly) constructed basis and its inherently wasteful character, war is universal in human culture. Warfare, therefore, brings together all that is confusing and contradictory in a human nature and society and pushes it to an extreme, presenting in stark contrast those social and cultural rules which were either invisible, or sometimes not yet existent, before—and as such it represents an invaluable lens for academic study.

The study of warfare has always represented one of the more popular areas of study in history, and particularly ancient history, albeit for a range of different reasons. Traditionally ancient warfare and battle narratives were studied as a way to improve contemporary armies, and the continued presence of lectures on ancient topics in modern military academies suggests that this is still seen as a viable avenue of study. Although technology has obviously moved on quite a bit since the days of Julius Caesar and Alexander the Great, the universality of war has always hinted that armies, past and present, may behave in roughly similar fashions and that there may be core principles used by successful generals in the past which can still prove useful today.[4] This scholarly mind-set can trace its roots back centuries, to Machiavelli and his contemporaries if not earlier, although it has not always been a healthy thing. In recent years, the long-standing, traditional focus on the practical details of combat and military organization has meant that ancient military studies have sometimes been relegated to the realm of ephemera within historical studies, as the social sciences more generally have moved on to investigating more communal principles and the cultural ideologies which underpin societies. In this changing scholarly environment, investigations into ancient warfare are sometimes seen as 'old-fashioned' or, as with catalogues and typologies of spear points or helmet types, worryingly esoteric.[5] But war's all-encompassing and pervasive nature means that it impacts and influences almost every aspect of a society, thereby providing insight on a number of often hidden facets within an ancient culture. War is also cross-cultural, in some ways by definition, making comparative studies not only possible but often extremely enlightening. And in contrast to other cross-cultural practices, warfare is also a typically well-documented activity, both by contemporary cultures and in the archaeological

---

3 Mead (1940): 402–405.

4 Bose (2004) 1–2; Brice and Roberts (2011a) 3–4.

5 Brice and Roberts (2011a) 1–10.

record, leaving behind military accoutrements, fortifications, burn layers, etc. Ancient military studies offer a myriad of different approaches, beyond the traditional practical models, for investigation.

In recent years, ancient military historians have increasingly moved away from the more practical or applicable (at least in a modern sense) aspects of ancient military studies and towards viewing warfare as a social construct and as a way to understand other, seemingly unconnected, aspects of the ancient world. From literature, to social bonds, to political structures and cultural norms, warfare is now understood to be an incredibly inclusive and valuable lens for viewing and understanding ancient cultures holistically. All human cultures practice warfare, making it useful as a cross-cultural tool, and yet each culture's approach to warfare, and indeed *modus operandi* within it, is arguably unique. As a result, by studying and comparing various societies' use of and reaction to warfare, an immense amount of information can be gleaned. The present volume touches on a number of these aspects and, adopting an overtly comparative approach, aims to illustrate both points of synergy and divergence between the cultures which ringed the Mediterranean in antiquity in the sphere of warfare. Each of the six themes discussed (narratives of war, the economics of war, military cohesion, military authority, irregular warfare, and fortifications) contains two studies looking at the same basic theme from the point of view of two different societies. The ultimate goal of the volume is therefore twofold, as it will serve to both demonstrate the various ways in which ancient warfare is currently being used in ancient world studies, and offer specific case studies and insight into various aspects of both ancient warfare and ancient societies more generally.

The first theme is that of military narratives, and Anthony Spalinger and David Nolan each offer examples of how military narratives—from very different cultures—can be used to investigate more complex social and cultural paradigms and how war was expressed in various genres. In his chapter, Spalinger demonstrates how comparing and contrasting the pictorial and literary records for an Egyptian Pharaoh's campaign in Asia reveals that each aspect of the record seems to have performed a distinct, but complementary, role in creating a single historical narrative which emphasized key aspects of the Pharaoh's persona. Spalinger's argument, that the Egyptian historical consciousness relied on both forms of media in order to construct its record, hints at the complexities of transmitting not only historical details, but also royal propaganda, in New Kingdom Egypt and how the entirety of the evidence needs to be considered when evaluating the various campaigns. David Nolan's chapter proposes to reconcile our developing understanding of literary *exempla* with Roman military history through an examination of combat anecdotes, and in particular the use of centurions, in Caesar's *Bellum Gallicum*. The discussion of centurions in this work has often been limited to a literal interpretation of their contributions to the Roman battle order, which has underestimated their literary value. Nolan's chapter argues that centurions were also used as *exempla* in the narrative to illustrate correct or incorrect behaviour, specifically regarding the conditions under which they were expected to fight. Centurions were expected to enter combat and bear the brunt of an enemy attack in order to restore order to a unit and were only supposed to engage in combat when a legion was in danger of collapse. Nolan, therefore, suggests that Caesar's *Bellum Gallicum* represents a far more complex literary work than usually thought, and that only through considering all of these aspects can a full appreciation of both the work and the war it recounts be attained. Taking both of these chapters together, the importance of understanding both existing literary genres and how warfare formed a unique subject within them becomes evident. The military accounts of both New Kingdom Egypt and late republican Rome both use existing forms of expression, but adjust and adapt them to express a particular, and unique, message. The way in which each culture approaches the existing genres and media, and the topic of war, puts aspects of each society into sharp relief—particularly the way they perceive of ideal behaviour.

The second theme discussed is the economics of warfare. In this section Nathan Rosenstein continues his work on the economics of Roman warfare in the middle Republic by arguing that existing financial resources were vital to Rome's military success during this period as the city's wars, contrary to popular opinion, rarely paid for themselves. Instead, the Republic financed these conquests through the *tributum*, a payment by its *assidui* to support the legionaries' *stipendium* and the other expenses of war. This chapter investigates how payment of *tributum* was distributed among the *assidui* in the later third century and then explores the value of the annual *tributum* compared to the cost of feeding a hypothetical family of five for a year in order to gauge its burden. Rosenstein then concludes with the suggestion that the need to pay *tributum* played a vital role in catalyzing widespread commercial development within the mid-republican economy and a general increase in the prosperity of Rome's citizens. Matthew Trundle's chapter then goes on to examine the role of money, and particularly coinage, in the navy of fifth century BC Athens. The absence of coin hoards within the Athenian naval *archê* has always raised questions concerning levels of monetization within the Athenian 'empire' itself. All the fifth century BC hoards we have come from regions outside the Athenian *archê*, mostly Egypt and the western Near East, and the majority of Athenian minting seems to have been aimed at producing large denomination coins for exporting the silver of Laureion. Indeed, even coins designed for payment of the fleet seem to have left the empire in great quantities as payments for food and other raw materials. However, in this chapter Trundle argues that coinage and its circulation within the Delian League played a crucial role in an increasingly mercenary Athenian navy (and an increasingly monetized economy), especially after the revolt of Samos in 440/39 BC. In short, by the 430s BC coins had become the central mechanism for the flow and management of resources, human and material, throughout the *archê* and as a result much of the eastern Mediterranean. Together, these two chapters clearly demonstrate the continued importance of economic considerations in any discussion of military action. However, moving away from the more traditional approaches to this topic, which generally focus on the elite, both Rosenstein and Trundle argue for the importance of broader socio-economic shifts as a result of warfare and military mobilization.

The third theme is military and social cohesion. Jeremy Armstrong's chapter focuses on the nature of military cohesion and morale in ancient armies, and argues against the idea that this was based solely on a pre-existing sense of community, citizenship, or a 'civic ethos'. Instead, using parallels and paradigms from modern military studies, Armstrong uses the case of early Rome to suggest that military cohesion in ancient armies may have been, at least in part, the result of 'task-based' cohesion instead. In this model, the importance of a goal (often economic) for the group is emphasized, and offers a solution to the previously problematic situations where cohesion is evident in armies which were socially, culturally, or politically heterogeneous and did not have a strong state-based mentality to rely on. Mark Hebblewhite's chapter focuses on the relationship between the Roman army and the emperor in the turbulent political period of AD 235–395. During this period, the *sacramentum militiae*, an oath sworn to the emperor alone which asked the individual soldier to express absolute fidelity to his cause, remained a constant feature of the relationship. Each soldier not only swore the *sacramentum* upon joining the army but also at the start of a new emperor's reign and then again at least once yearly for his entire reign. However, as Hebblewhite argues, the power of the *sacramentum* in maintaining the ongoing loyalty of the army must not be overestimated. Usurpations were common and each time a part of the army rebelled against the reigning emperor they were breaking the *sacramentum* they had sworn to him. Instead the actual effectiveness of the *sacramentum* ultimately depended on the existing attitude the army had towards the emperor, and represented but one part of a multifaceted relationship. Both Hebblewhite and Armstrong's contributions focus on the bonds which linked soldiers together, and to the state and/or commander, revealing that, despite the complex ideals and rituals often associated with membership in an army, the common soldier's motivation for fighting was, at its core, surprisingly simple and focused on the individual (and not necessarily the state) across antiquity.

The fourth theme is military authority. In this section James Kierstead examines the nature of power and control within Athens' Delian League. Kierstead proposes that the theory of collective action, where the Delian League is viewed in terms of a group of individual states seeking to provide themselves with the related public goods of security and a market, provides important insight into the inner workings of the group. As part of this model Kierstead eschews the traditional approach to power within the League, which generally depicts the relationship as decidedly one-sided—with Athens as the simple 'exactor of rents'. Instead, Kierstead argues that the Athenians did not set up the League with either exclusively altruistic motives or exclusively rapacious ones, and that the allies also contained a diverse range of positions. Rather, Kierstead argues that the internal dynamics of the League were shaped over time by the inherent nature of long-term collective action and that the change in the nature of the Delian League was produced by these principles and not necessarily by a consciously predatory hegemon. This chapter is followed by Ralph Covino's exploration of the Roman republican grant of *imperium*. While Kierstead's offering suggested that military power and command could be shaped by the unconscious dynamics of group action, Covino argues for the importance of conscious legal limits on authority. Although the various trials and speeches which occurred during the late Roman Republic, and particularly Cicero's *Verrines*, suggest that governors wielded substantial power and that control and oversight of provincial governors was a key issue in this period, Covino argues that the circumscription of *imperium* was actually a feature of the entire republican period. As Rome's empire became more pacified, and the importance of the military dimension of a governor's *imperium* began to fade, an ever-increasing set of laws, regulations, and customs were applied to restrict the power and ambition of the men charged with controlling the provinces. This chapter is focused on the fuzzy line between military and administrative power in the ancient world, and the often contentious relationship which existed between the centre and periphery within an imperial administration. Together, these chapters present two distinct aspects, or modifiers, of military authority in the ancient world—the influence of largely unconscious social developments and the conscious delineation and circumscription of power, largely in response to a changing socio-political situation.

The fifth theme is irregular warfare. Jeroen Wijnendaele's chapter builds on his previous work on 'warlordism' in late antiquity and seeks to highlight the dynamics behind the changing nature of the Western Roman Army in the first half of the fifth century AD. The term 'warlord' has often been used in a generic way in the field of history and only recently has 'warlordism' attracted the attention of scholars of the Late Roman Empire. However, these concepts have been vigorously developed in political sciences and third-world studies over the past few decades. Wijnendaele's chapter offers a survey of some of the most important sources in this area and then proposes a meaningful way to employ this model by giving an analysis of the career of the African field commander (*comes Africae*) Bonifatius. Special attention is given to the relationship with his warrior retinue—the so-called *buccellarii*—as a case study for the semi-privatization of the Western Roman Army. Bonifatius' rise to power witnessed the eclipse of traditional means to claim political and military authority through usurpation of the imperial office, whilst setting a precedent for future commanders to foster personal control over their armed forces. This chapter shows that middle-ranking commanders, such as Bonifatius and Aetius, broke the monopoly of violence hitherto exercised by the emperor and his court. This discussion is followed by Louis Rawlings' analysis of irregular warfare during Rome's first two wars against Carthage in the third century BC. In this chapter, Rawlings challenges the traditional assumption that large, set piece battles were the only, or indeed the normal, way in which armies confronted each other during this period. In contrast, the regular use of 'irregular actions' by both armies seems to have been a common, and intentional, strategy adopted by both sides. This suggests that ancient generals conceived of and organized their armies, even in this early period, in a far more fragmented and heterogeneous fashion then usually supposed—with a range of units and specialized troops being sent on various missions for limited goals. This reinterpretation of military norms in the third century BC has substantial implications, as it hints at a much broader strategic awareness and a more complex, diverse, and adaptive military structure than that explicitly presented in our extant sources. Together, Rawlings and Wijnendaele's chapters argue for the existence of multifaceted command structures existing, often largely undetected, within the larger military structures of ancient armies. The presence of warbands in the armies of the later Roman empire, along with the widespread use small, specialist units in the army of the early/middle Republic, hint at the incredibly complex network of relationships and structures which seems to have existed within ancient armies—despite the explicit assertions of many ancient writers.

The final theme discussed is that of fortifications and sieges. Brett Heagren starts this section with an examination of sieges and fortifications in Egyptian reliefs, and explores the complex interplay between depiction and reality in the Egyptian approach to this aspect of warfare. Heagren argues that although it is clear that Egyptians possessed a long and vivid history of 'siege warfare', it is entirely uncertain to what extent these images depicted historical reality—especially with those images and textual accounts dated to the New Kingdom. Fortifications seem to have been used to depict the strength or importance of a city even in cases where archaeology suggests that fortifications were absent—as at Meggido. As a result, Heagren suggests we can understand fortifications and siege warfare as representing literary and artistic tropes in military accounts, and as such they must be taken and understood in that context. This is followed by John Lee's study on the Achaemenid defenses of western Anatolia during the years 412–395 BC. Scholarship on this area has generally approached the matter from the Greek perspective, and only recently have scholars revealed what once seemed an elusive imperial presence in western Anatolia—demonstrating the overall, long-term effectiveness of the empire's defenses. But while this broad picture has been key in combating stereotypes of Persian weakness or decadence, Lee's chapter suggests that there is much to be gained from employing a short-term perspective that helps reveal some of the complicated realities and difficult choices involved in guarding the western frontier. In particular, Lee demonstrates how various developments in Persian politics and warfare, including the Athenian expedition of 409 BC and the civil war between Cyrus and his brother Artaxerxes II, changed the Persian approach to the defense of this vital region. These two chapters both demonstrate the psychological importance of fortifications in the ancient world, and particularly the eastern Mediterranean. Fortifications played a key role in how ancient empires and armies approached both the conquest and defense of various regions, which may have outstripped their actual, tactical importance on the battlefield.

Although each chapter in this volume focuses on a different period of history and, for the most part, a different culture, they are all linked through their use of warfare as a multifaceted tool for investigating various aspects of ancient societies. The resultant picture is still incomplete, and we (as always with ancient world studies) find ourselves frustratingly short on concrete answers because of the problematic and fragmentary nature of the evidence. But warfare can clearly be seen as a useful device for examining the ancient world in a holistic fashion. Additionally, it should also be noted that warfare is, obviously, not merely an ancient concern. Warfare is also something which is incredibly relevant in modern society, and the insights gained from the struggles of the ancient world can still resonate in today's world. Although it is often dangerous to draw parallels between societies, and particularly those separated by many thousands of years, the primitive and primordial aspects of warfare provide one of the strongest connections to antiquity which may exist. While it may be hard to put oneself in the place of an Athenian audience member at the theatre of Dionysius, with all of the cultural nuances lost, or a Roman senator arguing over legislation, where the motivations and mind-set are unknown, the base fears and reactions of a solider under stress or a society under attack are arguably easier to sympathize and connect with. As a result, while the old adage of 'learning lessons from the past' might not be entirely appropriate, warfare does provide one of the few very real connections where something resembling historical empathy is actually possible.

Military Narratives

∵

CHAPTER 2

Simple Words, Simple Pictures: The Link between the Snapshots of Battle and the War Diary Entries in Ancient Egypt

Anthony Spalinger

The inscriptional evidence for Egyptian warfare in the empire period has rarely been linked with the pictorial evidence, in view of their common narrative approaches. As an opening gambit, so to speak, along this line of inquiry, this discussion will center upon a working first-level approach that can help to explain further the issue of text in contrast to image in an ancient Egyptian context. In particular, I wish to adumbrate the communality of the simple narrative passages in the Egyptian war records with the frozen photographic presentations of the large New Kingdom war reliefs. By and large, the iconic nature of both means of recording the kings' manly and heroic deeds will not be covered here. Instead, I shall concentrate upon the coincidence of approaches, not to argue that one was the cause of the other but to demonstrate the parallelism of narrative techniques within a restricted corpus of two methods of recording history that were aimed at the same issue: the king at an Asian city.[1]

Let us begin with the famous and ever-present tomb biography of Ahmose son of Ebana.[2] Here, even though the literary composition narrating his career is not detailed and the aspect highlighted is not always pharaoh's prowess, we can nonetheless garner some basic insights concerning the role that Ahmose wished to signal to his later visitors who would read his account long after he had died. Right at the beginning, the doughty warrior pointed out that he "followed" (*šms*: an all too common verb in these private military biographies) his lord after "he perambulated in his chariot."[3] The word "to perambulate," *stwt*, simply indicates that Ahmose was not a chariot warrior, a point earlier

---

1 This study is thus a short follow-up to my larger analysis in Spalinger (2012).
2 The best study of his tomb and texts is by Davies (2009) 139–175. Earlier, one could rely upon Vandersleyen (1971). The texts still used by scholars are *URK.* IV 1–11.
3 Many of these common words and phrases were discussed by Lorton (1974), Spalinger (1982), and Grimal (1986).

© KONINKLIJKE BRILL NV, LEIDEN, 2016 | DOI: 10.1163/9789004284852_003

14 SPALINGER

stressed by Oleg Berlev.[4] Yet he also noted the ideal nature of his king: the commander in a chariot. It is as we were looking at an Egyptian ruler in one 'snapshot'. This image corresponds to the later pictorial reliefs of the Egyptian pharaoh, in which the chariot hero-nature of the king is always highlighted. Note further that Ahmose explicitly stated that he was brave "on his two feet," thereby indicating that he was not a very high-ranking military official in his lord's army.

Later passages in this biography that are worth signaling include Ahmose's references to the siege of Avaris as well as Sharuhen. On both occasions, the private account laconically noted the siege followed by the Egyptian plundering. The same may be seen with respect to some fighting, apparently serious, at the Djedku canal. Moreover, when we turn south to read of the warfare in Nubia, the sailing directions are, of course, given—a point that will be addressed in full below. No mention is given of the chariot nature of warfare. Pharaoh Ahmose in this case merely slaughters the southerners. Furthermore, when the ruler moved against the "rebel" Aata, the latter is described as having been simply "found" in a specific locality. How the king got there is left aside. Likewise, the crushing of the second rebel, Tetian, is presented as concisely as possible. When Amunhotep I, for example, also moved upstream, the chariot nature of the king or army is also not presented. Once more we note the sailing direction of the monarch. In that extensive southerly campaign, Amunhotep may have "searched" for the people and cattle of the Nubian ruler.

The personal deeds of Ahmose son of Ebana at the "upper well" are all therefore part of a marine-oriented voyage. The same may be said with regard to Amunhotep I's son, Thutmose I. He, as well, sailed south and destroyed. The distinct lack of any image of a chariot-hero cannot but be observed even though the king's role as an archer is then presented. Additionally, the presence, semi-mythically, of the flaming uraeus serpent, which by its nature must *precede* the king in firing against the enemy, can be read. Both images must therefore assume a chariot-based attack on the part of Thutmose I. But, as befits the geographic horizon of Nubia, the Nile, rather than a march across land, is stressed, if only by means of references to ships.

Let us now turn to the final event recorded in this biography; namely, the vast Asiatic movement northwards of Thutmose I. Although the success of this ruler in Syria is astounding to many scholars, the evidence is, in contrast, quite thin. Here we read of the king's "arrival" (*spr*) at the kingdom of Naharain. Note that a city is not mentioned. Thutmose I "finds" (*gmj*) the enemy leader ("the

---

4   Berlev (1967) 6–20.

SIMPLE WORDS, SIMPLE PICTURES

fallen one") and makes a "great heap of corpses," yet another common phrase.[5] Finally, we read of a chariot association within the Egyptian army:[6]

> ... while I was at the head of our army.
> And his majesty saw that I was brave.
> I fetched a chariot, a horse, and who was upon it as a living captive ...

Later historical texts, but those still dated before the Megiddo campaign of Thutmose III, barely help us in this analysis.[7] The biography of Ahmose Pen-Nechbet, for example, outside of repeating the concept of "following" in the king's footsteps and in "every place," is not helpful. Both Thutmose I's Tombos Stela,[8] and the war record at Aswan-Philae of Thutmose II,[9] are largely devoid of useful information for our scholarly mill. The first is an elaborate poetical rendition of the pharaoh and a great encomium for the newly ruling king, whereas the second narrates a Nubian war, first from the perspective of the belligerents and then turns to the army's participation in the campaign upstream. The only possible tidbit offered is that earlier under Thutmose I, the Nubians had fled (*wtḥw*) before him "on the day of slaughter." Hence, it is necessary to discuss more deeply the evidence from all of Thutmose III's military records.[10]

In the Megiddo account, the war diary provided an effective means of narrating the lengthy campaign northwards.[11] The signaling of specific localities, cities or towns in particular, by means of the original, infinitival, ephemerides account, highlights the issue of *topos* and icon; or in other words, when the king is in different places, things happen. For example, wherever he has arrived, the army diarist will have recorded some specific data: date or time of day, geographic locality (stream, ford, wood, city, or town), and what immediately transpired. It is to the latter that we must turn. In essence, the pharaoh will either have met opposition or not. In the pictorial representations, what will

---

5   *URK.* IV 9.10–13.

6   *URK.* IV 9.15–10.1.

7   For the Asiatic wars of Thutmose III see now Redford (2003).

8   Spalinger (1995) 271–281, and Goedicke (1996) 161–176; Egyptian text in *URK.* IV 82–86. The text was re-edited by Klug (2002) 71–78, and also by Beylage (2002) 209–219.

9   *URK.* IV 137–141; Klug (2002) 83–87; and Beylage (2002) 21–27.

10   Although now dated, the two basic studies of these diary excerpts located within the "Annals" of Thutmose II and later documents was compiled and explained by Grapow (1943) 156–174.

11   Classically, see Spalinger (1982), and Redford (1986).

be presented is a tableau of standard visual elements. These I have called icons. Specifically, the viewer will first encounter the king in his chariot, acting as the solitary archer-hero, shooting at a chaotic mass of enemy troops. This will almost always visually occur outside of a city or garrison. The latter is presented in the form of a fortress-citadel and clearly is modeled upon native Egyptian architectural forms.[12] The same aspect of presentation is found in the historical accounts of the New Kingdom Egyptian, if the author had chosen to record specific events at defined localities. Hence, the narrative use of the war diary of the army served the same role as the iconic underpinnings.[13] Namely, both present small, yet significant, events that are highlighted through the king's overt martial attitudes. Moreover, the original diary entries performed the same function, as do the standard snapshots that we can see on the temple walls of the New Kingdom. The terse adumbration of one specific event in written format parallels exactly the equally static and swiftly rendered pictorial representation. Both formats fit ideally into a schematic narrative progression—one that is staccato in orientation.

Before proceeding further with this aspect of the two, let us survey first the famous Megiddo campaign of Thutmose III in his "Annals" in order to determine the precise methods through which the royal presence as a chariot-hero was obtained. After the introduction we arrive at the commonplace ephemerides-based account. The first section is a simple one and was recorded with the date of the king's accession. "At the city which the ruler seized, Gaza ..." This is the standard war diary entry. If, however, there was to be resistance or a show of defiance, then the account would have been more detailed, of course. But—and this is the crucial point—if the latter occurred, then it is possible that the artistic representation would include the standard *topos* of the king versus city, and the pharaoh carved in his chariot attacking the enemy as an archer. If the visual narrative was felt to need addition or expansion, then we might see the monarch standing on the ground, again in the pose of an archer-hero. One could also have included a smiting scene. All in all, the "arrival" scene (*spr*), as some might call it, was the heroic image that was preferred above all. Its

---

12 Classically, see Oren and Shereshevski (1989) 8–22.

13 This is one of many reasons why the approach taken by Stevenson Smith (1965) or Gaballa (1976), remains simplistic. Neither felt the necessity of investigating the historical or time-oriented capacities and capabilities of the ancient Egyptians in order to develop their models of narrative art. As a result, both attempts were extremely formulaic repetition. See also the remarks of Eaton-Krauss (2000) 399, in her review of the third edition of Smith's work, following those of Wolf (1961) 18, in his review of the original edition.

SIMPLE WORDS, SIMPLE PICTURES

written format at first appears lowly, but was nonetheless of crucial importance. Whereas the "departure" diary account was infrequently represented visually, it was *de rigeur* in the diarists' journals.

However, during the march towards Megiddo, the presentation, though heightened by a personal *Königsnovelle* account, nevertheless eschews descriptions of Thutmose in a chariot until the actual clash of arms outside of the main city. Indeed, this is the first time that the pharaoh met strong resistance during his march northwards, or at least according to the account, a major force opposed to him. After commands were given to his army, Thutmose is then described as proceeding forth (*wḏ3*), "in a chariot of electrum, being equipped with his accouterments of war ..."[14] The dramatic nature of the encounter is heightened by this extremely common, yet strikingly visually oriented, phrase—one that acted as a *topos* of heroism. Absolutely nothing else in this narrative presents this aspect. Granted, one realizes that the march was with chariots, that those vehicles formed the heavy equipment for the elite troops, and that pharaoh naturally belonged to this thin superior group. Yet we encounter this simple phrase, partly sterile to be sure, but a highly effective background *topos*, only *at* the city of Megiddo, the goal of the campaign.

In the second half of his "Annals," those covering regnal years 29 and onwards there is once more no data. With the similarly organized war stele of Amunhotep II, however, the issue is otherwise. Let us survey both.[15] The following chart outlines the data presented by both inscriptions. I have included the key words, especially the verbs of motion that delineate the king's movements.

First Campaign of Regnal Year Seven

Memphis Stela	Karnak Stela
Departure (*wḏ3*)	
Arrival at Shamash-Edom (*spr*)	Lost
Crossing the Orontes (*ḏ3*)	Ditto, but "on a horse (team)"
King's "swoop" against the enemy	King "after them"

---

14   See *URK*. IV 650.3, 650.13, and 652.15.

15   Edel (1953) remains the standard edition; text in *URK*. IV 1301–1316.

18 SPALINGER

(cont.)

Memphis Stela	Karnak Stela
Killing enemy (with arrows)	King's use of axe (*mjnb*)
	Enemy leader brought back on the side of pharaoh's chariot
Enemy messenger brought back on the side of pharaoh's chariot	
	Turning back and south to Egypt (*ḥst* plus *ḥnt⟨jt⟩*)
Southward move (*ḥntjt*) to Ny (with *spr*)	Proceeding (*wḏꜣ*) "On a horse (team)" to Ny
Reaching (*spr*) Ugarit, Hatila; and Kadesh	
Proceeding (*wḏꜣ*) "on a horse (team)" to Khashabu	
Messenger brought back on side of chariot	*Ditto* (restored)
Proceeding forth (*prt*) to Egypt "on a horse (team)", (*msjbjn*) with a Maryannu on a "single horse"	*Ditto* (restored)
Arrival (*spr*) at Memphis	

Second Campaign of Regnal Year Nine

Memphis Stela	Karnak Stela
King on a "horse (team)"	
King's horse teams swiftly "fly like a star of heaven"	
King goes forth (*prt*) on his horse (team) at dawn	

SIMPLE WORDS, SIMPLE PICTURES

Memphis Stela	Karnak Stela
King goes forth (*prt*) at dawn, equipped with the expected panoply of war	
King reaches (*spr*) vicinity of Megiddo	
King arrives at (*spr*) Memphis	King leaves (*prt*) Perunefer for Memphis

The arrangements in which the king is highlighted are now simple to perceive. Both accounts reveals that Amunhotep may or may not be approaching a city "on a horse (team)," but he will be referred to in that way, especially when there is opposition. This is the key point. Bellicosity produced the champion. The pharaoh can go into battle after receiving the panoply (*ḥkrw*) of battle. In this case, the narrative did not have to present, in an automatic fashion, the chariot aspect. The key movement verbs, "to proceed" (*wḏꜣ*), "to reach/arrive at" (*spr*), "to go forth" (*prj*), and others that indicate northern or southern direction (*ḫdj* and *ḫntj*), with "to turn back" (*ḥsj*) likewise employed, are those most commonly used to refer to the king's physical activities. But as some of them are bland (e.g., *spr*) and others general in outlook ("to go north," or "to go south") it is understandable that no inherent heroic attribute is automatically indicated. Details of a specific nature, nonetheless, could be added, and they often indicate the chariot aspect of the monarch. See for example, the crossing the Orontes episode of Amunhotep II or, equally, his "swoop" down, as a bird of prey, against his opponents. Surely, the fast moving aspect of the chariot-hero was foremost in the mind of the writer who, of course, also implied the falcon's flight of attack. One can see how formal and regular these diary excerpts are. The infinitival use and its extensions remain, to no small degree, impersonal but informative. To some degree, this is what I was attempting to argue in my study concerning the Story of Sinuhe.[16]

In essence, these images—simple ones connected to the war diary—fit neatly with the visual presentations that are best known from the Ramesside Period. Other military-organized inscriptions, and I have in mind the Armant and Gebel Barkal Stelae of Thutmose III,[17] not relying upon an ephemerides-

---

16 Spalinger (1998).

17 *URK.* IV 1243–1247 and 1227–1243 respectively; see the re-edition of both by Klug (2002) 151–158 and 193–208; see also Beylage (2002) 157–203.

based approach, eschew such images even though it is reasonable to maintain that all and sundry knew their king's horsemanship and his skill in chariot warfare. Before departing form this empirical presentation, however, it will be necessary to trace the early Dynasty XIX war images of Seti I at Karnak in order to contrast that 'pure' visual outlook with the written ones so far presented.[18] By this means—using a series of visual snapshots of battle—we can compare them with the accompanying captions and determine the amount of emphasis that was placed upon the king's chariot role.

Register One on the east side contains, remarkably, nothing of importance. Of course the iconic images of king in chariot predominate, but that became the norm for the visual depiction of battles. Significantly, we might have expected a reference to the king in chariot in the 'snapshot' depicting Seti in front of the city of "Pa-Canaan." Perhaps it is not insignificant that the name of the king's horse team/main horse occurs only in that image.[19] This seems appropriate. After all, this event is the closest one in the first register that resembles the war diary accounts, and which reflect upon pharaoh advancing "on his horse (team)." The conclusive parallel is the earlier departure from an unknown city. In both cases it is the king in battle, and not the return to Egypt, which supply the information: setting and victory. The same may be said with respect to the return panels. In one of the dated rhetorical texts covering Seti's trip home a different horse team/horse name is given; instead of the earlier *Nḫt-m-Wȝst*, we read *Jmn-dj.f-pȝ-ḫpš*.

Even if the two Beth Shan Stelae conveniently avoid such common images, this may not be regarded in too serious a light. After all, both recount events that the king was not personally involved in. The same may be said with regard

---

18   Epigraphic Survey (1985).

19   Eshmawy (2007) 665–676, is one, among many, who have covered the issue. Earlier and more detailed discussion on the same topic will be found in Caritoux (1996) and (1998) 21–26, and Bouvier-Closse (2003) 11–37. Caritoux, by using one of the Seti I examples, noted two names associated with the chariot horses and concluded that we can read the principal and secondary names of the equids. With regard to the Kadesh data he added that "l'attelage porte le nom d'un des deux coursiers et attribue la précellence d'un nom sur l'autre au contexte iconographique." I mainly concur with this conclusion but do not necessarily believe that only iconographic reasons were the causes. The horses often would have been exhausted after battles and campaigns, especially the one undertaken by Ramesses II at Kadesh. The change of a name, therefore, could have had other, more mundane causes—death, for example. For the propitious nature, the context of victory, or the overt propagandistic nature of these names, see Fischer (1977) 177–178. Most recently, note Vernus (2009).

SIMPLE WORDS, SIMPLE PICTURES

to Seti's Nubian war accounts.[20] In those texts, the royal infantry and chariotry receive mention, as they do in the Beth Shan cases, but not the king. On the other hand, return 'snapshots' of war, such as the one in Register Two on the east side of the Karnak war monument, include the expected name of the horse(s) carved in the scene depicting Seti's attack at Yenoam, but also his subsequent return. In the Hittite depiction, this is made most clearly, quite possibly because the clash of arms was between two major powers, both of which possessed the advanced military technology of the day: chariots and horses. Note that the label for the horse team/horses' name is located within the scene of the triumphal return of pharaoh. The same may be said with respect to the second, or Libyan, register on the west side. Unfortunately, major portions of the third are missing. In summary, these are the cases:

	East		West
Shasu encounter	at town x at Pa-Canaan	Hittite (in field)	return scene
Yenoam	at city	Libyan	return scene
Lebanon	return scene	Kadesh	at city; other scenes partly lost
Lost			

The two "rules," if we can call them that, are thus self-evident. When the pharaoh is right at the gateway of a recalcitrant city, for example, this notation will be included. In an identical manner, when he is on the way back to Egypt the return scene *should* include the name of the horse team/horse. Therefore, additional evidence is proffered through a small identifying label to the pictorial representation, and if the campaign is the "standard" one (i.e., it is a true *topos*) then we should expect a label designating the pharaoh's chariot team. The only exception in these depictions is the rather cluttered one of Seti departing from Asia after completing his victory at Pa-Canaan. It should be added that the key 'snapshot' included many other details as well, such as the

---

20   *KRI.* I, 102–104; and Darnell (2011).

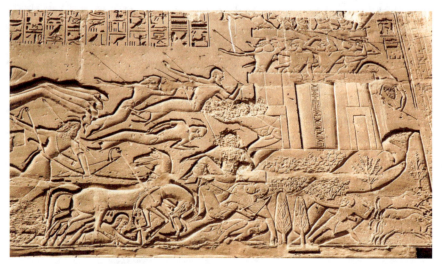

FIGURE 1   *Close up of the attack on Kadesh (Seti I), Northwestern sector of the great Hypostyle Hall, Karnak*
PHOTOGRAPH COURTESY OF PETER BRAND

welcome to the king by his officials. The depictions on the west side, though partially incomplete, further show that then a city is *not* present, the team of horses will be represented in the homeward bound images. We can see that the naming of his chargers reinforces the heroic image of the king as chariot-warrior. In the 'snapshot' of the king at a city, this is most usual and thus a strong parallel between the original war diary accounts and their later transformation into a literary narrative. The departure scenes, which appear to include these equid names as part of the written description, and thus perform the same role as those terse, originally daybook entries of departure, do in the narratives. In the latter case, the king's march away or home is often recorded, and the presence of an original ephemerides account behind the monumental one has to be recognized.

Therefore, we can set up a common formation for scenes of conflict, whether they are written or visual. Both are essentially historical in nature. There is no fantastic element, whether religious or otherwise. The visual heroic image of king in battle, acting as the solitary chariot-hero, may represent what the *aristeiai* of the greatest heroes of the Greeks performed in battle. On the other hand, these scenes are iconic presentations, or set pieces in writing/inscriptions. The questions pertinent to this first-level analysis concern the historicity of the accounts. The message sent out, through whatever of the two media, is sharp and exacting. The pharaoh must be presented in a heroically elevated position and, whatever exaggeration this entailed, his role needed to be overtly

SIMPLE WORDS, SIMPLE PICTURES

presented. Evidence of the latter included his expected supernatural size, but innovations in the New Kingdom were as follows: his solitary presence in a chariot, horses elevated in front, his use of the bow and arrow, the direction of movement aimed at one "living city." If there were two, then the lower one had to be devoid of people. The presence or absence of the famous blue crown can also be noted.[21] However, are these written passages highlighted earlier as well as the iconic presentations on temple walls type scenes or, in Kurt Raaflaub's terms, "formulae"?[22]

The analysis presented here in this short study belongs to a larger undertaking. For the moment, I simply wish to adumbrate the underlying connections between two graspable constructions, one pictorial and the other written. I presupposed a heroic image of the Egyptian monarch, its sketchy though powerful nature, and how it was transformed for anyone to read or see. Because the primary, as well as scholarly, evidence for this contention is well accepted, I reflected upon the underlying diary bases of many New Kingdom narratives of warfare. I then analyzed the visual snapshots, the ones that paralleled the ephemerides' accounts. By and large, in this study I have eschewed detailed observations on the *topoi* employed in both. Instead, I felt it necessary to link the bare, diary-like references in many empire texts with the arrival of pharaoh at a fortress-city and his departure. In both of these cases I observed that, combined with the oversized image of the ruler and his role as chariot commander-in-chief, one could still find other indirect reflections of this concept, especially within the distinct mention of pharaohs' proud and energetic horses. Here, I add, one can see the personal interest of an Egyptian warrior-king—a man, who when young, learned the art of horsemanship and shooting his arrows in a moving vehicle. Then too, is it not also the case that the same rulers not too infrequently placed their "pet lion" on the march or in camp?

The combination of standard images, such as the pharaoh trampling enemies on land whilst shooting at a citadel are thus identical, not merely in spirit but also in purpose, to the stock phraseology of the monumental narratives. Where divergences occur, and they can be great (e.g., the development of the story in the Kadesh "Poem") this combination no longer worked. Yet it is the case that as the basis of their military accounts, the Egyptians strictly depended upon a staccato-like progression through time.[23] In a written presentation, the

---

21 Davies (1982) and Hardwick (2003).
22 Raaflaub (2011) 1–34 and pages 14–15, in particular.
23 Compare the analysis of Hornung (1982).

war diaries formed the obvious basis for such an approach. In the visual, the snapshot presentation was equally useful. Both are similar and both represent an inherent embedded conception of progress through time.[24] We might call it paratactic, an earlier method of presenting a series of events embedded within a precise and progressive temporal framework.[25] In written presentations, no detailed, complex clauses were permitted. The pictures, equally, shunned subsidiary narrative elements—although they were not entirely absent. See, for example, the Syrian caught in a tree by a bear (Luxor) or the flight of the luckless Nubians homeward to their huts (Beit el Wali).[26]

In essence, this comparison of two distinct messages, visual and written, has been purposely limited to one major event and its immediate ramifications. The importance of the diary approach, archaic journalese one might amusingly call it, is a useful *point d'appui*. Those daybook entries were, by their very

---

24   See also the analysis of Loprieno (2003), depending in part upon Assmann (1975). In this case, note that Loprieno restricts himself to the "dialogue" between myth and history. Assmann's net is far wider but often too encompassing. I have addressed this in a forthcoming review to appear in *WZKM*. One needs to determine the chronographic aspects of ancient Egyptian time, the timekeepers' approaches, the instruments of telling time, and the workaday aspect of the diurnal, lunar monthly, and solar or annual cycles. Only then can we see how armies, literary men, and artists viewed the passing of their personal "narrative time."

25   The importance of parataxis, especially in connection with the development of embedded relative clauses, may be see in Probert (2006), Watkins (1976) 312–315, with his more important remarks on pages of Watkins (1970), 5–6, and Clackson (2007) 175–176, with a fine if abbreviated discussion of the later "embedded" relative clauses/sentences (versus the "adjoined" ones), which were not so at an earlier stage in the development of the languages. Watkins (1970) 6, stresses the presence "of an archaic paratactic Indo-European type which recurs most clearly in Hittite." He further adds that in such archaic syntax "the antecedent of the relative clause appears both in the relative clause and the succeeding main clause. (This type is also of interest for the general syntactical theory of the relative sentence.)" It is necessary to have a modern linguistic study of the Egyptian relative Clause. There is not even a compendium. Diemke (1934) unfortunately has been overlooked. His study was nonetheless discussed by Bühler (1934) which now may be read in an English translation (1990), 453–456. Both Bühler and Alan Gardiner had a close working relationship, albeit for a short period of time, and one that was mutually supportive. See Gardiner (1951) for his comments, as well as Bühler (1990) 32, n. 1.

26   Perhaps the easiest means of checking these well-known Ramesside images is to use Heinz (2001). The key ones mentioned in this paper, all of Ramesses II, in Heinz (2001) are to be found: 261 (V 6, Beit el Wali—Nubian flight), 272 (VIII 7, Luxor—birds departing from a deserted city), and 277 (Luxor, VIII 17—Syrian and bear in tree; VIII 18—the devastated landscape).

FIGURE 2   *Ramesses II in Syria against the Fortress-Citadel of Dapur, Ramesseum (Western Thebes) East Wall, Southern Half*
PHOTOGRAPH COURTESY OF PETER BRAND

nature, impersonal. Do they not reflect the bureaucratic mind?[27] Mahmoud Ezzamel, in fact, points out that the performative nature of accounting, and he solely discusses the ancient Egyptian practice, which depended on repetition and the need to create order from disorder.[28] As such, the use of the bare infinitive in accounting texts serves as a signifier of rationality and order, of regulation and control. The impersonal, agentless nature of the infinitive advanced these goals, and thus provided a ratcheting from one point in time to another, or within one time segment—a day, for example—or from one person or event to another.

Originating as brief remarks following a date, many of these infinitives were used within graffiti notations to serve as the equivalent to a "Kilroy was here" notation. See the classic example of the Amenemhet I's El Girgawi inscrip-

---

27   See Kemp (1989), in particular Chapter 3 ("The bureaucratic mind"). Helck's (1974) study is very important in this case. See now Ezzamel (2012). The impersonal nature of the journals' infinitival entries reflects the attitude, practice, and purpose of accounting. Ezzamel's analysis of the Egyptian bureaucratic records bears careful reading.
28   Ezzamel (2012), in particular, Chapters 3 and 4; see also the summary on pages 425–433.

tion: "We came to overthrow Wawat" ($jwt=n\ r\ s\underline{h}rt\ W3w3t$).[29] It did not matter whether the crucial jottings were in a military setting or not, the impersonal nature of the entry is what mattered. In fact, the action is what mattered, not necessarily the actor. Zbyněk Žába, for example, briefly discussed the impersonal nature of the syntax employed in these "simple" rock inscriptions; much more can now be added, including the ubiquitous employ of the impersonal suffix -tw.[30] Karl-Joachim Seyfried provided an additional and more helpful commentary later,[31] but we are now fortunate to have a recent detailed analysis by Julia Hsieh,[32] to which the work of Andréas Stauder can be added.[33]The latter scholar observed that the "active-transitive pseudoparticiple with events other than lexical statives is mainly confined to non-royal inscriptional registers," among the two being the genre of expedition inscriptions.[34] The four examples that Stauder refers to include one from Wadi el

---

29   I have cited this example owing to its importance at one time for the presupposed "coregency" between Amenemhet I and his son Sesostris I; see Simpson (1956) 215 (under "Korosko"); see now Zába (1974) 31–35.

30   Zába (1974) 253–255. See as well the comments in Sadek (1980) and (1985); add the commentary in Eichler (1993).

31   Seyfried (1981), *passim*, especially pages 132–134. This study replaces that of Sadek referred to in the previous note. There are some important comments on the vocabulary employed in these expedition graffiti by Luft (1988) 274–282, in his review of the work.

32   Hsieh (2012). She places some emphasis on the $sdm\ pw\ jr.n=f$ construction as "a grammaticalization of the narrative infinitive"; note as well her discussion of the prepositional phrase of location with r following the narrative infinitive. See as well Enmarch (2011) 97–121.

33   Stauder (2013). Once more, I am dependent upon this scholar's deep understanding of Egyptian syntax and its historical background. In his still very useful chapter on the Infinitive, Alan Gardiner stressed the wording wherein he maintained that the subject of the Infinitive was an agent with the help of the preposition in, "'by" and the object represented by the direct genitive, in Gardiner (1957) 225 § 300. Yet with intransitive infinitives the subject "can always be added as a direct genitive, whether noun or suffix", Gardiner (1957) 226 § 301. With a transitive verb, he claimed that the subject *and* object must be indicated. It is clear that much more needs to be written on this subject, especially from a diachronic point of view. Loprieno (1995) 84, indicates the "ergative-absolutive" coding in Egyptian where "the syntax of infinitives and of adjectival sentences displays 'absolutive' features: pronominal subjects are coded exactly like direct objects of transitive verbs (infinitive transitive $sdm=f$, 'hearing him,' vs. intransitive $prj.t=f$, 'his coming' ...)." In addition, "Logical subjects of transitive infinitives ... are all introduced ergatively by jn (Siut I, 68): $gmj.t=f\ jn\ \underline{h}m=f$ 'finding him by his majesty.' See also his comment son pages 64–65 on the particle jn, which may have been "originally a marker of 'ergativity'."

34   Stauder (2013).

SIMPLE WORDS, SIMPLE PICTURES

Hudi (I, number 14, 1.10: dated to the reign of Sesostris I)—"I brought it ...": *in.k(j)*. He also notes Hatnub Graffito 4.5 (probably dated to the reign of Pepi II) with the same method of presentation. However, perhaps we have diverged too far from the use of the narrative infinitive in such expeditional graffiti, even though Stauder's emphasis upon the textual scenario well represented in the Old Kingdom autobiographies needs remembering. As he concludes, such constructions had their origin "in the specific textual conditions of Sixth Dynasty *Ereignisbiographien*" and did not belong, as it is sometimes argued, "to the regular paradigm of Old Egyptian grammar."[35]

Thus, in a military setting, the entry "arriving at Kadesh" (*spr r Kdš*) merely designates the army's achievement of a goal, nothing more. Transform the entry into "His majesty arrived at Kadesh," with or without a *sḏm.n=f*, and the phrase becomes something else: either *spr ḥm=f r Kdš qr spr.n ḥm=f r Kdš*.[36] It is the living actor who is the crucial element of the accounts, just as in the cooperative communal "*We* came to overthrow Wawat" the body of troops was conceived as a unit and acted with one goal in mind. However, there remains a key difference between an infinitive plus suffix, where the intention is clearly close and identical to ownership, and an infinitive followed by the particle *jn* where the agent of the activity, an original verb transformed into a substantive by means of an infinitive, is employed thereafter. The latter situation places more emphasis on the maker or agent of the activity than the former, and thus is better suited for a literary narrative than a mere jotting on stone. The punctuated rhythm of such entries, even when they were used as the basis of a narrative, is identical in outlook to those visual 'snapshots' of the king in battle that we have already covered. Whereas the Gebal Barkal Stela may prefer a lengthy series of *sḏm.n=f* formations,[37] all reflecting the king's personal,

---

35   Stauber (2013).

36   I will avoid critiquing the important study of Kruchten (1999). I am not interested in the disappearance of the -*n*-. The use of a bare infinitive with suffix compared to the same infinitive, but with the agent particle *jn* + substantive and with a pronoun into the so-called *sḏm.jn=f* construction, heralds a greater emphasis upon the actor-agent.

37   Monumental hieroglyphic texts, themselves based on hieratic exemplars, therefore offer a wide range of possibilities for literary analysis. One may note them in so effusive a fashion as at Gebal Barkal by Thutmose III Stela; see Redford in Redford and Grayson (1973) 25–29. The Kadesh account of the Poem is also a good example because there are two extant hieratic copies in addition to the "officially recognized" hieroglyphic ones: Spalinger (2002) and von der Way (1984). As with the Berlin Leather Roll, one can see the coverage of the events contained with a specific literary setting. For this matter, note Spalinger (1997), and Müller (2011).

virile deeds in the first person, so too do the independent visual depictions of military campaigning in the Ramesside war reliefs serve to punctuate and delineate the narrative flow of events. In text, as well as in image, a relatively simple approach was taken in order to present history. In many ways both versions reveal a relatively unsophisticated approach to describing the past, one that was, nevertheless, not the be-all or the end-all of ancient Egyptian historical writing or visual projection. It is noteworthy that in the large number of undated war scenes of Ramesses II, the 'snapshot' of king versus fortress-city is almost always used.[38] The reason is again not hard to find: the entire image served as a frozen passage in time, just as the war diary's entry, "arriving at city x" fulfilled the same purpose—the journals of any official or accountant would do the same.[39]

Nonetheless, I feel that the elementary use of pictorial and written parataxis cannot be overlooked. In his linguistic analysis, Antonio Loprieno referred to the independent versus subordinate syntactic features of the clause.[40] By this opposition, he was referring to parataxis as opposed to hypotaxis, but he also considered "subordination" to be separate from these two, yet nonetheless independent in its own right. With respect to the narrative images discussed here, no subordination was present. It is as if we were retracing the well-known analysis of Robin G. Collingwood concerning the "scissors-and-paste history" so well analyzed in his *The Idea of History*.[41] In this case I am referring to the chronological placement of one event right next to another without any link. The simple paratactic grouped images or syntactical elements possess equal

---

38   See now Spalinger (2012), particularly Chapters 4, 6–8.

39   I am referring here to the well-known cases in the Late Egyptian Miscellanies: e.g., P. Anastasi III 6,1 and so forth: with *tst, spr, jj⟨t⟩*. For the accounts, see n. 27 above.

40   Loprieno (1995) 162–166 and page 163 in particular. Note, as well, the summary overviews on pages 189–190 and 225–226 in which parataxis, hypotaxis, and subordination are discussed. According to Loprieno, the "embedding situation" is a type of subordination without any morphological markers. With respect to our battle relief snapshots one much liken the familiar scene in which pharaoh is facing backward to the battlefield but whose foot in entering onto the flat platform of the cab of the chariot. The horses and war vehicle are thus facing in the opposite direction and geared forward, anticipating the return home, whereas the king is "betwixt and between." One might still say that this commonplace icon is still paratactic. See as well Junge (1989) 63–64 concerning "figure" and "ground." Does not this simple mode of linguistic analysis also serve us well in discussing the icon of pharaoh superimposed on the background of a spaceless zone but fighting a mass of enemies in front of a closed city?

41   Collingwood (1946), in which pages 33 and 269–270 can be signaled. See also Strauss (1952) 582.

SIMPLE WORDS, SIMPLE PICTURES

weight, although not indicating equal dramatic intent.[42] Erich Auerbach has also noted the same dramatic context of paratically ordered sentences.[43]

Further research on this theme will demand a more sophisticated approach, but one, I feel, that is dependent upon these simple cases, all of which can be exploited by modern scholarship in order to explain further the ancient Egyptian uses of historical methods. Pascal Vernus, for example, has stressed within a linguistic study that the term "paratacic" is questionable.[44] Still, one can argue for an extensive use of embedded visual elements within those paratactically oriented visual scenes—see the Syrian in a tree with a bear, the *topos* of the Nubian flight which is beautifully rendered at Beit el Wali, and some despondent and fleeing Asiatics. By and large, however, these small portions of major 'snapshots' do not advance the temporal narrative much, though I definitely would set aside Ramesses II's Nubian depictions at Beit el Wali.

Prolepsis seems to be far removed from either method of narration. The expectation of the future is placed in a secondary position, even if the first person narrative "lends itself better than any other to anticipation."[45] Yet at this point have we not reached the crucial difference between the first person pronoun(s) and the other two? Émile Benveniste, in his well-known study concerning the linguistic aspects of the three personal pronouns, stressed the various grammatical interpretations associated with each of them. He desired to prove that the only "correct" or "true" persons of the verb are the first and second.[46] Subsequently, Romana Lazzeroni attempted a refinement of Benveniste's position by proposing that there is a major opposition between the first person and the remaining two; namely, a "correlation of subjectivity," to quote Maurizio Bettini.[47] How this affects the narrative infinitive use in

---

42   Dramatic intensity is also discussed in Spalinger (2012).

43   Auberbach (1953) 71. Of course, parataxis can be an excellent method of presentation an oration, and was widely used by Winston Churchill. On the other hand, as Cicero's orations indicate, hypotaxis also has its own method of control.

44   Vernus (1990) 72, n. 71.

45   Genette (1980) 67.

46   Benveniste (1966a) 251–257, originally published in 1956. As Bettini (2010) indicates, the first person is "he who speaks" and the second is "he to whom one addresses one's self", which are strongly different from the third which is "he who is absent." In Egyptian, I can refer the reader to the remarks of Loprieno (1988) 62, wherein he stresses the first person's "higher topicality"—"predictability as Topic of the discourse portion"—than that of the third person.

47   Bettini (2010) 22, n. 69; with Lazzeroni (1994) 267–274.

Egyptian texts will be interesting to evaluate in more detail than presently offered, but the well-known first person approach of some war records, outside of performance speeches (e.g., the Gebel Barkal Stela of Thutmose III) needs additional commentary.

Concerning narrative and its grammatical formation, Robert Kawashima, continuing from Benveniste, has offered an interpretation with which some may disagree.[48] Yet, for an Egyptologist, I can cite his useful analysis of the use of the *qatal* form in Classical Biblical Hebrew, which "occurs primarily in direct discourse in biblical narrative,"[49] and his evaluation of the "converted imperfect" (*wayyiqtol*) formation. The difference, states Kawashima, is close to that between modern French's *passé composé* and its *passé simple*. The Hebrew perfect is mainly used in direct discourse whereas the other formation will be found mainly in narration. Thus, through "particularized speech," a term that Kawashima employs, two different methods of narration are presented.[50] With respect to Egyptian, however, none of his categories—and I am eliminating his further discussion on narrative—are appropriate. Hence, we must turn to other discussions in Egyptology regarding narrative.[51]

Recently, I have used the modern background scholarship concerned with narrative art in order to explain further the prevalent attitudes of the Egyptian who established their military reliefs in the New Kingdom. It will be sufficient to quote part of that study:

> The individual scenic events were discrete building blocks, all interconnected, and all contributing to certain dramatic events within a war or a campaign. Yet the attempt to represent "continuous narrative" by means of a representation of successive events within one scene, was never present. Instead, we might borrow Franz Wickhoff's other term, "distinguishing style," in order to define, more narrowly, the Egyptian approach to narrative. Yet the as-pect of climax or zenith (in an illustrated campaign, for example) must be analyzed independently from the type of narrative organization employed by the artists, and perhaps the modern term "monoscenic" is more appropriate in the context of pharaonic mil-

---

48   Kawashima (2011) 341–369.

49   Kawashima (2011) 349.

50   Kawashima (2011) 351.

51   Tait (2011) 397–411, does not hit the mark. Johnson's (1980) preliminary and all-too-short analysis is closer to my aim. A study subsequent to her scintillating presentation was that of Junge's, referred to above in n. 40. I adhere nonetheless to the short presentation of Loprieno, also cited earlier in n. 40, and that of Vernus (1978) 100–102.

SIMPLE WORDS, SIMPLE PICTURES

itary reliefs, for at least we have one moment in time placed within one space and centered upon one picture per one event.[52]

In this work I referenced Chikako Watanabi and Zainab Baharani.[53] Perhaps I should have chosen investigations of another kind, such as those in film and even the closely related analyses of Karl Bühler with respect to language and film.[54] After discussing the Egyptian organization of relative sentences and parataxis, Bühler continued his theme by addressing the flow of images in a (silent) film. He noted that "with respect to imagination-oriented deixis, the film is more closely related to epic than to drama."[55] Point of view changes are achieved with a jump, he stated further, being displacements that move, such as from a long shot to short, or from one side of an object to another. Hence, Bühler refers to similar jumps in dimension from Book 21 of the Odyssey, when Homer traces Penelope's walk on the way to the city's treasure vault. "Continuous transitions and jumps in the range of magnitude ... are favored techniques of representation in film."[56] Surely, the parallel to Egyptian snapshot montage is self-evident here? Indeed, I can cite Sergei Eisenstein's still pertinent remarks concerning silent film: "The earliest conscious film-makers, and our first film theoreticians, regarded montage as a means of description by placing single shots one after the other like building-blocks."[57] This is not very different in approach from the infinitival system of the skeletons used in the construction of the written narrative war records of New Kingdom pharaohs, or for that matter Goethe's pregnant words: "Die Baukunst ist eine ertarrte Musik."[58] Yes, one is frozen for a moment, be it short or long in duration. Yet the visual tableau of war or the transformed diary entry serves to enhance the dramatic nature of the event. The mental halt, so to speak, allows the viewer, listener, or reader to absorb the detailed information *as a unity*. Hence, subsidiary dependent

---

52   Spalinger (2012) 14.

53   Watanabe (2004) 105, and Bahrani (2008) 32. See also Kemp (1996) 63 who provides a more useful summary of the viewpoints of Franz Wickhoff whose ideas are pertinent to my study. Kemp observes that the concept of a "distinguishing style," which establishes a 1:1 equivalence between unit of time and a unit of narrative, is appropriate here. He further remarks that "time past, time present, and time to come must all be implicitly present." In essence, this is exactly what our Egyptian military scenes/snapshots provide to the viewer.

54   Bühler (1990) 444–448.

55   Bühler (1990) 445.

56   Bühler (1990) 448.

57   Eisenstein (1949) 48. See also Arnheim (1957) 87–89.

58   Eisenstein (1949) 48.

events, or clauses, hypotactic connections are shunned, and the anaphoric system of the visual art, as with the written format, connects one scene to the next. For example, the chariot role of the king in pictorial representation or a repetition of words in the literary presentation provide identical means of linking one snapshot image or one simple verbal entry to the following one.

The close association of Egyptian infinitival diaries and their transmutation into some type of paratactic narrative should now be evident, even if I have purposely relegated the artistic evidence to the New Kingdom and selected only those reliefs associated with military affairs.[59] Note that I am not arguing for direct influence. Of course, one can trace back in time the development of narrative art, but that was not my purpose. This aspect of analysis was only partly studied in my study *Icons of Power* and it remained beyond the parameters of that study. However, is not enough to state whether Egyptian "historical" writing was "literary" or not, or that its art was essentially atemporal, mainly static, and elementary in historical consciousness.[60] The issue is how the ancient Egyptians approached their past by means of their artistic and literary sensibilities, and how the joining together of brief written or visual images lay at the heart of their temporal thinking.[61]

---

59   One can turn to the compendia of William Stevenson Smith or G.A. Gaballa if a need for non-bellicose scenes is desired. Their works are referred to in note 13 above.

60   See Eyre (1996). The question is ill-proposed. Most monumental texts, for this is what surely Eyre was aiming at, were publicized. They were drawn up extremely carefully as they had to have, ultimately, pharaonic approval. The drafts would have been on soft copies, papyri, and constantly refined before the final "proof" was accepted. The various copies were made to be sent out to officials at certain places, the temple of Amun, for example. Ultimately, hard copies on stelae or walls would have been expertly carved and the master copy, perhaps on a leather roll, kept somewhere. Surely, this explains the reuse of Ramesses II Festival Calendar at Medinet Habu, the mortuary temple of Ramesses III. For a useful recent discussion, see Haring (1997) *passim*, especially chapter 1.

61   The Eyre analysis, therefore, blurs the issue by separating monumental historical texts from others on soft copies. If the encomia can be easily cut out of the former material, so can the *Königsnovelle* accounts as well as the poetry. The latter is best seen in the famous Poetical Stela of Thutmose III, an account that was present later in the Ramesside Period. For example, see Grapow (1936). There is an excellent discussion of the material by Kitchen and Gaballa (1970), following upon the commentary of Edgerton and Wilson (1936) 119–120. Additional texts can be brought into discussion such as the Blessing of Ptah and the like. There, the reader is recommended to turn to Kitchen's comments, in Kitchen (1999) 159–163, with the remarks on the "pairing-off" with the First Hittite Marriage Inscription. Indeed, one can use the First Hittite Marriage text of Ramesses II for the evidence of a hieratic *Vorlage* that was literary, and one that must have circulated outside

of the temple environment. Vernus (1990b) 35–53 is useful to consult on this matter. The life-long brilliant work of Henry Fischer, though subsumed by an overwhelming mass of data, small or large, must form the basis for a deeper exploration of Egyptian time consciousness as evidenced in the narrative art. That issue was covered, albeit in a brief fashion, and with reference to Fischer, Ludwig Morenz and other scholars interested in the *Königsnovelle* in Spalinger (2011).

CHAPTER 3

Caesar's *Exempla* and the Role of Centurions in Battle*

David Nolan

The role of literary *exempla* in military history is a highly problematic one, particularly for the characters recorded in Caesar's *Bellum Gallicum*, as the figures appear under extraordinary circumstances and often have a close association with the literary objectives of this work. E.L. Wheeler noted this troublesome aspect of *exempla* in Caesar stating "... emphasis on combat anecdotes ignores both ancient historic penchant for the dramatic effect of such *exempla* and the possibility that the exceptional of such anecdotes may not be typical."[1] This comment typifies the problems associated with using *exempla* in military history, as their appearance in the texts is determined by a range of criteria that are not necessarily related to the normal circumstances of combat. However, exceptional behaviour does not make these figures useless in determining roles on the battlefield, and it is the purpose of this chapter to demonstrate that behind the exemplary activity that these characters engage in are a set of fundamental values and responsibilities that give important insights into actual Roman combat. Even though the veracity of the described events cannot be confirmed, centurion *exempla*, even as literary figures, have value for what information they can provide about battle, as they serve in a summary capacity to support Caesar's presentation of a battle. It is also possible to use *exempla* to define the role of centurions in combat as they act against a background of responsibilities when they appear in the text. Through an analysis of the broader literary context in which the centurions are used, it is possible to identify these responsibilities and show that exemplary combat at the front of the cohort was not part of their normal duties. Rather, Caesar expected centurions to involve themselves in combat only when a legion was on the edge of collapse, with the specific aim of restoring the situation. This is most evident in the manner that the centurions Lucius Vorenus and Titus Pullo

---

* The author is indebted to Dr Jeremy Armstrong of the University of Auckland for his support, encouragement, and for the opportunity to contribute to this work.
1 Wheeler (1998) 645.

© KONINKLIJKE BRILL NV, LEIDEN, 2016 | DOI: 10.1163/9789004284852_004

are paired with the defeated commanders Q. Titurius Sabinus and L. Aurunculeius Cotta, in which the pairing helps to determine how centurion roles should be understood. These boundaries on expected behaviour are best illustrated through an examination of the account of the battle of Gergovia, where Caesar makes use of *exempla* to explain why control was lost on the battlefield, and in the process of defending his own conduct, provides information on his general expectations regarding the rank. The account of Gergovia is particularly illustrative of how an analysis of the *exempla* and the purposes of the text can yield rich rewards in understanding the role of centurions in combat.

Exempla and Roman Battle Narrative

The traditional use of *exempla*, particularly in the works of authors such as Livy, is generally associated with 'moralising' and a desire to instruct, and there is a long Roman tradition of using moral *exempla*.[2] Livy states that such characters are "worthy of imitation" or are used to illustrate behaviour that should be avoided (Livy *Praef.* 10).[3] Oakley even argued that Livy made some important factual changes to historical accounts in order to moralise, such as to inflate the *superbia* of Tarquin, or direct attention to the *clementia* of Scipio Africanus.[4] The extent of exaggeration may be unknown, but the emphasis on traits and the moralising element is an important aspect in such accounts.[5] Nevertheless, where they appear in combat the exemplars in Livy are used to make general points about battle according to his overall schema. This is most evident in the duel of Manlius where the attributes of the opposing nations are expressed in terms of the two combatants; Manlius representing the Roman values of *pietas*, *virtus*, and *disciplina*, the Gaul representing characteristics such as size,

---

2 For more comprehensive examinations of *exempla* in Roman culture refer generally to Roller (2004); (2009) and Chaplin (2000) 11–31. For general information on *exempla* see also Price (1975) and Kornhardt (1936). See also Oakley (1997) 114–116 on *exempla* and moralising.

3 See Moles (1993) for an overall examination of Livy's preface. See also Cic. *Arch.* 14. Wiseman (1979) 37 notes that Cicero regarded *exempla* as the main value of history in guiding individual conduct.

4 As Oakley (1997) 115–116 states, several episodes are shaped so as to reveal particular moral qualities. For example Manlius Torquatus is revealed to have *pietas*, *clementia* and *disciplina* at Livy 7.9.6–7.10.14. Oakley notes that Manlius even calls himself an *exempla* self-consciously to his son at Livy 8.7.17.

5 See also Oakley (1997) 115–116 for examples such as the Volscian campaign and *temeritas* of L. Furius Livy 24.9, and the *temeritas* of the Manlii at Livy 6.30.2–8. At Livy 7.5.2 Manlius rescues his father and is described as exemplary.

36   NOLAN

boastfulness, and lack of self-control (Livy 7.9.6–7.10.14).[6] According to Livy's account, these attributes were directly associated with the result of the battle, so that the vignette of the duel has direct connection with victory.[7] While the emphasis in this use of *exempla* is on the moral dimension, it is important to note that these figures have a causal relationship to their battle narrative, as they can be linked with the result or be expected to influence events.

More particularly, *exempla* had a role in oratory and it is as part of supporting an overall statement about a battle or a related issue that the figures in Caesar's *commentaria* should primarily be considered. *Exempla*, like other precedents, were used to support arguments in rhetoric, and this skill was common to aristocratic training (Cic. *De Or.* 1.18, 256, *Orat.* 120; Quint. *Inst* 10.1.34, 12.4.1–2).[8] There is evidence to suggest Caesar understood this concept, as the notorious remark about Sulla attributed to him suggests (Sall. *Cat.* 51.4–6, 28–34, 37–42).[9] The *exempla* as they appear in the *Bellum Gallicum* may therefore be regarded as serving as a literary method of conveying and supporting the objectives of the author.[10] The flexibility of the figures in their rhetorical context also

---

6   See Oakley (1997) 116 who examines national characteristics as an aspect of battle in Livy. Walsh (1963) 200 also notes that the duel between champions motif demonstrates "which race is superior in war." Walsh (1963) 255–256 describes how the duel of Manlius and the Gaul are transformed into examples of cultural contrasts. Note how the actions of the *exempla* are given causality and made the reason the Gauls withdraw at Livy 10.9.1.

7   Chaplin (2000) 3 states that Livy's *exempla*, "turn out to embrace practical matters as well as moral concepts," such as military strategy, and constitutional procedures. When such concepts appear, even though they may be stated using individuals or literary dichotomies, they nonetheless express those military principles as understood by the author. See Chaplin (2000) 55 regarding Livy 38.11 on caution and the Fabius approach to war, see also Livy 42.4 on Paullus and foresight.

8   Cicero in *Orat.* 120 states that *exempla* gain credibility for the speaker and delight the audience. See Cic. *De Inv.* 1 29.46 on *confirmatio* and 1 .49 for his definition, which is that *exempla* are types of comparisons that strengthen or weaken a case by the authority/precedent or experience of a certain person or event, as discussed in Price (1975) 104–105. See also Price (1975) 97, 120. As Chaplin (2000) 13 notes, young men were expected to have a 'firm control' of *exempla* for speeches.

9   According to Sallust, Caesar was aware of value of precedents, and the saying that evil precedents originated from good intentions is attributed to him, see *Cat.* 51.27. Refer to Chaplin (2000) 26–27 for a discussion of Caesar and *exempla*. Chaplin cites Suet. *Iul.* 77.1 and Caes. *B Gall.* 1.33.4, 1.40.5 as examples of Caesar's use and understanding of *exempla*. See also Morgan (1997) 35–39 and Nordling (1991) for further reading on *exempla*, rhetoric and Caesar.

10   As Price (1975) 89–90 notes *exempla* can have two interrelated functions, ostensibly as *exornatio*, but also in order to strengthen the proof of an argument (*confirmatio*). The

CAESAR'S EXEMPLA AND THE ROLE OF CENTURIONS IN BATTLE 37

means that there is precedent for Caesar to fit his *exempla* to the case he presents in each battle narrative, and he is not burdened with a historical tradition for the figures he uses.[11] For example, the nameless *aquilifer* who inspires the men in the first invasion of Britain is no historical figure, however his speech and presence in the battle reinforces the interpretation of the event as a great venture for the commander and Rome (Caes. *B Gall.* 4.25.3–5).[12] Caesar was skilled in oratory, with a reputation that even Cicero admired, so it is not surprising to find evidence of such training in the *Bellum Gallicum* (Cic. *Brut.* 252).[13] This is particularly so in the use of *exempla* as interpretive tools, and their presence in support of self-promotional objectives for each battle narrative.

Caesar makes use of non-aristocratic *exempla* in marked contrast to an author such as Livy, who primarily expresses values through the actions and words of aristocrats.[14] For instance, in Livy's account of Cannae and Trasimene Roman values are expressed through the fate of commanders like L. Paullus and C. Flaminius, who behave in a particularly Roman manner in defeat (Livy 22.49.1–13).[15] Caesar certainly includes aristocratic figures in his accounts, with P. Crassus, T. Labienus and other subordinates described in terms of their contributions, and the text even includes historical figures where they are

---

   fundamental idea that these figures assist in the presentation of argument is an important aspect of how Caesar uses them to support his interpretation of battle.

11   Chaplin regards the roles of *exempla* as becoming more fixed by the time of Livy, in contrast to Cicero who recycled and reinterpreted his *exempla* to fit the context in which he was using them. This chapter proposes that Caesar also had a flexible approach to *exempla*, at least in the *Bellum Gallicum*. See Chaplin (2000) 170; Roller (2004) 7. For Cicero see Schönenberger (1911) 19 and Rambaud (1953) 46–50.

12   See Nolan (2014) 258–260 for an analysis of the first invasion landing. The use of individuals in this way also applies to the *Bellum Civile*. Consider the case of Scaeva and the shield pierced by many arrows, which Caesar uses to encapsulate the ferocity of the assault on the defensive position at Caes. *B Civ.* 3.53.5. See also the speech of Crastinus in Caes, *B Civ.* 3.91.1–4, where Crastinus summarises an interpretation of Pharsalus as the final confrontation of the Civil War.

13   See Nordling (1991) 2–8. Nordling provides a summary of Caesar's rhetorical training and facility according to the sources.

14   As Roller (2004) 6 notes most, but not all *exempla* tend to be aristocrats. A notable exception is Livy's description of the centurion who placed the standards in the forum, demonstrating that in one instance where when a centurion appears, he does so in the context of discipline and holding ground. See Chaplin (2000) 87, and Livy 5.55.1–2.

15   Paullus clearly expresses aristocratic values in this account. See also Livy 22.6.1–2 for the redemption of Flaminius through combat.

relevant to the context of the battles.[16] However, Caesar primarily uses non-Roman or non-aristocratic *exempla* in the *Bellum Gallicum* and these are often the most memorable figures in his narratives. For example Gallic auxiliaries may be used to represent qualities such as *pietas*, as in the case of two brothers who die trying to save each other in a skirmish against the Usipetes and Tencteri (Caes. *B Gall.* 4.12.4–6).[17] Among Roman forces, centurions are particularly important and are mentioned alongside senior ranks to demonstrate correct Roman behaviour. This is the case with Baculus, who exemplifies courage and tenacity in combat, or the two centurions casualties mentioned in a defeat alongside their aristocratic commander (*B Gall.* 2.25, 5.35). While the use of centurion *exempla* in Caesar is not completely unique, the frequency and emphasis on non-aristocratic over aristocratic is certainly notable.

Centurions and Battlefield Conditions

Caesar's understanding of battle is conveyed through his use of centurions as interpretive tools, most importantly in their representative capacity to summarise, through anecdote, the status of the Roman forces. While Caesar may show indulgence to his centurions by including them in his account of battle, it appears likely that he would present them according to the expectations of their rank and a correct knowledge of battlefield conditions in order to establish his own understanding of war.[18] As Kraus states, centurions "focus

---

16   For Caesar's reference to his officers see Welch (1998). Welch recognises the credit given to officers, noting however at 102 that they 'are not in the end the really memorable heroes or characters who emerge from the pages of Caesar's commentaries'. See also Caes. *B Gall.* 1.12.5–7 and 2.29.4–5 where Caesar cites historical figures. In the case of the slaughter of the Tigurini at 1.12 history is used to show that the battle is the enactment of vengeance. The Aduatuci are associated with the Cimbri and Teutones at Caes. *B Gall.* 2.29.

17   The two brothers demonstrate *pietas* in their support for each other. See Nolan (2014) 311–313.

18   See Welch (1998) 85–86. As Welch states, Caesar's *commentaria* were probably in part an answer to the military prowess of Pompey. Consequently, we may expect him to convey his understanding of battle correctly, including the general responsibilities and roles of subordinates. It may be expected that he might exaggerate or outright manufacture particular episodes, but he would nevertheless do so within the confines of what was regarded as correct or incorrect behaviour from within the ranks of Roman forces. *Exempla* are also likely to be described with strong reference to their correct duties due to Caesar's desire to associate his own activity with state interests and sanctioned military behaviour. This is certainly evident in his identification with the *Populus Romanus*, as noted by Ram-

CAESAR'S EXEMPLA AND THE ROLE OF CENTURIONS IN BATTLE

action ... are granted rare direct speech, and generally serve as the stylised representatives of [Caesar's] legions."[19] This is evident at the battle of the Sabis River, where Caesar utilises Baculus as an indicative example of the state of a legion:

> *Caesar ab decimae legionis cohortatione ad dextrum cornu profectus, ubi suos urgeri signisque in unum locum conlatis duodecimae legionis confertos milites sibi ipsos ad pugnam esse impedimento vidit, quartae cohortis omnibus centurionibus occisis signiferoque interfecto signo amisso, reliquarum cohortium omnibus fere centurionibus aut vulneratis aut occisis, in his primipilo P. Sextio Baculo fortissimo viro multis gravibusque vulneribus confecto, ut iam se sustinere non posset ...*[20]

From encouraging the 10th legion Caesar set out to the right wing, where he saw his men driven back with the standards gathered in one place. He saw that the soldiers of the Twelfth, clustered together, were themselves an impediment to fighting. He saw that all the centurions of the 4th cohort were slain and the standard bearer had been killed. The standard was lost and nearly all the centurions of the remaining cohorts were either wounded or killed, among whom was chief centurion Publius Sextius Baculus, the bravest of men. He was exhausted with many grievous wounds, so that by this point he was not able to stand ...

CAES. *B Gall.* 2.25.1

This passage illustrates how Caesar uses an *exemplum* to reinforce his interpretation of events on the battlefield. The artifice involved is evident as the

---

age (2002) 132–134. Feldherr (1998) 107–108 notes something similar with regard to Livy. Unsanctioned combat, such as that of the younger Manlius does not reinforce state legitimacy and aims. See Livy 8.7.8–12, 8.7.15–19.

19 Kraus (2010) 56. This aspect of the work has been omitted in some studies of Caesar's battle narratives. For instance see Lendon (1999) 318 for an analysis of the Sabis River battle that does not include Baculus in his representative role. Kagan (2006) 122 views anecdotal information like the *pila* volley against the Helvetii in Caes. *B Gall.* 1.25.2–1.25.5 as 'subtactical' elements often added for colour" or with a direct causal role and does not analyse the role that *exempla* play in such narratives, particularly in the analysis of the Sabis River battle at 128–136, even though Baculus has such an important position in Caesar's text.

20 Editorial note: Given the overtly literary approach adopted by this chapter, the original Latin text (taken from the 1905 Teubner edition) has been included for all the passages discussed in detail.

narrator's perceptual framework presumes he can see all the centurion losses as he arrives, drawing on casualty figures to describe the danger of the situation even though it is unlikely such detail was fully apparent at the time.[21] Caesar also moves from a broad description of the circumstances regarding the whole legion to the casualties among the fourth cohort centurions and other centurions of the legion, to finally a description of Baculus. Consequently, the passage is structured towards an increasing level of casualty detail that culminates with the description of Baculus, who appears as representative of those casualties within the struggling legion. The manner with which Caesar moves from the general to the specific using various artifices shows how Baculus is being used in a summary capacity; as an anecdote that serves as an illustrative example of the listed casualties. Furthermore, the presence of Baculus in the narrative and the details given about him demonstrate how Caesar uses the individual to summarise the overall state of affairs. The injured state of Baculus is representative of the situation, as like him the legion is barely able to hold itself in fighting condition. As Caesar states:

> ... *reliquos esse tardiores et nunnullos ab novissimis desertos proelio excedere ac tela vitare, hostes neque a fronte ex inferiore loco subeuntes intermittere et ab utroque latere instare et rem esse in angusto vidit neque ullum esse subsidium, quod submitti posset ...*

> ... the rest were very slow and some from the rearmost ranks were deserting the fighting, drawing away and avoiding the missiles. He also saw that in front the enemy were not ceasing their advance from the lower ground and that they were threatening on each side. The matter was in dire straits, and there was no help to be sent in ...

> CAES. *B Gall.* 2.25.1–2

The parallel between the individual and the unit shows that the appearance of Baculus in the text is more subtle than a simple tale of courage or great deeds. While the description does hint at previous combat, Caesar does not specifically state any such thing, choosing to only describe the man's wounds and difficulty in standing. In contrast, Caesar is clear about the overall situation in this passage, that it is *angustus*, or close to disaster. The appearance of Baculus,

---

21   While the loss of the standard would probably have been apparent at the time, it is difficult to see how, in the chaos of battle, Caesar could be so sure of the centurion casualties even if he took some time to assess the situation.

CAESAR'S EXEMPLA AND THE ROLE OF CENTURIONS IN BATTLE   41

and the parallels between his and the legion's status seem deliberately designed to emulate each other, as both are in a state between combat effectiveness and collapse.

Caesar later closes the battle narrative with a description of the *virtus* of the enemy and the credit he attributes to his own army (Caes. *B. Gall.* 2.27). This is directly relatable to the tale of Baculus who epitomises the valour of an army that struggles to succeed under difficult circumstances. These parallels suggest that the centurion is being used to reinforce the more complicated description of the situation occurring among the men, as one individual appears to represent generally what was happening.[22] This emphasis on the injured state, and the tenacity of the man, rather than the deeds that brought them about, demonstrates how Caesar had a representative role in mind when including the character. Indeed, Baculus is used in such a representative capacity on several occasions.[23] In the description of the defence of Quintus Cicero's camp against a Germanic attack in Book Six, Caesar describes the state of the defenders as near breaking point, which he then follows with a description of this centurion:

> *Totis trepidatur castris, atque alius ex alio causam tumultus quaerit; neque quo signa ferantur, neque quam in partem quisque conveniat provident. alius castra iam capta pronuntiat, alius deleto exercitu atque imperatore victores barbaros venisse contendit ... Erat aeger in praesidio relictus P. Sextius Baculus qui primum pilum apud Caesarem duxerat, cuius mentionem superioribus proeliis fecimus, ac diem iam quintum cibo caruerat. hic diffisus suae atque omnium saluti inermis ex tabernaculo prodit; videt imminere hostes atque in summo rem esse discrimine ...*

> Through the whole camp there was fear, and each queried of the other the reason for the tumult. They did not see to where standards might be borne, nor to which quarter each might gather. One announced that the camp was already taken; another contended that the army and gen-

---

22   Campbell (1987) 23 notes that the audience may be uninterested in the technical details of combat, something that may explain the use of this *exemplum*. Note also the difference between Caesar and Livy and the latter's description of several duels, where a direct link is established between the *exemplum* and the outcome of battle. Caesar associates the actions of this centurion with the progress of the greater battle, rather than creating a direct causal link. See Feldherr (1998) 93–94 and Livy 7.10.7–13, 7.26.8–14.

23   See Nolan (2014) 106–107 for the purpose behind the appearance of Baculus in Caes. *B Gall.* 3.5.

eral were destroyed and that the barbarians came as victors ... There was one Publius Sextius Baculus who was ill and left behind with the baggage. He whom we have mentioned in previous battles as first spear had commanded for Caesar and by this point had been without food for five days. Having despaired for the safety of himself and that of everybody he came forth unarmed from the tent. He saw that the enemy were at hand and that the matter was at a severe crisis point ...

CAES. B Gall. 6.37.6–6.38.2

Baculus, like the legion, was in poor health and had lost hope. While he later took action instrumental in restoring the situation, at this point in the passage there is a clear parallel between the morale of the army and that of the individual as both observed the dire nature of their situation.[24] The manner with which Caesar presents Baculus is therefore indicative of the representative role the character plays in summarising or clarifying circumstances through use of a centurion, and his use is consistent as a representative figure.

Exempla and Behavioural Expectations

Beyond the summary capacity of such figures and their role in the literary objectives of the work, Caesar's use of *exempla* from among the ranks of the centurions is particularly useful as it enables the commander's behavioural expectations to be determined. The episodes in which the centurions appear are all clearly demonstrative of correct or incorrect conduct. Even if the activity itself is spurious from an historical perspective, the underlying principles for correct behaviour are less contentious as Caesar's knowledge of battle, and his understanding of the workings of his army, seem less subject to dispute.[25] The

---

24   Centurions such as Lucius Vorenus and Titus Pullo are utilised in a similar manner in a later account, where they display exceptional self-confidence that echoes the general feeling of defiance of an entire camp. This is apparent as the two men appear in battle immediately after a description of centurions taunting the enemy. See Caes. *B Gall.* 5.43.6–5.44.14.

25   The two extremes of the arguments regarding the veracity of Caesar are represented by Rambaud (1966) and Collins (1972). For more recent analysis see Welch and Powell (1998) for articles that examine various aspects of Caesar's military reporting. Why Caesar used centurions is unclear. It may be that he had genuine affection for the men, or that they were simply useful tools for illustrating certain aspects of battle. See de Blois (2000) 23–25 for some aspects of the relationship between Caesar and his centurions, but the

CAESAR'S EXEMPLA AND THE ROLE OF CENTURIONS IN BATTLE 43

activity may be exceptional but the actions occur in relation to a set of norms, and a close examination of each episode can reveal what Caesar expected of these men.

The behaviour of Baculus is important as it reveals some information about Caesar's expectations, such as the idea that when the fighting was hardest, the centurions were expected to act in a way that was not only critical to the state of the legion but that led to heavy casualties among their rank.[26] This is evident in the battle of the Sabis River, where Baculus is first described. Here Caesar makes it clear that the casualties among the centurions matter the most in this account, describing how the majority of the legion's centurions have been killed or wounded, with the fourth cohort's centurions entirely lost (Caes. *B Gall.* 2.25.1). While it is possible that the use of centurion casualties is merely an artistic or literary method of stating the dire circumstances, any exaggeration is nevertheless framed around the importance of the centurions at this time and their activity and the situation have a close association. The importance of the centurions is supported as there is no mention of general casualties, or even a description of the state of the *tribunes militum,* so it appears that Caesar regards the centurions as best suited to describing a desperate situation. Caesar's description of the crisis with specific reference to the centurion casualties therefore suggests they had a particularly important and dangerous role at this particular time, even if there is an artistic purpose in using them in the passage.

The responsibilities that form the background to the appearance of Baculus shed some illumination on what was happening. Baculus was clearly attempting to carry out his duties at this time, and although his wounds were exceptional, he was trying to stand and presumably continue on with the duties he and the other centurions were expected to perform. There is an implied expectation in his behaviour that, when a legion was on the edge of breaking, centurions took up a position that placed them in particular danger. Their role was so critical that even severe wounds would not stop a particularly brave centurion from performing his duties.[27] While at this time it is not specified what this activity involved, it seems clear that centurions were expected to stay in that

---

idea of favour is in part derived from the simple fact that Caesar included them in his commentaries.

26 Ward (2012) 57, 81. While Ward states that centurions generally led from the front, he also notes that they consistently appear in the texts when a situation is dire.

27 See Lee (1996) 211. Lee draws the conclusion that centurions bear a heavy cost among the casualties, but does not assess the implications of such a link between casualties and particular conditions in a battle, and does not examine Caesar's *exempla*.

position if they were able. We therefore have in the description of Baculus an extreme example of the idea that centurions took up a highly dangerous position at the point when a legion was in danger of collapsing, and were expected to continue in their duty while the situation was critical.

Caesar gives further insight into what this activity might be through a description of his own intervention in combat, which seems to take on some of the attributes of centurion actions due to the number of casualties among that rank. Caesar indicates that his normal command responsibilities were unable to function when he states that he was not able to send help to the struggling legion. (Caes. *B Gall.* 2.25.1). His actions are consequently the last resort of the commander, taking on some duties of the lower ranks in order to restore the battle. While Caesar mentions that the men were restored in hope due to the presence of their commander, unlike his personal intervention at Alesia, where his red cloak was enough to inspire the men, the details of his intervention are much more synchronous with the actions of the lower ranks (Caes. *B Gall.* 7.88.1).[28] With the casualties so high among the centurions and the absence of effective command at this level Caesar advanced to the front of battle as follows:

> ... *scuto ab novissimis uni militi detracto, quod ipse eo sine scuto venerat, in primam aciem processit centurionibusque nominatim appellatis reliquos cohortatus milites signa inferre et manipulos laxare iussit, quo facilius gladiis uti possent.*

> ... with a shield taken from one of the last soldiers in line, because he himself had come without any shield, he [Caesar] proceeded to the front of the line. He called out the centurions by name; he encouraged the remaining soldiers and ordered that the standards be advanced and the maniples relaxed so that more easily was it possible to use swords.
>
> CAES. *B Gall.* 2.25.2–3

The idea of what a centurion might have been expected to do at this point in the battle seems implicit in this description.[29] A centurion advanced to

---

28   Note that in the Alesia passage emphasis is placed on the sight of Caesar's cloak, and his arrival with reinforcements rather than any personal action. At the crucial moment of the great confrontation with Vercingetorix, it is not surprising to find Caesar emphasising his commander's responsibilities, whereas at the Sabis River, where the emphasis is on surprise, he focuses on his behaviour under pressure and his ability to adapt.

29   It is less clear what is expected of the *optiones*, and how this might differ from centurion

CAESAR'S EXEMPLA AND THE ROLE OF CENTURIONS IN BATTLE   45

the front at this time and stabilises the men by supporting other centurions, exhorting them, and getting the *milities* to form up properly and in fighting order, all of which occurs in close proximity to the enemy and with a great deal of risk involved. Caesar specifically mentions the taking up of a shield, an indication that he was going into a form of danger that, as commander, he was not expecting until now and is demonstrative of how he was switching roles. Importantly, Caesar's choice to relate the shield incident may be indicative that this role was not so much about the offensive capabilities, as he makes no mention of his sword, but more of a bolstering action while the fighting capability of the legion was restored. The details are important, as the nature of Caesar's self-promotion, and the possibility he was serving as a centurion analogue provides some insight into what was might have been expected.

The concepts inherent in Caesar's self-promotion are supported by the other major instance of battle in which Baculus is involved, where further clarification of how and why a centurion intervenes are given (Caes. *B Gall.* 6.36–6.42). In the following passage, Caesar describes how Baculus performed in preventing an enemy break in to a camp, again illustrating how centurions should behave when a legion was at the point of breaking:

> ... *videt imminere hostes atque in summo rem esse discrimine; capit arma a proximis atque in porta consistit. consequuntur hunc centuriones eius cohortis quae in statione erat; paulisper una proelium sustinent ... hoc spatio interposito reliqui sese confirmant tantum, ut in munitionibus consistere audeant speciemque defensorum praebeant.*

> He [Baculus] saw that the enemy were at hand and that the matter was at a severe crisis point; he took up arms from those closest and stood at the gates. Centurions of the cohort on guard followed him as a group; little by little they held up the attack ... with this break made the rest strengthened themselves so much that they dared to stand among the fortifications and to offer the appearance of defences.

> CAES. *B Gall.* 6.38.2–5

Baculus certainly showed extraordinary courage as this situation occurred after he rose from his sickbed (Caes. *B Gall.* 6.38.1–2). In spite of this exceptional status, an examination of his behaviour provides important clues to a centurion's

---

activity, as Caesar does not mention them in his accounts. For *optiones* see Cowan (2007) 5.

duty at this time. Baculus, in response to the fear and trepidation throughout the camp came out from his bed. He observed that the enemy was threatening and that the matter was at a crisis point, similar to Caesar's observation in the Sabis river account. It is at this point that Baculus took up arms and moved to the front gate. Importantly, the other centurions followed him, the use of *consequuntur* suggesting that what he initiated was a group activity. While Baculus is exceptional in his personal action, his deeds were also part of a desire to perform his duty alongside his fellow centurions, just as he tried to do at the Sabis River. The consistency between the two battles, and in the behaviour of Baculus and Caesar, suggests that behind the unusual actions of the exemplar is the same set of expectations regarding centurion duties—that they advanced to hold the unit together when a legion is in trouble.

Further details that clarify the purpose of such activity are evident in the details of Baculus' actions in this second passage. Baculus received many wounds, acting out of a sense of obligation to his duty to hold position and give the men time to recover (Caes. *B Gall.* 6.38.4). More importantly, Caesar specifically states that the actions of Baculus were successful in holding up the enemy assault, and that due to the time given, the legion was able to restore itself. As with the Sabis River account, while other men from the cohort on guard may have stood with the centurions, Caesar chooses to describe the action only in terms of centurions, so that in both cases the action is defined in terms of their responsibility. While the specific instances described are unusual, the underlying behaviour is based on a consistent set of rules in each account. The expectation was that centurions moved forward as a group, to hold the line together while the legion either corrected itself or was restored by their activity.

Defining Behaviour through Literary Context

In order to understand the relationship between the actions of the centurion *exempla* and the historical expectations of the rank, it is critical that the literary objectives of each passage be examined to determine the context within which Caesar utilises these figures, and the message that he intends to portray through their use. Examination of the literary objectives and context in which the figures appear shows that in some cases the behaviour described was not actually heroic action, but reckless, selfish or irresponsible. When the literary context is considered a more precise definition of the responsibilities of the rank may be determined.

The best example of a literary context that affects the interpretation of centurion behaviour is a contrast of the centurions Lucius Vorenus and Titus Pullo with the legates Q. Titurius Sabinus and L. Aurunculeius Cotta, all of whom appear in Book Five. The association between episodes provides important context for understanding how the centurions' activity should be understood and that exemplary combat in front of the other troops was not part of a centurion's duties. Vorenus and Pullo appear in the defence of a camp, where they advanced beyond the fortifications, engaged in close combat with the enemy then retired to safety. (Caes. *B Gall.* 5.44.1–14). While the Vorenus and Pullo episode has been viewed as a form of *aristeia*, with the two centurions performing great deeds and inspiring their compatriots, there is evidence that the incident was not designed to be completely positive.[30] An examination of this instance of ostensibly heroic combat indicates that when examined as part of the literary schema, it is apparent that the exemplary combat of the two centurions was not part of their duties, and may even be proscribed by the context it appears in.

The defeat of Sabinus and Cotta and the incident with Vorenus and Pullo have numerous parallels that shape an understanding of the two battles. As has been noted by scholars such as Brown, the two centurions are paired with the commanders in a manner designed to evoke contrasts between the episodes and the behaviour of the combatants.[31] This is evident in the basic sequence of actions performed by the requisite pairs. Both were involved in the defence of a camp, and both pairs left the camp only to find themselves in difficulty (Caes. *B Gall.* 5.31.3–5.37.7, 5.44.4–14). In the case of Sabinus and Cotta, they failed to support each other, Sabinus giving himself up to the enemy with the eventual result that both legates and their command fell (Caes. *B Gall.* 5.36.1–5.37.3). Vorenus and Pullo present a contrast, as they provided mutual support to each other, and were able to return to camp safely (Caes. *B Gall.* 5.44.13–14). Consequently, there is an association of the two accounts at a structural level, where the similarity of initiated action, followed

---

30 See Welch (1998) 97 for the idea of an *aristeia*. Welch notes that the two centurions are inspirational, and certainly their courage and survival are to be admired. This does not mean that the initial behaviour in advancing beyond the defences is entirely sanctioned, and in view of the parallel to the Sabinus and Cotta episode, it seems problematic at the least. Goldsworthy (1996) 269–270 and 179 has trouble interpreting the episode and notes that the actions of Vorenus and Pullo are 'for no apparent purpose'.

31 See Brown (2004) 307. Brown's interpretation of centurions' contest is that it is to show the expiation of Sabinus episode with *virtus militum*, with Caesar in a divine beneficence role. Rambaud (1966) 231 and (1974) 40–41 states that Vorenus and Pullo are an antithesis to Sabinus and Cotta, and Rasmussen (1963) 27–29 regards the speech of Pullo as a specific reminder of Sabinus' words.

by different activity and contrasting fates, is not only critical of the defeated commanders but holds important implications for the roles of centurions.

The structural similarities suggest that the centurions should not have advanced into single combat in the manner they did, and that exemplary displays were not their normal duty in battle. Caesar was highly critical of the advance beyond the fortifications in the Sabinus and Cotta affair, as the commanders did not follow his instructions. Caesar's opinion of the better course of action is summarised through the speech of Cotta:

> *L. Aurunculeius compluresque tribuni militum et primorum ordinum centu-riones nihil temere agendum neque ex hibernis iniussu Caesaris disceden-dum existimabant; quantasvis, magnas etiam copias Germanorum susti-neri posse munitis hibernis docebant;*

> Lucius Aurunculeius and many of the tribunes of the soldiers and centurions of the first rank thought nothing must be done rashly and they must not leave winter quarters unordered by Caesar. They explained that in winter quarters it was possible to hold off the forces of the Germani no matter how great they were.
>
> CAES. *B Gall.* 5.28.3–4

As Caesar states through Cotta, the correct course of action was to stay within the fortifications and defend the camp, so, prior to the Vorenus and Pullo episode, Caesar provides an important qualification on actions that involved leaving the safety of a camp. This is particularly evident as, like the commanders, the two centurions quickly found themselves in a great deal of trouble once they did advance beyond the fortifications (Caes. *B Gall.* 5.32.1–2). The episode was not designed to be a positive reflection on all the centurions' actions, and there is an implication that in leaving the safety of the camp, they abandoned established orders or doctrine in the face of the enemy. The association of the two episodes in this regard does not support the idea that heroic individual action by centurions normally occurred in front of the rest of the men. Instead there is a strong implication from the previous episode that such an advance was reckless and dangerous. Furthermore, the motivation behind the advance of the two centurions is not unequivocally positive, and there is a strong suggestion that their actions derived from the wrong attitude towards competition. A purely positive interpretation of their enthusiasm is tempting, however the two centurions were rivals and, like Sabinus and Cotta, were mentioned in the context of a dispute prior to the episode (Caes. *B Gall.* 5.44.1–3). While such rivalry might seem healthy, there is nothing to suggest it

CAESAR'S EXEMPLA AND THE ROLE OF CENTURIONS IN BATTLE   49

was friendly, and in the context of the earlier episode it may even have been as bitter as the arguments between Sabinus and Cotta. In the case of Sabinus and Cotta, the council they were involved in nearly came to blows; such was the hostility and intensity of the dispute, so the association of the episodes appears to encourage a negative interpretation of such rivalries and disputes (Caes. *B Gall.* 5.30.1–5.31.2). Caesar even recognises the hostilities inherent in the rivalry, as Vorenus and Pullo are described as *inimicus*, having ongoing disputes with each other regarding seniority (Caes. *B Gall.* 5.44.9).[32] This rivalry suggests that the initial advance beyond the fortifications to settle a dispute was, like the actions of Sabinus and Cotta, not to be interpreted in a completely positive manner.

It should also be noted that when Vorenus and Pullo left the fortifications, they did so for highly selfish reasons, which represents yet another reason their activity should be questioned. This is evident in the speech of Pullo:

> ... *Pullo, cum acerrime ad munitiones pugnaretur, 'quid dubitas' inquit 'Vorene? aut quem locum tuae probandae virtutis exspectas? hic dies de nostris controversiis iudicabit.' Haec cum dixisset, procedit extra munitiones quaque hostium pars confertissima est visa, inrumpit. ne Vorenus quidem sese tum vallo continet, sed omnium veritus existimationem subsequitur.*

> While it was being fought most bitterly by the fortifications Pullo said 'why do you hesitate Vorenus? What place for showing your courage do you await? This day will be the judge of our disputes.' When he said these things he proceeded beyond the fortifications. He saw which part of the enemy was most densely packed and charged in. Nor indeed did Vorenus hold himself on the ramparts, but fearing the judgement of all he followed closely.

> CAES. *B Gall.* 5.44.3–6

The two men proceeded beyond the fortifications to decide who was better suited for promotion, not for the good of the defence, as at this time the fighting was fierce but there is no indication that the defences might fall, or that the legion was in trouble.[33] In fact, other centurions had been taunting

---

32   The use of this word as Vorenus rushes to the aid of Pullo seems deliberately designed to evoke a contrast with Sabinus and Cotta, where Sabinus abandons his rival in order to surrender.

33   Caesar describes the fighting as *acerrime* in Caes. *B Gall.* 5.44.3, but there is no indication

the enemy and acting in a confident manner, so the idea the two centurions had to engage in exemplary behaviour to inspire confidence is not evident in the account (Caes. *B Gall.* 5.43.6–7). While Caesar might have admired their courage, it is important to note that such activity was fundamentally selfish and unnecessary, so the idea that this exemplary conduct should be unequivocally admired is not suggested by the manner in which the incident is presented.

While the two centurions returned to praise, this was largely for the redemptive behaviour shown through combat. It is only through mutual support that the centurions were able to recover and return to the fortifications, a marked contrast to how Sabinus abandoned Roman values and surrendered in the earlier massacre. Caesar's closing remarks on the Vorenus and Pullo episode are important in this regard:

> *huic rursus circumvento subsidium fert Pullo, atque ambo incolumes compluribus interfectis summa cum laude intra munitiones se recipiunt. sic fortuna in contentione et certamine utrumque versavit, ut alter alteri inimicus auxilio salutique esset neque diiudicari posset, uter utri virtute anteferendus videretur.*

> Pullo in turn bore back help to him [Vorenus] who was surrounded, and having killed a few both men retreated beyond the fortifications unharmed and to the highest of praise. So fortune in competition and in contest changed hands, so that that one man as an enemy was help and safety to the other, and it was not possible to decide who was before the other in courage.
>
> CAES. *B Gall.* 5.44.13–14

The emphasis here is on the idea that through co-operation and an abandonment of the attack, the two centurions returned to the ranks, and Caesar appears to be praising the men, not for advancing beyond the lines, but for helping each other under difficult circumstances.[34] He even specifically makes

---

the defence was about to collapse. Note also the emphasis on selfishness in the earlier account of Sabinus and Cotta, such as the soldiers' concern for their baggage at *B Gall.* 5.31.4, 5.33.6. This is overtly contrasted with the overall lack of such motivation in the defenders of Cicero's camp at *B Gall.* 5.43.4–5.43.5. An examination of self-interest is a feature of both battle narratives and such initial motivation in Vorenus and Pullo dangerously associates them with the earlier massacre.

34 This approach is different to Gilliver (2007) 137 who states that centurions led from the front. Sabin (2000) 11 also states that centurions continually fought in the front rank. This

CAESAR'S EXEMPLA AND THE ROLE OF CENTURIONS IN BATTLE

the point that neither won the initial dispute as both showed equal valour. The basic principles of co-operation and mutual support at a crisis were more important than glory hunting, as the emphasis is on what was required to return safely to the ranks.[35] Through an understanding of the literary context of this passage, it is possible to determine that, while this is an exceptional episode, it adheres to a basic principle Caesar established regarding centurion behaviour; that there were strict limitations on when their duties involved personal combat or aggressive singular action.

Consideration of Caesar's overall structure in his battle narratives also reinforces the idea that the entry of centurions into combat was unusual as the use of unequivocally positive centurion *exempla* is specific to situations where a legion was in trouble. Caesar's manner of describing generally successful action does not involve the use of centurions when the legion was performing as expected.[36] For instance, in the battles against the Helvetii and Germani in Book One, Caesar described instances of intense, but generally successful fighting as follows:

> *Ita ancipiti proelio diu atque acriter pugnatum est ... nam hoc toto proelio, cum ab hora septima ad vesperum pugnatum sit, aversum hostem videre nemo potuit.*
>
> CAES. *B Gall.* 1.26.1–3

---

instance from the *Bellum Gallicum* would indicate otherwise, as centurions only seem to appear in combat for close fought encounters. The number of centurions in a cohort is key here as there are only six per cohort and as fighters centurions would have had limited battlefield functionality in normal combat circumstances. In identifying faltering lines and shoring them up, centurions could play a more precise role in combat with such numbers, especially with the help of other experienced soldiers. The approach of this chapter also challenges Marshall based theories that propose centurions were the main instigators of aggressive action. See Goldsworthy (1996) 219, 222 who cites Marshall. Wheeler (1998) 648 challenges the Marshall approach.

35 See also Josephus *BJ* 6.81–90. In this passage a centurion charges to rally men and dies, something that leads to a rout. While Josephus was not a contemporary of Caesar and describes events from AD70, the continuity of such a portrayal is interesting for later studies of centurion roles.

36 For instance, there are no centurion casualties mentioned in the massacre of the Usipetes and Tencteri at *B Gall.* 4.14.1–4.15.5. Similarly, in the first invasion of Britain, Caesar describes difficulties but not the idea of a legion about to collapse. See *B Gall.* 4.24.2–4, 4.26.1–4. This is not to say that the centurions were not leading the men or near the front line, at such times, only that they were not expected to be closely engaged, and not expected to suffer unusual casualties.

So it was fought in a battle wavering for a long time ... in this whole battle, although it was fought from the seventh *hora* to evening, no one could see the back of an enemy.

Relictis pilis comminus gladiis pugnatum est. at Germani celeriter ex consuetudine sua phalange facta impetus gladiorum exceperunt. reperti sunt complures nostri, qui in phalangem insilirent et scuta manibus revellerent et desuper vulnerarent.

With *pila* dropped, it was fought hand to hand with swords. The Germani made a phalanx according to their own custom and swiftly took up the attack of swords. Many of our soldiers were found who would jump onto the phalanx, tearing shields from hands and wounding from above.

CAES. *B Gall.* 1.52.4–6

Both instances describe combat that was hard fought, either through the duration of the battle or the intensity of the fighting. Nevertheless in both cases the action is not described as in doubt, and the legions involved were not near breaking point. In both cases Caesar uses descriptions of the enemy, or the general *milites*, to support his interpretation of these hard fought but ultimately successful actions, and centurions are not featured in the combat anecdotes. This suggests that within the overall literary structure of the battle narratives, the same set of rules applied regarding the appearance of the centurions, and that there was a background set of historical values that guided their appearance.[37] Caesar does not use them as anecdotes in successful encounters as centurions were generally only involved in combat when the situation was in grave doubt and the order of the legion was under threat.

---

37 There is one possible exception to this in the *Bellum Civile* at the battle of Pharsalus, where Crastinus appears to lead an attack. However it should be noted that his rank at the time is unclear, as he was a first spear centurion in the previous year. He is also mentioned as a casualty separately (*etiam*) to the centurions, further suggesting he may not be of that rank. Also note that in structuring the overall context of the battle, Caesar may have regarded the whole event as a final conclusive encounter, and the mention of centurion casualties in this account may be included for its symbolic relevance to the context of the whole battle as a last ditch confrontation. See Caes. *B Civ.* 3.91.1 for Crastinus and Caes. *B Civ.* 3.99.1–3.99.2 for the casualties. See also Caes. *B Civ.* 1.46, 1.80 for further examples. See also Scaeva at Caes. *B Civ.* 3.53 and Fabius at Caes. *B Civ.* 2.35. Note how instances where advancing to attack leads to a centurion's death.

CAESAR'S EXEMPLA AND THE ROLE OF CENTURIONS IN BATTLE   53

Throughout the work centurion casualties that would indicate their participation in high risk situations are only mentioned when a unit breaks, or comes close to collapse or annihilation. For example in Book Six a group of centurions are described as holding their men together in order to achieve safety, something that comes at the cost of their own lives. (Caes. *B Gall.* 6.40.7–8).[38] Caesar even attributes such action as directly responsible for the survival of the other soldiers, indicating the direct link to the safety of the unit (Caes. *B Gall.* 6.40.8). This behaviour is consistent with Caesar's general comments, as while in the *Bellum Civile* he states that soldiers must be cultivated with fighting zeal, he also states in the *Bellum Gallicum* that such spirits must be moderated, so that the *virtus* of centurions was clearly a more complex affair than directed aggression.[39] Caesar's presentation of centurions overall through the *Bellum Gallicum*, and the values he ascribes to them matches the expectations he ascribes to individual *exempla* in combat.[40] The use of centurions

---

38   See also Caes. *B Gall.* 7.12.4 where a group of centurions hold a gate allowing their men to retire safely. See also Caes. *B Gall.* 5.35. As these instances show, when centurions are mentioned it is always in regard to a defensive or holding position.

39   Contrast the following from Caes. *B Civ.* 3.92.4–5 "... *est quaedam animi incitatio atque alacritas naturaliter innata omnibus, quae studio pugnae incenditur. hanc non reprimere, sed augere imperatores debent ...*" with Caes. *B Gall.* 7.52.4 "*non minus se in milite modestiam et continentiam quam virtutem atque animi magnitudinem*" The statement from the *B Gall.* provides qualification to the more open statement of the *B Civ.*, and while it is most likely the result of Caesar wishing to cover himself for the loss of control among the men at Gergovia, it still raises the possibility that the centurions were responsible for ensuring that '*modestiam et continentiam*' were maintained. This approach qualifies Lendon's definition of centurion qualities, as Lendon regards their role as more aggressive by the time of Caesar. See Lendon (2005) 218. For a contrary view see Brice (2011) 36. The approach of this chapter is consistent with the observation of Polybius that experience, more than aggression, was an important quality in centurions, and that there is a continuation of responsibility through to the late republican period, where centurions were foremost expected to hold position and control the men. See Polyb. 6.24.

40   The qualifications regarding centurion casualties are supported by other contemporary sources where centurion casualties or combat are mentioned. See [Caes.] *B Alex.* 43, although note how other casualties are given in this instance. See also Livy 27.12.16, 35.5.14. In the *Bellum Hispaniense* two centurions cross a river to restore a failing unit by entering combat, and it is specifically mentioned how one fought to hold back the enemy line ([Caes.] *B Hisp.* 23.3.3). While centurions do receive mention as leading attacks, it is interesting to note these instances can include the qualification that they grab standards rather than engage in combat. See Livy 10.36.10 where a centurion seizes the standards at critical moment rather than fight, and 25.14.71 for Titus Pedantius who takes a standard into an enemy camp and men follow the standard. See also 26.5.12.3 for Quintus

only under particular circumstances suggests that Caesar's approach is based on consistent behavioural expectations, and that centurions were to engage in combat only when their unit was in danger of collapse.

Gergovia and Centurion *exempla*

The interplay of literary objective, *exempla* and battlefield responsibilities can best be illustrated through an examination of Caesar's account of the battle of Gergovia. The picture revealed is consistent with others in the *Bellum Gallicum* and demonstrates how the use of *exempla* in Caesar can be an important tool in reconstructing behaviour of the Roman military in combat. Caesar makes use of both general combat anecdotes and *exempla* to summarise the situation at various points in the battle, and to reinforce the interpretation provided through a speech to the men after the battle:

> *Postero die Caesar contione advocata temeritatem militum cupiditatemque reprehendit, quod sibi ipsi iudicavissent, quo procedendum aut quid agendum videretur, neque signo recipiendi dato constitissent neque a tribunis militum legatisque retineri potuissent ... quantopere eorum animi magnitudinem admiraretur, quos non castrorum munitiones, non altitudo montis, non murus oppidi tardare potuisset, tantopere licentiam arrogantiamque reprehendere quod plus se quam imperatorem de victoria atque exitu rerum sentire existimarent:*

> On the next day Caesar called an assembly and reprehended the desire and temerity of the soldiers because they had judged for themselves where to go and what seemed must be done, and with the signal for holding back given they did not stop nor could they be held back by the tribunes of the soldiers and the legates ... As much as he admired their greatness of spirit, who were not checked by the fortifications of the camp, the height of the mountain nor the wall of the town, so much he rebuked their lawlessness and insolence because they thought to think about victory and the result of affairs more than the commander.
>
> CAES. *B Gall.* 7.52.1–4

Naevius. See also 34.46.12, 39.31.9. The emphasis on centurions seizing standards, while patriotic, suggests that the men normally advanced according to the movement of the standards, rather than by centurion-inspired combat.

CAESAR'S EXEMPLA AND THE ROLE OF CENTURIONS IN BATTLE 55

Caesar's interpretation of the battle is clear in this summation which explains that the men lost control and advanced too far out of daring and desire.[41] This explanation forms the context in which he presents the event, and the narrative is designed to support this interpretation.[42] He uses anecdotes as part of an interpretive repertoire, describing centurion activity in a manner that illustrates what would normally be expected of the position in battle, and how they failed in this regard.

Caesar's narrative technique is apparent in the general anecdotes he uses to illustrate that the situation was out of control. The first occurrence is when the men advanced too quickly, and Caesar states of the speed of the attack '... *ac tanta fuit in capiendis castris celeritas, ut Teutomatus, rex Nitiobrogum, subito in tabernaculo oppressus, ut meridie conquieverat, superiore corporis parte nuda vulnerato equo vix se ex manibus praedantium militum eripere'.* ("... and so great was the speed in taking possession of the camp that Teutomatus, the Nitobrigian King, beset suddenly in his tent as he had taken a midday nap, shirtless on a wounded horse scarcely tore himself from the hands of plundering soldiers," Caes. *B Gall.* 7.46.5). While an event worth telling due to the high rank of the fleeing king, this anecdote raises two ideas that Caesar wishes to convey. The first is that the men attacked so quickly that they were able to catch the enemy leader unawares, demonstrating that the rush was much faster than even an enemy located in a higher position could expect.[43] Moreover, Caesar specifically mentions that the king barely escaped from the hands of plundering soldiers, thus illustrating the idea that the men were seized with a desire for booty and were not under the control of their leaders.[44] The use of *praedantium* is a small, but important qualification as it is illustrative of the concept that the men were out of control, and the anecdote is a vivid reference to Caesar's overall explanation for the battle.

Caesar provides another important anecdote soon afterwards, to emphasise his point about how thoroughly control was lost. As he states:

---

41 See Nolan (2014) 116–123 for an analysis of the battle on which this chapter is based.

42 Kagan (2006) 163–180 accepts this explanation by Caesar; however, she relies heavily on the battle narrative itself to build her case, using information that is highly problematic considering Caesar's facility in the use of anecdotal detail to support his interpretation.

43 While such information helps Caesar to exonerate himself from the defeat, by explaining how quickly control was lost, it is also an example of how such anecdotes support his overall interpretation of the battle summarised at *B Gall.* 7.52.

44 Sall. *Cat.* 11 provides one aristocrat's written view on such avarice.

matres familiae de muro vestem argentumque iactabant et pectore nudo prominentes passis manibus obtestabantur Romanos, ut sibi parcerent neu, sicut Avarici fecissent, ne a mulieribus quidem atque infanti bus abstinerent; nonnullae de muro per manus demissae sese militibus tradebant.

Mothers of the household threw from the wall clothing and silver, and hanging over with naked breast and outspread hands beseeched the Romans to spare their families, unlike Avaricum, where they [the Romans] did not hold back from the women or even from the infants. Some descended from the walls by hand and were giving themselves over to the soldiers.

CAES. B Gall. 7.47.5–7

This colourful depiction of the activity on the walls allows Caesar to show how utterly chaotic the situation had become, and he describes in detail aspects of the assault that represent the powerful temptations that overwhelmed the normal discipline of the *milites*. Details of women removing clothing and even handing themselves over to the soldiers evoke the idea that some men were succumbing to their lusts, and may have lost all ability to perform in a soldierly fashion. Following on from the description of the Nitobrigian king, these details illustrate how Caesar provides such incidental detail to reflect the state of the legions.[45] This information supports his overall contention regarding the progress of the battle, and the anecdotes fall into a stylistic pattern in which individuals and vignettes served a representative role in summarising battlefield conditions and interpreting conditions for the audience.

As with other battles, the centurions serve a representative role. However, there is an important qualification to make as Caesar is not using the anecdotes to attribute blame to the individual centurions, but to illustrate the general circumstances of battle. A combination of terrain and the enthusiasm of the soldiery led to the loss of control, and it is important to note that Caesar clearly exonerates the tribunes and legates who attempted to follow orders. (*B Gall.* 7.47.2–3). In contrast, the centurions are included with the general *milites* and Caesar is assigning blame to all the lower ranks rather than a few individuals. The action took place across a frontage three legions wide so it also seems unlikely that Caesar would have attributed the loss of control across that entire

---

45   Note how the enemy later gain the upper hand and Caesar again uses the women as an anecdote in order to illustrate how dramatically the situation has changed. He describes how they encouraged their men, capturing the sudden reversal of conditions and the mental state of the defenders. See 7.48.3–4.

CAESAR'S EXEMPLA AND THE ROLE OF CENTURIONS IN BATTLE 57

front to the actions of a few individuals of lesser rank.[46] Individual centurions should instead be viewed as representative examples of the attitudes and motivations of the men of their rank. Based on the size of the forces, and the general loss of control that Caesar wishes to communicate, it seems most likely that centurions described are used to support the general contentions regarding loss of control, rather than to exist as figures of individual blame. One of these representative figures is Lucius Fabius, a centurion of the Eighth Legion, who appears in the account to exemplify the loss of control of the army. Caesar states of this individual:

> *L. Fabius centurio legionis VIII, quem inter suos eo die dixisse constabat excitari se Avaricensibus praemiis neque commissurum, ut prius quisquam murum ascenderet, tres suos nactus manipulares atque ab his sublevatus murum ascendit, eos ipse rursus singulos exceptans in murum extulit.*

> It was established among his men that Lucius Fabius, a centurion of the Eighth Legion, had said on that day that he was excited by the rewards of Avaricum and would bring it about that no one would ascend the wall before him. He obtained three of his own company who lifted him up. One at a time in return he then bore them onto the wall.
>
> CAES. *B Gall.* 7.47.7

The reference to Avaricum and booty ties the actions of Lucius to the plundering motives of the men, mentioned earlier through the anecdote of the Nitobrigian king.[47] Lucius Fabius and the men are thus associated through traits that exemplified the reasons for the loss of control.[48] The behaviour of

---

46   Kagan (2006) 167 states that Lucius Fabius, one of the centurions involved, is blamed. According to Caesar, Lucius is certainly culpable, however, based on the number of men involved it is more likely he serves to represent the problem rather than take the entire blame for the defeat.

47   In her analysis of Gergovia, Kagan (2006) 167 does not address the representative role of the Nitobrigian king, demonstrating how such anecdotes can be overlooked in an analysis of Caesar's battle narratives. Kagan takes the incident with the king literally and suggests it may even be directly responsible for the loss of control. While this is possible, the illustrative purpose seems more suited to Caesar's likely aim of exonerating himself in this account.

48   How such a view of Lucius is reconciled with the awarding of the *corona muralis* requires further study, as centurions could be the first to ascend the wall. One example in Livy involves a dispute at New Carthage over who ascended first, a marine or centurion (Livy

Lucius is representative in illustrating why the men sought to storm the town, providing an example of the general overconfidence and desire for booty. He is demonstrative of the general conditions of battle, in the same manner that Caesar uses centurions in other accounts.

In his representative role Lucius also provides important information about centurions and their role, as criticism is enunciated in terms of negative traits and bad behaviour. The over eagerness of the man is evident in his boastfulness prior to the battle, a problematic characteristic that is evident in Gallic and Germanic adversaries before they receive their comeuppance.[49] He also displays a selfish attitude, being motivated by self-interest similar to the way Vorenus and Pullo initially behaved. His action in climbing up onto the wall is not to be regarded as heroic or particularly courageous due to the manner in which he is characterised. Centurions had a specific responsibility to control the aggression of the men, and to ensure they obeyed the commands passed through via the standards and senior officers rather than charging off seeking glory, something Lucius fails to do in this account.

The boundaries on centurion behaviour in combat are further addressed through the next *exemplum*, the centurion Marcus Petronius, who is used to contrast good conduct with that of the overzealous Lucius Fabius. The actions of Marcus support the idea of limited personal activity in combat by centurions, and that they should only advance into the frontline of combat in order to restore their unit to fighting capability.[50] As Caesar states of Marcus Petronius:

> ... *L. Fabius centurio quique una murum ascenderant, circumventi atque interfecti de muro praecipitabantur. M. Petronius, eiusdem legionis centurio, cum portas excidere conatus esset, a multitudine oppressus ac sibi*

26.48.6). It seems apparent that centurions could be first on the wall, but perhaps there was a hierarchy of responsibilities that enabled Caesar to show such an action as reckless in the circumstances at Gergovia.

49 As Rosenstein (2007) 141 notes, moral character is what counted in heat of battle according to the sources, supporting the idea that any inferences regarding character in a battle narrative should be given particular attention. See also Caes. *B Gall.* 1.32.4–1.34.5 for the arrogance of Ariovistus, and 4.7.4–5 for the Usipetes and Tencteri. See also 2.30.4 and note how such contempt is quickly overturned in the text.

50 Kraus (2010) 57, n. 55 cites Cato *Orig.* HRR F83 and states that Marcus is the spiritual descendant of Cato's military tribune who, with a small group, sustains the charge of enemy allowing the group to escape. Kraus regards the role as one of "sacrificial glory." Nevertheless the speech of Marcus includes an admission of error which is critical for a definition of responsibilities.

CAESAR'S EXEMPLA AND THE ROLE OF CENTURIONS IN BATTLE   59

desperans multis iam vulneribus acceptis, manipularibus suis, qui illum erant secuti 'quoniam' inquit 'me una vobiscum servare non possum, vestrae quidem certe vitae prospiciam, quos cupiditate gloriae adductus in periculum deduxi. vos data facultate vobis consulite.' simul in medios hostes inrupit duobusque interfectis reliquos a porta paulum submovit. Conantibus auxiliari suis 'frustra' inquit 'meae vitae subvenire conamini, quem iam sanguis viresque deficiunt. proinde abite, dum est facultas, vosque ad legionem recipite.' ita pugnans post paulo concidit ac suis saluti fuit.

Lucius Fabius the centurion and those who had ascended the wall were surrounded, killed and thrown from the wall. Marcus Petronius a centurion of the same legion, when he had tried to cut down a gate was beset by a multitude. Despairing for himself with many wounds received he said to the men of his maniple, who had followed him, 'seeing that I am not able to save myself along with you, at any rate let me see for your lives, whom, having been drawn by a desire for glory I led into danger. With opportunity given look after yourselves.' At the same time he burst into the middle of the enemy and with two of them killed he moved the rest a little from the gate. With his men trying to bring aid he said 'in vain do you try to come to the aid of my life, which is already failed by blood and strength. Therefore go, while there is an opportunity, fall back to the legion.' Thus fighting before the gate for a little while he died and saved his men.

CAES. *B Gall.* 7.50.3–6

As indicated in this passage, Caesar gives the death of Lucius, the exemplar of bad behaviour, close proximity to Marcus in the text. The close pairing of the two centurions is important, as it suggests Caesar is drawing a deliberate contrast between their actions.[51] The role of Lucius as a bad *exemplum* is complete, and Caesar indicates his opinion of the figure through a death absent of any redemptive qualities. By contrast Marcus is given direct speech in which he enunciates the codes of expected behaviour and acknowledges the error of his ways, making his presence important for an examination of how a centurion should behave.[52] The message implicit in this pairing is an important contrast that allows a close understanding of expected conduct.

---

51 See Kraus (2010) 55 on the literary aspect of doubling.
52 See Kraus (2010) 57. Kraus addresses the question of why Marcus, not Lucius, is given the speech. Based on Caesar's contrast of good and bad behaviour, Lucius may not be granted direct speech as he died and failed to save anyone, but the success of Marcus'

The picture that emerges is consistent with the expectations regarding centurions as they appear elsewhere in the *Bellum Gallicum*; that centurions were expected to control the men rather than lead attacks. This is evident in the glory-hunting behaviour that Marcus repents of, stating '*cupiditate gloriae adductus in periculum deduxi*' (Caes. *B Gall.* 7.50.4–5). His selfishness and failure to his rank are also evident as he was personally attempting to break down a gate.[53] His recognition of this suggests that when there was a breakdown in responsibility at a centurion level these men lost control along with the men. They could be found acting as individuals, seeking out glory and fighting in the front ranks rather than acting as part of the disciplinary structure of the unit. Through the structure of Marcus' speech and actions, once he realises his error, the correct principles of centurion conduct are elaborated, and these are also consistent with how Caesar presents centurion roles throughout the work. Already wounded, Marcus engaged in extreme behaviour in this instance, as he burst in among the enemy. The reason he did so is nevertheless consistent with those of the positive figure of Baculus. Marcus held the position and allowed his men to recover, supporting the idea that when a unit was in trouble, it was only then that a centurion should enter the fighting to allow the men time for correction.[54] In this case, Marcus' speech is an important enunciation of this responsibility, stating "*vos data facultate vobis consulite*," that he was giving them an opportunity to save themselves. (Caes. *B Gall.* 7.50.5) He engaged in combat so the unit could restore itself and Caesar even closes the vignette by stating "*post paulo concidit ac suis saluti fuit*," directly linking the centurion's sacrifice with the safety of the soldiery (Caes. *B Gall.* 7.50.6). The successful retreat and return to fighting order shows that Marcus' sacrifice was successful, and is an implicit affirmation of his actions (Caes. *B Gall.* 7.51.1–4).[55] The exemplary actions of Marcus are consistent with the expectation that these

---

objective in ensuring the escape of his men is inherent in the survival of the legions. The circumstances of Marcus' death are much more appropriate for an enunciation of correct behaviour through direct speech.

53   The implicit condemnation of this action suggests that centurions, while close enough to react to problems at the front line, were not fighting in the front rank normally, as they would find it difficult to implement commands if usually engaged in hand to hand combat or other intense physical activity.

54   Note also the similarity to Vorenus and Pullo, in that an initial transgression is redeemed through more appropriately motivated combat.

55   Note that the retreating legions are able to reform once they reach the plain, suggesting that while Caesar states they were cast from the hill, this was in some sort of order rather than an outright rout.

men entered combat only under specific circumstances, to achieve the objective of restoring order to a legion and ensuring the safety of the men.

This view of centurion roles is supported not only through the activity of Marcus, but in the general centurion casualty figures of the battle. Caesar states after the death of Marcus that another forty-six centurions were killed in the fighting (Caes. *B Gall.* 7.51.1). While Caesar later gives the overall casualties among the men, the losses among the centurions are mentioned at the time of the retreat, suggesting a close association with the retirement of the legions.[56] It also seems likely that the losses were due to deliberate behaviour, rather than simply a random result, as the legions would have been in some disorder when they reached the walls. *Milites* would have been milling around, receiving women or attempting to break through, and it seems likely that for such a relatively high number of centurion casualties to occur they must have deliberately placed themselves in danger during the chaos once they saw the danger, in a manner similar to how Marcus behaved when he came to his senses. The figure of Marcus may be presented in a highly literary manner with direct speech and a heroic death, however, the manner in which he is closely associated with general casualties suggests that his actions were meant to serve as an example of how a centurion should behave. Marcus was a model that conformed to the expectations of his rank, and provides important information regarding the expectations of his commander once his actions have been placed into their literary context.

Conclusions

The use of the centurion *exempla* in the account of Gergovia, and throughout the *Bellum Gallicum*, demonstrates that these characters served in a representative capacity designed to reflect Caesar's interpretation of events and the messages he wished to communicate about battle. The figures are included in the text for literary purposes which are independent of their contribution to the historical conflict, in particular to represent battlefield circumstances in a summary capacity. They also conform to some consistent standards of behaviour that provide important information regarding activity at the front line of Roman combat. Beneath these exemplary conditions are the expectations of their aristocratic commander whose behavioural standards were consistent across the work. The absence of centurion *exempla* during successful

---

56   See Caes. *B Gall.* 7.51.4 for general casualties.

advances, and the problematic nature of their presence when in front of the men, suggests that these men did not normally engage in personal combat unless absolutely necessary, and were primarily expected to follow orders and keep discipline within the ranks. When a legion came under serious threat, they were expected to advance and hold position, ensuring that the legion had time to correct or extricate itself from trouble. This conclusion, drawn from the structure of Caesar's accounts and his use of *exempla*, has significant implications for discussion about the nature of legion combat in the late republic, as it was perceived by one of Rome's most accomplished commanders.

The Economics of Warfare

∵

CHAPTER 4

Coinage and the Economics of the Athenian Empire[*]

Matthew Trundle

In a country as poor as Greece, war became a principal way for individuals and states to increase their revenue. Homeric heroes acquired property and wealth through conquest.[1] Greeks saw spoils of war as the victor's property (Andoc. 3.11, Xen. *Cyr.* 7.1.44). Herodotus (e.g. 5.77, 8.123, 9.81), Thucydides (e.g. 1.11–13, 2.13, 7.52), Plato (*Phd.* 66c, *Resp.* 372e–374a), and Aristotle (*Pol.* 1.8, 1255b37, 1256b1, 1256b23–27, 1333) each knew the material benefits of a successful campaign. Ancient literary evidence attests, therefore, that plunder produced wealth on a far greater scale than other economic activities. Modern scholarship has agreed.[2] The poverty of many Greek states, and the dearth of food resources especially, made wars of acquisition more common in the fifth and fourth centuries BC.[3] In this period, states, like Athens, became more cohesive and more able to coordinate their resources to steal those of others. This chapter explores the acquisition, redistribution, and circulation of resources, primarily in the form of coinage, which accompanied the rise and dominance of the Athenian Empire in the fifth century BC. It seeks to follow the trail of money and show that Athens was as much an economic as a political imperial society and empire and the important role played by economic and therefore monetary 'flow' in understanding the rise and fall of the Athenian *archê*.

Fifth century Athens reflected the economic necessities of antiquity. The *archê* that the Athenians created in this period sought to control grain resources within the eastern Mediterranean. What follows demonstrates the economic significance of Athenian efforts and, most importantly, the role that

---

[*] I would like to thank David Rosenbloom, now of the University of Maryland, Baltimore County, for the many discussions and stimulating ideas about the Athenian Empire. His influence permeates this essay. I would also like to thank the editor of this volume and the outside reader for their comments and suggestions. Naturally, all the errors are my own.

1 van Wees (2004) 34–43; Rihll (1993) 92–105; Rawlings (2007a) 151–157; Trundle (2010) 227–253; Millett (1993) 177–196.

2 Rihll (1993) 92–105.

3 All dates are BC unless otherwise stated.

© KONINKLIJKE BRILL NV, LEIDEN, 2016 | DOI: 10.1163/9789004284852_005

coinage played in framing the Athenian Empire. The efforts of the Peisistratids in the late sixth century to secure the Hellespont, Chersonesus, and other areas of the northern Aegean further developed Solon's earlier policies to control food supplies within Attica at the start of the sixth century. Economic interests underpinned Athenian imperialism well before the imperial period of the fifth century. During the fifth century, Athenian imperialism feels ever more economic, with all manner of techniques to control resources within Attica and to control food production centers like the Black Sea, Egypt, and later Sicily.[4] Silver, in the form of Athenian coinage, became the mechanism by which such resources might be controlled. Silver coins flowed out of Attica from Laureion, around the empire, the Aegean Basin, and back again, but, as many hoards illustrate, it also flowed out of the empire.[5] Understanding this money 'go-round' is not as simple as it theoretically might appear. The flow of coinage, the fiscal focus of many inscriptions, and Thucydides' clear emphasis on the importance of money illustrates much about the Athenian Empire's economic basis.

Two sets of premises are important in this discussion. The first set connects naval warfare, money, and empire just as Thucydides (1.11) did at the start of his history. Here he links naval power with money and with longer and more aggressive wars. It is no surprise that wealthy regions and states produced the first large navies: Phoenicia, Egypt, Miletos, Samos, Corcyra and Corinth, and then, finally, Aegina and Athens.[6] Thucydides clearly recognized the significance of navies on the one hand and money on the other.[7] Thus, the Athenian naval *archê* rested on the production and redistribution of coined silver. This silver paid for the ships, the oarsmen, and specialist crews that defended the empire. Tribute in coin, after the initial investment from a windfall find at Laureion, became a significant part of a redistributive cycle of resources flowing from the allies, as *phoros* and other mechanisms of financial exploitation, to

---

4 See for example the Phaselis decree, Meiggs and Lewis (1989) 31.6–14, regarding grain supplies and other traded goods to Athens. For further evidence see Thuc. 3.86.4, 115.3–5 and Arist. [*Ath. Pol.*] 51.3. For discussion and a contrary view see Garnsey (1988) 127–128 who observes the absence of official controls of grain supplies.

5 See for discussion Howgego (1995); Kallet-Marx (1993); Kallet (2001); Samons (2000); Samons, (1993) 129–138; Mattingly (1996); Meiggs (1972); Meritt et al. (1939–1953). Most recently, the important contribution of Kallet (2013) 43–60 which argues, I think rightly, for an economic basis from the start of (and even before) the Athenian control of the Delian League.

6 For an overview see van Wees (2004) 199–231; Gabrielsen (2001) 72–98; Gabrielsen (2008) 46–73; Haas (1985) 29–46; Rawlings (2007a) 107.

7 Thucydides' obsession with money has been well noted, see for example Kallet-Marx (1993); Kallet (2001).

COINAGE AND THE ECONOMICS OF THE ATHENIAN EMPIRE

Athens that then financed the ships and men that maintained the empire. Finally, the costs of maintaining the empire were vast. Revenues fell short of expenditure in times of military activity, as they did, for example, in the first years of the Archidamian War. This meant that Athens had no choice but to expand in order to replenish resources for maintaining the empire, which in itself required more investment in empire. Money, therefore, necessitated imperial expansion in a vicious circle of resource redistribution and acquisition.[8]

The second set arises from Thucydides' (1.11.1–2) statement that wars in his own day, as opposed to the past, were greater, not due to lack of manpower, but due to lack of *chrêmata*. *Chrêmata* in earlier sources referred generally to wealth but, by Thucydides' day, *chrêmata* specifically meant coined-money. He suggests that soldiers in previous wars had to spend a good deal of time gathering provisions, farming the surrounding land, and plundering rather than fighting. The role of wealth and its coordination in Greek military engagements feeding, paying, and redistributing resources in the classical period remains central to understanding how Greek warfare changed from small-scale raids and border skirmishes to the great wars of the later fifth and fourth century.[9] Thucydidean *chrêmata* had transformed the way in which many soldiers received remuneration instead of redistributed plunder. Coins substituted for the redistribution of cumbersome slaves, animals, and metals and enabled commanders not only to centralize supply redistribution, but also to attract suppliers to armies more effectively. Thus, coins became the mechanism to centralise the logistics of warfare. The introduction of a widely accepted measure and means of exchange gave military organisation a centralized internal economy that could also engage with the outside world of traders and markets. Commanders captured goods, all of which they could now sell in return for coins. These they could keep for long periods of time, leave at a designated place, or carry with them more easily than wagon-loads of plunder. In the past, as Thucydides said, in a world before coinage, men had to spend their time in the countryside gathering food, now the food came to them. In short, money enabled the efficient centralisation of economic exchange for armies and navies in the field. Hence wars became greater and campaigns longer in the classical period.

---

8   As I have argued elsewhere. See Trundle (2010) 227–253.

9   On this subject in some detail see Trundle (forthcoming). On the culture of early land warfare see Hanson, (1998); Hanson, (1995); Schwartz (2010); van Wees (2004) 26–30; Pritchard (2010a), especially 7–15. On naval raiding prior to the Persian Wars see Haas (1985) 29–46; Tandy (1997); de Souza (1998) 271–294; finally most recently see van Wees (2013), especially 44–75.

As you will notice, both of these sets of premises involve the use of coined-precious metal (silver). Significantly, the Greek word for money was *chrêmata*. The word comes from *chrêma*, meaning a tool, a useful thing, the plural *chrê-mata* therefore means tools or useful things. Literally, *chrêmata* were means to an end. Money facilitated and coordinated wars, especially naval wars, and money bought the resources with which wars might be waged by attracting merchants to a central point, either Athens itself or the military forces, both of which commanded these resources. The purpose of these wars was, ironically, additional resource acquisition. Wars and money both became means to ends in a resource poor environment. Both became a part of the same vicious circle.

As many have pointed out, the Athenian Empire was 'an empire of the owl'.[10] Athenians had economic interests overseas even before the Persian Wars and the Athenian naval *archê*. Aristagoras' attempts to persuade both Sparta and Athens to join the Ionian revolt illustrate this well (Hdt. 5.97). He failed to persuade Cleomenes of Sparta, but the Athenians were deceived into joining the Ionian cause. Herodotus (5.97.3) criticized democracy and how many can be persuaded into a bad decision more easily than one man. Athenian greed for money lay at the root of their decision to follow Aristagoras and aid the Ionians. Soon after Marathon, the Athenians gave Miltiades money (*chrêmata*) and men in hopes he might enrich (*kataploutizô*) them from a place supposedly full of gold—*chrysos* (Hdt. 6.132.1). In 483 Themistocles convinced the assembly to use their new found Laureotic wealth for a fleet of ships that would effectively become a piratical operation in the Aegean. He extorted "large amounts" of money from Andros, Paros, and Carystos after Salamis (Hdt. 8.111).[11] After the defeat of the Persians, the Athenians imposed a tribute (*phoros*) upon the allies within the new Delian League (Thuc. 1.96, Plut. *Arist.*).[12] Even under Aristeides, in the wake of the creation of the Athenian *archê*, Athenians reaped the economic benefits of the league. Aristotle (*Ath. Pol.* 23.5–24.3) paints a vivid picture of a city into which the resources of the Aegean and the allies flowed and from which the increasingly urbanised citizens of the Athenian-*polis* benefited economically in food (*trophê*) and resource redistribution (if not yet coined-money). Several early Athenian-led campaigns focused specifically on resource rich mining regions in the northern Aegean, including those most

---

10 As noted in the title of Samons (2000); see also Blamire (2001) 99–126; Kallet-Marx (1993); Kallet, (2001); Mattingly (1968) 450–485; Meiggs (1972); Meritt et al. (1939–1953); Samons (1993) 129–138.

11 van Wees (2004) 260, n. 32 suggests such demands were "common practice." Samos had demanded 100 Talents from the Siphnians (Hdt. 3.58).

12 Jackson (1969) 12–16.

COINAGE AND THE ECONOMICS OF THE ATHENIAN EMPIRE 69

famously against Eion and Thasos.[13] According to several accounts Pericles ultimately won over the poor through payments of wages (*misthos*) for military and jury service (Plut. *Per.* 9.1, *Cim.* 10.1; see also Theopompus, *FrGH* 115 F89; Cic. *Off.* 2.64; Nep. 5; Plut. *Cim.* 4.i–iii). The expenses of the Archidamian War led the Athenians to raise the amount of tribute significantly c. 425/4.[14] Money (*chrêmata*) had become an essential component of war and empire.

In practical terms, the Athenian naval empire sucked money and resources from the islands and cities of Thrace, the Black Sea, and Asia Minor. The precedents of tribute came from earlier imperial models—the Lydians (Hdt. 1.6.2, 27.1) and naturally the Persians (Hdt. 3.89–96, 6.42.2)—just as coinage itself had come from the Lydians (and had been adopted by both Persians and Greeks). By the 430s almost all the allies, save for Chios and the cities of Lesbos, contributed money rather than ships and men to the Athenian league. The annual amount of money ranged from 460 Talents (Thuc. 1.96.2; Plut. *Arist.* 24.2) to 600 Talents (Thuc. 2.13.1) to 1000 talents (Xen. *An.* 7.1.27) and even 1300 (Plut. *Arist.* 24.2), each figure provides an "on-paper" amount paid by the allies at some point in the fifth century. As noted above, the level of tribute was increased in the 420s to build up exhausted reserves. Decrees of this time illustrate well this increase as much as they also show the meticulous care and painstaking enforcement of the financial contributions of the allies within the empire.[15] By 425, the Great Panathenaea had become a festival associated with the assessment of the tribute.[16] So too the City Dionysia, which displayed the wealth of the city and the gifts of the allies both in the *pompe* and in the theatre. This was an imperial, theatrical, and economic spectacle. Several inscriptions and other evidence attest the bountiful produce provided at these festivals by the allies. This includes a Cow and a Panoply, both economically valuable, that were brought to Athens along with a Phallus for the Dionysia (*IG* 1³ 46; *IG* 1³ 71,

---

13 See Kallet (2013) 43–60 for the most recent discussion of the economic origins of the Athenian alliance in the Aegean, especially relating to Eion, Thasos and Thrace and the mineral resources of these regions.

14 For debates on the dating of these events for the 420s see Mattingly (1996); Samons with Fornara (1991) 98–102; Samons (2000) 329–331; for the 440s, see Walbank (1978), especially Chapter 2; Figueira (1998) 431–465; and, for 414, see Kallet (2001) 205–225 and most recently the discussion in Rhodes (2008) 501–506.

15 See Cleonymus in 426/5 BC (Meiggs and Lewis (1989) 68 = *IG* 1³ 34 = Fornara (1983) 133), Kleinias in 425/4 BC (Meiggs and Lewis (1989) 45 = *IG* 1³ 34 = Fornara (1983) 98) and a decree requiring the allies to use Athenian weights and measures and silver coins (Meiggs and Lewis (1989) 45 = *IG* 1³ 1453). See also Blamire (2001) 99–126, especially 111.

16 Meiggs and Lewis (1989) 2 65.5–9, 29–32; 69.26–33.

ll. 57–58).[17] We know that cakes and bread and other food were also distributed at these festivals to the people (Ath. 3.111b; Poll. 6.75) and, according to Photius (*Lexicon* s.v. *obelias artos*; see also Pollux 6.75), bread described as *obelias artos* or "penny bread" circulated. Aristophanes' *Knights* (51, but see also 715–718, 789, 904–905) parodies food's connections with money in what is very much a *trope* of his comedies: "here's something to nibble, wolf down, savour," says the Paphlagonian "a three-obol piece." Coinage and empire appear symbiotic.[18]

Chrêmata's centrality to naval *archê*—empire—and to war was undeniable by 431. Pericles allayed Athenian fears that the Spartans might take the Delphic funds at the start of the Great Peloponnesian War and use them to finance their own fleets (Thuc. 1.141–144). He also reassured Athenians of their own resources in coinage and precious metals and their own skills as seamen (Thuc. 2.13). Thus, the allies paid 600 talents each year, and the Athenians possessed 6000 coined talents on the acropolis, along with a further 500 talents of un-coined gold and silver, with even more in other sanctuaries.

The war, of course, cost Athens dearly.[19] In the hard times of the 420s, we know of the money-collecting ships (*argyrologoi*) pressing the allies for cash. Thucydides (2.69, 3.19, 4.50.1, 4.75) presents four sailings in 430/29, 428/27, 425/24, and in 424 of such *argyrologoi* during the Archidamian War. The Sausage-Seller and Demos allude to such ships collecting tribute to pay *misthos* to soldiers and rowers in the *Knights* (1065–1066, 1070–1072). Xenophon (*Hell.* 1.1.8, 1.12, 4.8.30, 4.8.35; see also Plut. *Alc.* 35.5–6) described missions to extort money with threats of violence late in the Peloponnesian War. Not only the allies, but Athenians felt the pinch of war, as some two hundred talents were raised by the citizens of Athens in a tax—an *eisphora*—during the financial crisis of the siege of Mytilene as early as 428.[20]

Athenian awareness of financial matters meant they made their wars pay for themselves in the fifth century. Samos had to refund Athens 1200 Talents after its defeat in 439.[21] But Athens also knew their limitations and Blamire suggests

---

17   See also Rhodes and Osborne (2003) 146–149, n. 29 for the Athenian Decree concerning Paros and the Second Athenian League dated to 372.

18   See Rhodes (2006) 41–53; Meritt et al. (1939–1953); McGregor (1987); Meiggs (1972).

19   On finances in the Peloponnesian War especially see Blamire (2001); Kallet-Marx (1993); Kallet (2001); Mattingly (1968) 450–485; Meiggs (1972); Meritt et al. (1939–1953); Samons (1993) 129–138.

20   Thuc. 3.19.1. For discussion see Blamire (2001), at 110 he thinks that Mytilene required additional expenses from the allies.

21   Isoc. 15.111; Diod. 12.28.3; see Meiggs and Lewis 55; *IG* 1³ 363; Fornara 113 for figures of 1400 Talents spent on the war, see also Thuc. 1.117.3.

COINAGE AND THE ECONOMICS OF THE ATHENIAN EMPIRE

the Athenian decision to put aside about 1000 Talents from the funds of the Acropolis after the Peace of Nicias was an illustration of their fears regarding the importance of money in warfare (Thuc. 2.24.1).[22] Finally, and significantly, came the so-called twentieth, or *eikostê*, by which a 5% tax on the use of harbours across the empire replaced tribute in 413 (Thuc. 7.27.4). Kallet sees this as central to her argument that the Athenians moved from a political conception of empire to an economic model, and that the Athenians saw money and its extraction as more important than the need "to gain a stronger hold" of their subjects within the empire.[23] She also attempted to link this *eikostê* to the Athenian imposition of common standards, weights, and measures. In doing so, Kallet, alone of ancient historians, sought to date the standards decree to the 410s and not the more regularly accepted 420s.

Thucydides (7.27) detailed both the reasons for the imposition of the *eikostê*—that "their expenditures (*dapanai*) were not the same as before, but were far greater, in as much as the war was greater"—and the motives the Athenians had for it—"thinking that more *chrêmata* would come into their hands." Most scholars see this *eikostê* as a failure, with tribute reinstated in 410, but it is possible that the tax continued in addition to the tribute.[24] Aristophanes (*Frogs* 363) refers to an *eikostologos* in 405. Certainly, other measures ran parallel to tribute and tithes at moments of financial disaster, as Xenophon (*Hell.* 1.1.8–23) records a specific *dekatê* on grain imports through the Hellespont and Aristophanes jovially highlighted old and outlandish taxation schemes in the *Ecclesiasusae* (823), which seems to lampoon efforts and inflated percentage taxation programmes.

Athenians justified the tribute they took because they did the fighting. Property captured in warfare was justification for ownership in itself. Thus, Pericles could divert 1/6oth of the tribute to beautifying the city by stating that Athenians did the fighting, while the allies contributed "not a horse, nor ship, nor hoplite, but only money, which belongs not to the givers, but to the takers" (Plut. *Per.* 12.2–3). Thucydides (1.99; see also Plut. *Cim.* 11) explains the transformation of the Delian League into Empire as the result of Ionian reluctance to fight. They

---

22   Blamire (2001), especially 109.

23   Kallet (2001) 196. See for discussion also 197–205. On the *eikostê* see also Kallet (1999); Meiggs and Lewis (1989) 349; McGregor (1987) 158.

24   Regarding the majority viewpoint: Xenophon *Hellenica*, 1.3.9, notes that Chalcedon paid the normal tribute when recovered in 409 BC. See also Kallet (2001) 223; Meiggs and Lewis (1989) 438–439; especially Lewis on *IG* 1³ 291; Meiggs (1972) 369. As always Mattingly (1967) 13–15 and Mattingly (1979) 320–321 holds a different position on dating. For the viewpoint that tax continued in addition to the tribute, see Merrit et al. (1939–1953) 148 (vol. 3).

paid tribute rather than contributing ships and men and their excuse was that they did not want to spend time away from home. Thus, they added to Athenian power in this way by funding the Athenian war machine and when they came to revolt the battle-hardened Athenians easily overcame any islands' forces. But this historical process produced an ironic result as many Ionians did fight for the League, but on Athenian ships for Athenian wages in Athenian coin.

It is clear then, that the economic aspect of empire cannot be overstated. Thucydides (1.122.1, 143.5; 2.13.2, 3.46.3) referred to money's importance as a symbol of Athenian power (see Thuc. 1.121.3; 2.13.2; 6.17.7; 7.66.2). The money go round is also a central premise of Aristophanes' *Wasps* (672–679, 684–685, 1099–1101, 1114–1121). In Aristophanes' *Knights* (1365–1367), Demos asserts that he will ensure payment to rowers who return to port. The *Knights* even has triremes that are called *misthophoroi* (554–555), wage earners, a term usually referring to mercenaries.[25] Poor Athenians hoped for years of *misthophoria* or wage-earning from the Sicilian expedition in 415 (Thuc. 6.24.1–4; see 6.46.3–5).

The financial aspects of empire made Athens vulnerable. Maintaining the empire alone cost money, revolts cost more, and the Peloponnesian War much more. The siege of Samos cost 1200 Talents, but at least the Athenians made the Samians pay for the siege after their surrender. The siege of Potidea, on the other hand, derived no recompense and cost about 2000 Talents (Thuc. 2.70.2; Isoc. 15.113). In other words, Potidaea absorbed three years of imperial income in one siege. The Sicilian expedition represents the pinnacle of spending and, of course, lost revenues, running to at least 150 talents per month for the fleet, crews, and army, perhaps more than 3,500 talents in coin alone.[26] The losses were ridiculous, and the consequences are well known, as the last years of the war drained Athenian resources further.[27] The appearance of the "wicked little bronzes" of Aristophanes' *Frogs* in place of good silver shows the plight of the Athenian cause at the end of the war. The resources of Persia proved too much and Aegean sailors and rowers found new paymasters in the Spartan Lysander and the Persian Cyrus. As allies deserted the Athenian cause, less tribute flowed into the Athenian treasury and the money-go-round collapsed in on itself and Athens lost the war.

---

25   Trundle (2004) 15–21.

26   Rhodes (2006) 93. A fragment of an inscription (Meiggs and Lewis (1989) 78 = *IG* 1³ 93 = Fornara (1983) 146) has 3000 talents set aside for Sicily in 416/5 BC and further investment of about 500 Talents followed (in three instalments, thus 300T = Thuc. 6.94.4; *IG* 1³ 370; 120T = Thuc. 7.16.2; and an unspecified amount = *IG* 1³ 371).

27   Gabrielsen (2001) 72–98; Kagan, D. (1974) 37–40 thought the war would have cost about 2000 Talents a month.

COINAGE AND THE ECONOMICS OF THE ATHENIAN EMPIRE 73

The literary sources, therefore, illustrate clearly the significant role of money in the Athenian Empire's rise and fall. The numismatic evidence, however, paints a slightly different picture in the context of this Athenian money-go-round. The evidence for coinage within the Athenian imperial sphere is sparse when compared to coinage found in parts of Asia Minor and Egypt, outside of the Athenian naval *archê*. Good evidence suggests that one of the original purposes of coinage was a way of exporting silver at a profit.[28] In support of this is, an often quoted line from Xenophon's *Poroi* (3.2), written in the middle of the fourth century, that states that "wherever they sell it (i.e. the silver), everywhere they take more than the principal." Xenophon's treatise directs the fourth century Athenians to exploit the mines at Laureion through trading silver. Kraay thought that the Athenians had always minted coins for export and this explained the distinctive markings of Athena and the Owl on their coins, almost as a trademark for an industry of money.[29] But if the purpose of Athenian coins was export, rather than empire, then this challenges the image outlined above that money acted primarily as a part of a hegemonic, political, or even an economic mechanism for controlling the allies of the Delian League.

Kroll has argued that large producers of coinage like Athens, Aegina, Cyzicus, Phocaea, and Mitylene profited greatly from minting coins for export regionally, and on occasion well beyond their own regional dominance.[30] As he states, Athenian coinage became "preferred monetary specie" and "among certain distant populations, Athenian owls and acceptable money became virtually synonymous."[31] Several major hoards outside the Athenian empire, alongside the large quantities of imitation Athenian coins found in Egypt, the Levant, and Syria, not to mention in smaller numbers further east, attest to the export and validity of Athenian coinage as silver beyond the *archê*'s borders.[32] A majority of the coins in hoards in Egypt and Syria are high denominations, *dekadrachms*, which again associates them closely with export of bullion rather than export of useful or usable coined-money.

One of the arguments as to why Athenian coinage in particular became so popularly imitated and so sought after beyond the limits of the Athenian

---

28 For discussion of this see Bissa (2009) 92 who actually challenges the notion of the export of coinage to Egypt based partly on the fact that bars and ingots are also found in large quantities there and even in hoards (IGCH 1636, 1637, 1639, 1640, 1644, 1645, 1649 and 1651).

29 See Kraay (1956); Kraay (1964) especially 82.

30 Kroll (2011) 27–38.

31 Kroll (2011) 32.

32 See Flament (2011); Elayi (1992).

Empire was quantity.[33] The Athenians produced enormous numbers of coins in the fifth century. Estimates range to as many as nine million coins minted by the mid-fifth century.[34] Thus, supply provides part of the answer for Athenian dominance. But we might also note demand for their coinage. The quality of Athenian Laureotic silver was extremely high. This explained the acceptability of Athenian coins across the eastern Mediterranean and Near East and naturally why they travelled so far and were imitated so often.[35]

But, and this has become a contentious question, where did Athenian coins predominantly go? Flament has recently illustrated the massive flow of funds to the military from both the treasury and the allied *phoros*.[36] At the start of the war, this flow took 1300 talents from the central treasuries and 393 talents from the tribute, thus the tribute represented a relatively smaller percentage of coins than those from the central treasury destined for Athenian military commitments. But he went on to show that only half of this money returned to the Athenian marketplace, while half went into foreign markets in expenditures for resources. Thus, a very high number of coins disappeared from the empire altogether. He then examined the likely course of funds in the middle of the 420s. By that decade, the military received 130 talents from the treasury, and a much increased 587–790 talents from the *phoros*. A similar amount as at the start of the war, about half, then went into non-Athenian markets again. The seepage of money out of Athens and Athenian control appears disturbing to those who would argue, like myself, for a highly monetized Athenian imperial economy.

Hoard evidence supports the disappearance of coins out of the Athenian imperial sphere. Konuk has discussed the fact that Athenian Owls are not found in hoards inside the *archê* at all. Asia Minor represents a case in point, as hoards of Athenian coins only occur outside of the imperial sphere in Asia Minor.[37] Hoard evidence also demonstrates this phenomenon elsewhere in the fifth century. Christopher Howgego, discussing the Northern Aegean at the time of Athenian dominance, identified the same problem in this region, as he wrote:[38]

> There is practically no hoard evidence to show that Athenian coin played an important role in the North Aegean at any period in the second half

---

33   Kroll (2011) 31.

34   Kroll (2009) 198.

35   We ought to note here the particularly high quality of Athenian silver.

36   Flament (2011).

37   Konuk (2011).

38   See Carradice (1987).

COINAGE AND THE ECONOMICS OF THE ATHENIAN EMPIRE   75

of the fifth century. One would have expected the contrary if Athenian coin had been the only currency for a significant period. This absence of evidence may not, with total confidence, be taken as evidence of absence, but it is worrying.[39]

We must, however, draw a distinction between the way that large, as opposed to small, denomination coinages circulated and how each related to our hoard evidence. Hoards, wherever they are found, tend to consist of large and impressive high-value coins. Hence the prevalence of *dekadrachma*-coinage in several hoards found outside of the empire. As I, and others elsewhere, have noted, this suggests these heavyweight coinages were minted specifically for export rather than circulation, let alone spending money for ordinary soldiers, sailors, or public servants of the empire. In addition, valuable coins tended to be saved and so hoarded, hence their prevalence in coin-hoards. Smaller coins were spent, circulated, and easily got lost. In addition, much recent work has revealed increasing numbers of small denomination coins circulating in closed and local economic zones, as for example on Aegina and at Abdera.[40] It would be dangerous to conclude that the Athenian imperial economy did not rely heavily upon the circulation of small denomination silver coinage based on the relative lack of such coins found in hoards within the imperial sphere. Indeed, the silver flowed out to the periphery and returned to the centre in a cyclical process, via, as illustrated in the earlier discussion, a series of heavily controlled Athenian legal, economic, and political mechanisms. The coins that stayed within the imperial system are more likely to have not found themselves in hoards because of their constant use within the system. Coins that did leave the empire also left the system as a whole and so ended in collected hoards and, naturally, out of circulation.

The internal arrangements of the Athenian imperial system regarding coins had significant effects on the allies and their own coinage production. Many member states curtailed or even stopped minting. Thus, Picard clearly identified changes in Thasian minting habits after the island's incorporation into the Athenian imperial sphere in 465–463.[41] The Thasians began minting staters of 8.6 *grammes* exactly half the Attic *drachma* standard, at the same time as the decline in Thasian coinage generally after the mid-460s. As soon as the Athenian empire collapsed within the region of Thasos in 411, the Thasians minted

---

39   Howgego (1995) 48.
40   For Aegina, see Kim (2001); for Abdera, see Kagan, J. (2006).
41   Picard (2011) 79–109.

as they had before. It seems probable, therefore, that imperial monetary controls in this instance and in this region transformed with the ebb and flow of power, and surely this demonstrates the significant role that coins played in that control. Kroll chose to highlight the several avenues open for the coins of Athens, from imperial use and circulation to private export. As he wrote:

> What this means for the Athenian mint is that it was but one element in what can be called an Athenian industry of money, much of it struck of course for state use (including, especially fifth century, large overseas military and naval expenses) and as a currency in the local economy, but much of it too—namely, the privately owned surplus—as profit for private individuals who invested in the movement of the silver.[42]

Bearing this sensible observation in mind, then, let me conclude this discussion with an analysis of the main recipients of the money of empire in the latter part of the fifth century. Military and political demands illustrate that coins were a significant part of the Athenian imperial experiment, as suggested in the opening discussion of this discussion. As Athenian citizens became increasingly involved in political life and the redistribution of imperial resources within Athens, allied mercenary oarsmen served in increasing numbers on Athenian ships. Thus, the Athenian navy became an increasingly mercenary one and the Athenian *archê* was increasingly an economic empire, concerned as much with coined money and its control and circulation as with political control. A case in point is the introduction of the *eikostê* of 413, whereby goods were taxed at one twentieth of their value as they entered the Athenian sphere. This brought non-Athenian allies and non-tribute payers engaged in commerce within the *archê* into the Athenian economic sphere. Perhaps it became a mechanism to bring resources back into the empire that had flowed out of it and illustrates Athenian awareness of this phenomenon. The empire constantly pursued money, which played an ever-increasing role as an intermediary between the state and the fleet, and which is further reflected in the fleet's increasingly mercenary nature.

Other naval powers must have utilised non-citizen oarsmen to man their ships. Thus, even in the Persian wars, states like Aegina must have utilised their non-citizen populations to fully compliment their approximately forty five ships.[43] The 9,000 men required for such a fleet cannot have come from

---

42   Kroll (2011) 33.

43   18 ships at Artemesium (Hdt. 8.1) and 30 at Salamis plus others at Aegina (Hdt. 8.46).

COINAGE AND THE ECONOMICS OF THE ATHENIAN EMPIRE

citizen oarsmen alone.[44] Thucydides (1.55.1) stated that 800 of 1050 rowers on Corcyraean ships captured at Sybota were slaves (*douloi*). As for the Athenian system, by the 430s only Chios and the cities of Lesbos provided ships and their own citizen crews to the Athenian military machine. The rest of the allies provided money alone, but their men must have served in some capacity on the Athenian ships. This is alluded to clearly in several passages of Thucydides. At the start of the war Athens' enemies hoped that they might draw non-Athenians away from the fleet with offers of money (Thuc. 1.121). Pericles, as noted above, allayed fears that the Spartans might appropriate the funds of the Delphic sanctuary also to attract the *xenoi* who served in the Athenian fleet into their service. Thus, non-citizen-Athenian manpower played a significant role in the Athenian navy even in 432/1. Later in the war, slaves played their role as oarsmen as well.[45] Recent scholarship has supported the notion that slaves and mercenary crewmen made up at least part of the Athenian navy's oarsmen.[46] Money must have acted as a mechanism of remuneration from which, we can assume, Athenian slave owners gleaned wealth from their payment, in much the same way as slave-miners enriched their owners at Laureion or slave-workers did in other industries. Money then made citizens rich at the expense of the labour of others. Aristophanies comedies of the 420s allude to the role of money as a means to remunerate citizens for explicitly political rather than military service. Thus, it is possible to conclude that a main objective for Athenians was civic sinecure, rather than military risk.

Three examples from Aristophanes illuminate this situation. The first comes from The *Wasps*. Cleon-Hater points to those Athenians whose goal was jury-service rather than military service.

> There are drones sitting among us who have no stingers, who stay at home and feed off the fruits of the tribute without toiling for it. And we're nettled

---

44   For full discussions see Hansen (2006a); Kirsten (1964) 160; Beloch (1886) 123 where the figure of 80,000 is arrived at through addition; Figueira (1981) 22–64. For further discussion with examples see Hansen (2006b).

45   See for examples and allusions, Thuc. 1.121.3, 143.1–2. Thucydides 7.63.3 has Nicias in Sicily address the non-Athenian allies who seem to make up a large number of the forces present. Slaves too rowed in number. *IG* 1^3 1032 notes at least 20% of naval crew named as slaves and possibly a lot more.

46   For discussion of this whole topic see Hunt (1998) 83–101. For support of the non-Athenian presence amongst oarsmen see Graham (1992); Graham, (1998); Jordan (2001); Krentz (2007), especially 150. See also van Wees (2004) 210–212. For the traditional view that only Athenians crewed Athenian ships see Sargent (1927).

if some draft dodger gulps down our pay, when in defence of this country he's never raised an oar, a lance, or a blister. No, I think that from now on any citizen, bar none, who doesn't have a stinger should not be paid three obols.

ARIST. *Wasps* 1114–1121

The second from The *Knights* shows that pay for jury service trumped military concerns for assembly goers, as the Sausage-Seller speaks:

Yes, by Zeus, you were (to Demos asking 'am I that stupid'). If two orators spoke up, one proposing to build long ships for war and the other to spend the same amount to pay off certain citizens, the one who spoke of pay would always go away victorious over the man who spoke of war ships.

ARIST. *Knights* 1350–1354

And the last also from the *Knights* requires a little more explanation to show that jury service was the goal of military expeditions and imperial control. The Cleon speaks:

I did that so Demos might rule over all the Greeks—for the oracles declare that one day he must sit in judgment in Arcadia at five obols a day, if he bides his time. At any rate, I will feed and care for him and use fair and foul means to see to it that he receives three obols every day.

ARIST. *Knights* 794–800

Arcadians were ubiquitous mercenaries by the 420s.[47] This passage clearly shows the juxtaposition of the mercenary Arcadians who took wages for their military service happily versus the litigious jury-serving Athenians who sought political pay and avoided military service—and of course five obols a day for political or military service was an enormous sum at that time.

By the time of the Peloponnesian War, Athenians preferred political service and its rewards to military campaigns. Further references illustrate this fact. Nicias' letter to the Athenian Demos from Syracuse demonstrates the numbers of slaves and foreigners in his fleet, not to mention the role that money and mercenary activity played in their behaviour (Thuc. 7.13.1–7.14.2). In addition, his speeches at the end of the campaign try hard to rally a losing cause and

---

47 On Arcadia as a source of mercenary service see Parke (1933); Griffith (1935) 237–238; Fields (1994); Fields (2001); Roy (1999); Trundle (2004), especially pp. 52–54.

COINAGE AND THE ECONOMICS OF THE ATHENIAN EMPIRE

aim as much at the allied crews as to the Athenians (Thuc. 7.61–64; 7.77–79). Xenophon provides telling information on the make-up of the crews of Athenian triremes. Slaves and free men together manned 110 triremes to rescue Conon blockaded at Mitylene, admittedly in what appear as desperate times, but still a major campaign (Xen. *Hell.* 1.6.24). Of some significance, the Spartans captured (and executed) only 3000 Athenian citizens at Aegospotami from a fleet of around 171 ships (Plut. *Lys.* 11, 13.1; see Xen. *Hell.* 2.1.20, 28). If fully manned, this equated to 36,000 men. Admittedly, some Athenian crewmen were killed in the capture of the vessels themselves and Xenophon also alludes to empty and undermanned ships, but on these figures then Athenians made up only 9% of naval personnel or a staggeringly low number of seventeen men per ship. Of course, we have no means of knowing how many ships were specifically Athenian at the disaster, but their numbers, more than likely on the evidence, represent a relatively small minority of men actually in service.

Are we to conclude then that Athenians served in fewer numbers than they had in the past in these latter years of empire? It is tempting to see triremes manned by status, even in the glory days of the earlier years of the empire, in which Athenian citizens sat on the upper bench, as *thranitai*, paid more than those of the lower benches, the *zygioi* and *thalamioi*, who perhaps came from amongst the metics, and the foreign-subjects and even slaves (Ar. *Ach.* 161–162; *Frogs* 1074). The very fact that the flagship of the Athenian fleet, the Paralus, had to be rowed only by Athenian citizens strongly suggests the rest of Athens' fleet did not and, therefore, it was not a citizen navy (Thuc. 8.73.5 and 8.74.1; Aeschin. 3.162, Poll. 8.116). Thus, the money of empire had produced a mercenary navy.

To conclude then, the Athenian empire required payments from the allies in silver and in Athenian coin. Often lacking mines and mints, and eventually direct access to valid Athenian currency, the allies needed to acquire Athenian silver coinage to pay the *phoros*. The mechanisms by which they might achieve this included selling their goods, especially food, to the city of Athens and the Athenian fleets in the field directly, both of which had the money and controlled the money supply. Plenty of evidence demonstrates that the Athenians often diverted grain from their allies in the Aegean as part of their imperial demands. In addition, military service, by which individuals received payments in Athenian coinage, must have been another means of such silver acquisition. The Athenians provided the ships, possibly a large number of specialists on board and some of the oarsmen, but the allies—non-Athenians—provided many of the rowers for the fleet in the latter part of the imperial period—as we have seen even in the 430s. Coined money mediated this mercenary relationship and paid for the distribution of resources in the empire and ensuring that food flowed into Athens.

CHAPTER 5

Tributum in the Middle Republic

Nathan Rosenstein

This chapter examines the financial resources that supported Rome's military operations during the middle Republic. It argues 1) that the wars Rome fought between the late fourth century and 167 BC were for the most part not financed through spoils and indemnity payments. Instead, 2) *tributum* payments from Roman citizens constituted the principal means of meeting the costs of war. The chapter next 3) offers a sophisticated model for understanding how *tributum* payments were apportioned among Roman citizens and 4) endeavors to model the relative financial burden the *tributum* payments would have placed on citizens. Finally, 5) the chapter concludes by considering the effect that the citizens' need to make regular payments of *tributum* might have had on the broader economic development of mid-republican Rome.

Money was a mainstay of the Roman Republic's military success. With it Rome and its allies were able to pay their soldiers as well as furnish their armies with food, equipment, transport, and support personnel. That material support in turn enabled the legions and their allied contingents to remain in the field for months and even years on end. Long service together, and the training they underwent during that time, allowed legionaries and *socii* to develop a skill at arms, unit cohesion, and ability to maneuver under the stress of combat far superior to those of their opponents. Armies unconstrained by the demands of the agricultural calendar and the need to return men to their farms every fall to plant next year's crops gave their generals great latitude in conducting campaigns. And monetary strength allowed Rome to field several armies simultaneously and pursue the kind of coordinated grand strategy that not only allowed it to fight the war of attrition that finally overcome Hannibal but then to go on to conquer the entire Mediterranean world within the span of a generation.

The importance of money in creating and sustaining Roman military power makes understanding its source of surpassing importance. One might assume, along with many scholars, that it came mainly from the very same conquests that it made possible. As one eminent historian has put it, "[E]xpansion before the Second Punic War had greatly increased public revenues without a comparable increase in regular liabilities. Once the war was over, the impression must have returned to senatorial minds that in general both war and expan-

© KONINKLIJKE BRILL NV, LEIDEN, 2016 | DOI: 10.1163/9789004284852_006

TRIBUTUM IN THE MIDDLE REPUBLIC

sion were profitable to the state."[1] Yet surprisingly, although a few rich victories during the early second century BC yielded spoils that offset the expenditures involved, in most cases the spoils from the army's victories rarely equaled the money spent in winning them. Where the evidence allows for quantitative analysis—Livy's reports of the sums carried in the triumphs celebrated during the early second century and Polybius' evidence on legionary pay scales—it is clear that while a few spectacular victories brought enormous riches into the treasury, only about half of the campaigns that culminated in triumphs produced spoils in excess of the *stipendium* paid to the soldiers involved.[2] Were it possible to include the costs of equipment, food, and transport in the reckoning, income from these victories is unlikely to have exceeded expenses in more than a handful of cases. Still, that handful of victories paid for far more than just their own costs. Between 200 and 167 Rome fielded an average of 8.7 legions per year. Their *stipendium* cost the Republic a little less than 171,400,000 denarii. The wealth gained in the wars these legions fought amounted to more than 267,850,000 denarii, a profit of about 96,500,000 denarii over expenditures for the armies' *stipendium* (although this last figure would shrink somewhat if other expenses could be factored in). However this profit was produced entirely by the multi-year indemnity payments levied on Carthage, Philip V, Nabis of Sparta, Antiochus the Great, and the Aetolians under the terms of the peace treaties they struck with Rome following their defeats. Together these combined to bring into the treasury somewhat more than 140,200,000 denarii. Without these payments, booty actually carried in triumphs amounted to only about 127,600,000 denarii, and the cost of *stipendium* exceeded spoils by some 12,600,000 denarii—a figure that would only increase if the other expenditures these wars entailed were included.

In view of the critical difference these indemnity payments made to the Republic's military balance sheet, it is important to note that they were very much the exception rather than the rule in Rome's conquests between the Latin Revolt and the Hannibalic War. The only clear examples are those levied on Hiero in 263 and on Carthage in 241.[3] The Republic apparently imposed

---

1   Harris (1979) 68–69; see now Kay (2014) 25–29.

2   Rosenstein (2016) on what follows. All dates BC.

3   Carthage: Polyb. 1.62.8–63.3, 88.12; Hiero: Polyb. 1.16.9, although the balance of Hiero's indemnity was forgiven at some point in gratitude for his assistance in Rome's war against Carthage: Zon. 8.16 with Walbank (1957–1989) 1.68–69. Some type of payment may also have been demanded from the Illyrian kingdom in 229 and 219 as well: Rosenstein (2016) n. 34. How much the tithes from Sardinia and Western Sicily were worth must remain an open question,

indemnities only on those enemies whose finances enabled them to meet the heavy, long-term obligations they entailed, and these foes were few prior to the second century. There is little reason to think that very many if any of the Republic's other wars between the later fourth and the early third centuries produced similar sorts of continuing money payments. Even though the Republic's military costs prior to 218 were lower because normally only four legions were in the field every year, the enemies these armies fought were unquestionably poorer and their economies less well developed than those of the Hellenistic kingdoms and states.

The conclusion seems inescapable that Rome's wars down to c. 167 ordinarily cost more to win than they returned in spoils and that their funding came primarily from the *tributum*. *Tributum* was paid by those Roman citizens, known as *assidui*, whose wealth qualified them for legionary service but who were not conscripted to serve in any given year. For that reason, a clearer picture of how the *tributum* worked and the extent of the burden it placed on the *assidui* is essential to an understanding of how the Republic funded its remarkable military success in this period. As is well known, *tributum* was technically not a tax but rather a kind of compulsory loan that could on occasion be repaid.[4] In 187 the senate instructed the quaestors to use the enormous booty from Cn. Manlius Vulso's campaign in Asia Minor to repay the arrears in *tributum* still owing to the citizens (Livy 39.7.4–5). Nearly a century earlier, C. Duilius proudly recorded on his monument the sums he had handed back to the citizens following his naval victory in 260 over the Carthaginians in Sicily, apparently a similar instance of returning *tributum* to the *assidui*.[5] A few other cases are also on record, yet their scarcity suggests that instances in which *tributum* payers got their money back were rare occurrences, occasions to be celebrated and memorialized. Ordinarily the *assidui* paid their *tributum* year after year with little or no expectation that they would ever see their loans repaid.

Tenney Frank, following De Sanctis, held that *tributum* was a flat one mil (one tenth of one percent) levy on the wealth an *assiduus* declared at the quinquennial census.[6] However Claude Nicolet argued convincingly that *tributum* was not levied at a fixed rate but varied according to the Republic's annual

---

but what evidence we have suggests they did not provide enough food to meet all of the needs of the Republic's armies even after the Second Punic War—see below for further discussion.

4   Nicolet (1976) 19–26; cf. Nicolet (1980) 149–169.

5   *CIL* I².2.25 line 17; further discussion of the authenticity of this inscription: Rosenstein (2016) n. 18.

6   Frank (1933) 66, 75, 79; De Sanctis, (1907–) 3.2.626.

TRIBUTUM IN THE MIDDLE REPUBLIC

military needs. The senate in his reconstruction calculated what military operations would cost at the outset of the year, then divided that sum by the total wealth of the *assidui* to find the rate at which *tributum* was to be paid. That rate was then multiplied by the wealth of each *assiduus* to determine the sum to be handed over.[7] Thus

$$\frac{\text{Anticipated Military Expenses}}{\text{Total Wealth of Assidui}} = \text{rate} \times \text{Wealth of Assiduus} = \text{Tributum}$$

For example

$$\frac{3,000,000 \text{ denarii}}{5,000,000,000 \text{ denarii}} = .0006 \times 100,000 = 60 \text{ denarii}$$

However it is possible to go considerably beyond Nicolet's simple formula. Based on what is known about the size of the Roman citizen population in the later third century and the distribution of wealth within it, the operation of the *tributum* can be modeled more precisely. The former is known from Polybius' report of the Republic's military resources in 225, on the eve of the great Gallic invasion, which he found in Fabius Pictor, a contemporary of the event.[8] The figures Polybius records are not without problems, and the question of the size of the Roman population at this date has in recent years excited considerable controversy, but for the purposes of this illustration Peter Brunt's reconstruction and interpretation will be followed, which is the lowest estimate.[9] The implication of using interpretations that put the number higher can be considered later. Brunt reckoned the total number of free adult male Romans at about 300,000. We also know something about the distribution of wealth among these Romans. In the same passage, Polybius reported that Rome at that time could call on a total of 23,000 citizens for cavalry service. This group was primarily composed of the *equites equo privato*, those citizens whose wealth qualified them to serve in the cavalry rather than the infantry and who supplied their own mounts when they did so.[10] Only a very small number were members of the centuries of *equites equo publico*, whose horses were furnished at public expense. The *equites equo privato* represented a select group within the first census class, the remainder of whose members served in the infantry. Hence the first census class itself must have been at that date larger

---

7 Nicolet (1976) 35–46; Nicolet (1980) 156–164.
8 Polyb. 2.24.14; Pictor as the source for Polybius' figures: Eutr. 3.5; Oros. 4.13.6.
9 Brunt (1971), 44–60, esp. 54.
10 Polyb. 6.20.9; McCall (2002) 2–7.

than 23,000. An incident from 214 also tells us something about the number of Romans at the bottom of the census hierarchy. In that year Livy reported that the senate determined to increase the Republic's naval forces by 100 ships. However, because free oarsmen to man them could not be found, the senate resorted to conscripting slaves (Livy 24.11.5–9). The inability to recruit enough citizens for the new ships reveals a lack of *proletarii*, the very poorest citizens who occupied the lowest census category. Their sole military obligation was to serve as rowers on Roman warships. Brunt has calculated that at this point in the war, the Roman fleets already at sea would have required the services of a total of about 20,000 oarsmen.[11] That figure makes it possible to arrive at an estimate of the size of the total number of adult male *proletarii*.

A couple of additional facts about how the *tributum* operated are also important here. The first is that *tributum* was paid by those *assidui* who were not serving as legionaries for the benefit of those who were. The latter therefore paid no *tributum* since otherwise they would have been subject to a kind of double taxation, once in money and a second time through their military service. Secondly, while *tributum* was at least in theory paid by every *assiduus* who was not conscripted for service in the legions on the basis of his declared property, the adult sons of *assidui* represented a special case. They remained in their fathers' *potestas* until he died, and hence they could own no property since they were not *sui iuris*.[12] Technically at least they were *proletarii* and could not be held liable for *tributum*. However an adult son of an *assiduus* could substitute for his father and fulfill his obligation for military service (e.g. Livy 42.4.12).

The foregoing helps refine our estimate of who might have been paying *tributum* in the later third century and so how much might have been collected from those *assidui* in each census category. All of that affects the denominator in the equation sketched above. However the numerator is far more problematic, representing as it does the total estimated military expense of the year. Without it, no calculation of *tributum* is possible, and we have no way to ascertain the costs of all the equipment and food Roman armies required or the numbers and pay of the support personnel who accompanied them—to say nothing of the expenses that might have been incurred in transporting all this material. Estimates of the numbers of mules alone that would have accompa-

---

11 Brunt (1971) 64–66, 417–420.

12 A father with adult sons registered them along with himself at the census: Livy 43.14.8 and Festus 58 L s.v. *duicensus* cf. Brunt (1971) 22 n. 1.

TRIBUTUM IN THE MIDDLE REPUBLIC

nied the legions and were required to move food and equipment from depots in the rear up to the front reach into the thousands.[13] Further, the costs of warships and their crews are completely unknown. And although Rome obtained much of the food for its armies and fleets after 241 through taxation in kind from Sicily, Sardinia, and eventually Spain, the amounts arguably were never sufficient to meet all of the military's needs during the third and second centuries.[14] The only cost we can be relatively confident about is the *stipendium* itself, based on Polybius' report (6.39.12) that in his day (that is, the midsecond century) an ordinary infantryman earned two obols a day, a centurion twice that, while a cavalryman received a drachma a day. Dominic Rathbone has convincingly argued that the correct conversion of Polybius' Greek currency into its Roman equivalents produces rates of three, six, and nine asses for each rank.[15] These pay rates probably had been established at the time of the monetary reforms of c. 212/211, which introduced the denarius system of coinage together with the bronze sextantial as of two ounces. They remained unchanged until Julius Caesar doubled them.[16] More importantly, the value of the legionaries' *stipendium* prior to 211 is likely to have been about the same as the later rates of pay. This is because, even though the soldiers were paid fewer asses, these coins weighed ten ounces rather than two. The total weight of bronze that a legion's *stipendium* represented before c. 212/211 would therefore have been more or less equal to its weight after that date. On that basis, the Republic's yearly outlay for *stipendium* c. 225 can be calculated using Polybius' post-211 rates as a substitute, since amount of bronze expended would have been roughly the same. Even though that figure will not represent the total cost of Roman warfare, *stipendium* undoubtedly represented its principal ongoing expense, probably well over half of the cost in most years. Therefore it can stand as a rough proxy for the actual numerator in the equation, bearing in mind that the actual figure will undoubted have been higher, perhaps considerably higher. Hence, a model for annual *stipendium* can be formulated as follows:[17]

---

13   Roth (1999) 79–91, 115.

14   Erdkamp (1998), 89–90; Rosenstein (2016) 116–117.

15   Rathbone (1993) 151–152. See Rosenstein (2016) n. 15 for further discussion.

16   On the reform and its date, Crawford (1985) 55–56 with further references; Caesar's increase in military pay: Suet. *Iul.* 26.3.

17   Note: 10 asses = 1 denarius after the reform of the coinage c. 211 down to the mid-second century.

86   ROSENSTEIN

(3 asses / day × 4,140 legionaries) + (6 asses / day × 60 centurions) +
(9 asses / day × 300 cavalrymen)
× 355 days of a Roman year = $\frac{5,495,400 \text{ asses}}{10}$ = 549,540 denarii × 4 legions =
2,198,160 denarii

Figuring the denominator is more complex since it requires solving two problems: how many citizens paid *tributum* and how much each *assiduus* paid. The first problem in turn involves establishing three things: how many citizens were *proletarii* and so exempt from paying *tributum*; how many of the rest were *in potestate* of their fathers; and how many of the remainder were serving in the legions. Subtracting each group from the total number of citizens should tell us how many *assidui* remained to pay *tributum*.

A population of 300,000 adult male citizens 17 years of age and older implies that the total male population was roughly 476,000, assuming that their age distribution corresponds to Coale-Demney[2] Model West 3 Female (life expectancy of e_0 = 25 and a rate of reproduction of r = 0.0). To estimate how many were *proletarii* we can assume that rowers, like legionaries, were drawn mainly from men between the ages of 17 and their early 30s.[18] Men in this age range constituted about 30 percent of all males of any age in a Model West 3 Female population. If we know that the 20,000 *proletarii* serving as rowers in 214 represented this 30 percent of all male *proletarii*, then we can extrapolate from that number that about 66,600 of all males of any age were *proletarii*. Finally, 63 percent of those 66,600 *proletarii* of all ages would have been 17 years of age or older: 42,000.[19] For the purposes of this exercise, we can assume that the number of *proletarii* ten years earlier would have been about the same.

To estimate how many of the rest would have been *assidui sui iuris* we can use the tables that Richard Saller has developed that estimate the percentage of males who would have had a living father in 5 year age cohorts and in a model population with the age distribution of Coale-Demney[2] Model West 3 Female.[20] Those percentages can then be multiplied by the numbers of males in our population of 476,000 less 66,700 *proletarii* to establish how many of them would have been *in potestate* and how many would have been *sui iuris*. The results are set out in the following table:

---

18   Rosenstein (2004) 82–86.
19   100 / 30 = 3.33 × 20,000 = 66,600; 66,600 × .63 = 41,958.
20   Saller (1994) 48–65.

TRIBUTUM IN THE MIDDLE REPUBLIC

TABLE 1   *Distribution by Age of 409,400 Roman* Assidui, *225*

Age	% at Age	# at Age	% with a living father	# with a living father	# without living father
0–1	0.0321	13,142			
1–4	0.0953	39,016			
5–9	0.1053	43,110			
10–14	0.1	40,940			
15	0.01892	7,746			
16	0.01892	7,746			
17	0.01892	7,746			
18	0.01892	7,746			
19	0.01892	7,746			
17–19	0.05676	23,238	0.63	14,640	8,598
20–24	0.0881	36,068	0.51	18,395	17,673
25–29	0.081	33,161	0.39	12,933	20,228
30–34	0.0736	30,132	0.28	8,437	21,695
35–39	0.0662	27,102	0.17	4,607	22,495
40–44	0.0591	24,196	0.09	2,178	22,018
45–49	0.0522	21,371	0.04	855	20,516
50–54	0.0452	18,505	0.01	185	18,320
55–59	0.0375	15,353			15,354
60–64	0.0291	11,914			11,915
65–69	0.0203	8,311			8,311
70–74	0.0123	5,036			5,036
75–79	0.0058	2,375			2,375
80–84	0.0019	778			778
85–90	0.0004	164			164
Totals		409,400			195,475

Of the roughly 258,000 *assidui* between the ages of 17 and 90 therefore only 195,475 will have been *sui iuris*. Not all of them will have been liable to *tributum* however. Those serving in the legions or whose sons were doing so were exempt from payment. During most of the third century the Republic normally fielded four legions every year each containing 4,500 infantry and cavalry, a total of 18,000 *assidui* or their sons. These must be subtracted from the 195,475 potential *tributum* payers to yield a final total of 177,475 *assidui* who actually had to pay.

TABLE 2A *Distribution of Roman citizens by wealth ca. 225, Hypothesis I*

1st class (100,000 asses)	25,000
2nd class (75,000 asses)	25,000
3rd class (50,000 asses)	50,000
4th class (25,000 asses)	65,000
5th class (11,000 asses)	93,000
proletarii, etc	42,000
Total	300,000

The total wealth of the *assidui* liable to *tributum* is much more difficult to estimate than their numbers. The only attempt to do so, De Sanctis' calculation of the total wealth of all Roman citizens around 200, is vitiated by his assumption that *tributum* was a flat one mil levy.[21] A different approach is to begin with what we know about the size of the census classes and extrapolate from there. We know that in 225 about 23,000 Romans were *equites equo suo*; hence the first class was greater than 23,000 although by how much is unknown. We have also estimated the number of adult *proletarii* 17 years of age and older at the same date at about 42,000. And we know that the total number of adult citizens to be distributed among the various census classes was 300,000. Because we know the size of at least part of the first class and the number of *proletarii*, the remainder of the 300,000 citizens must somehow be fitted within those parameters. This information enables us to establish two limiting hypotheses, extreme cases neither of which is likely to represent reality but which together can suggest the boundaries within which reality ought to lie. The first assumes that the distribution of wealth in 225 was steeply pyramidal with the first class only slightly larger than the number of cavalrymen and the second through fifth classes increasingly large.

The second hypothesis assumes, on the contrary, that the first class is significantly larger than its 23,000 cavalrymen and that the distribution of wealth resembles an inverted pyramid.

Although Hypothesis II may seem counter-intuitive—and indeed wildly improbable—in view of the usual assumptions about how wealth was distributed among the population of middle republican Rome, its plausibility is

---

21   De Sanctis (1907–) 3.2.623 cf. Frank (1933) 125–126.

TABLE 2B   *Distribution of Roman citizens by wealth ca. 225, Hypothesis II*

1st class	75,000
2nd class	57,000
3rd class	51,000
4th class	40,500
5th class	34,500
proletarii, etc	42,000
Total	300,000

at least somewhat strengthened by the likelihood that the *equites equo publico* did not constitute all but a handful of the members of the first class. They were an elite corps among them and so probably would have constituted a minority. In addition, there is the possibility that the allocation of centuries in the *comitia centuriata* was not so heavily gerrymandered to favor the well-off as is often thought. It may to some extent have corresponded to the numbers of citizens in each category. Hypothesis II allots about a quarter of the citizens to the first census class in view of the fact that the *comitia centuriata* apportioned about 46 percent of the centuries to the first class along with the six centuries of *equites equo publico*.

Finally we can make one more simplifying assumption, namely that for the purposes of *tributum* each *assiduus* was assessed only at the minimum amount of wealth necessary to qualify him for inclusion in his census class so that, e.g., every member of the first census class was assumed to have property valued at 100,000 (post-211) asses, even though other, higher levels of wealth are known (see e.g. Table 6 below). In this case the assumption may be fairly close to what was in fact the case. Citizens declared the value of their property to the censors (or perhaps their representatives), and it is difficult to believe that they did so on the basis of an exhaustive inventory and valuation of their possessions.[22] In all likelihood a citizen placed himself (within limits) where he believed he belonged in the census hierarchy, balancing the increased *tributum* paid by those in the upper census classes against the greater honor that membership

---

22   Ulpian, *Digest* 4.932 (Watson): *Omnia ipse qui defert aestimet*. Note, too, Cic. *ad Brut.* 26 (1.18 = s.b. 24).5: *impudenti censu locupletum*. cf. App. *B Civ.* 4.32.

in those classes bestowed.[23] And indeed some such system of regularizing the declarations of citizens must have been used to facilitate arriving at a total for the wealth of all *assidui*, since otherwise the task of adding up the nearly 200,000 different figures collected from individual citizens would have been overwhelming.

On the basis of the foregoing, the following two models of how the *tributum* might have been distributed and how much each *assiduus* might have had to pay can be offered. The figures in each table have been established by first assuming that the proportion of men *sui iuris*, and so liable to pay *tributum*, was the same for each census class of *assidui*. Likewise it is assumed that the proportion of *assidui* who themselves, or whose sons, were serving in the legions and so were exempt from paying *tributum* was the same for all census classes. Subtracting them from the number of all *assidui sui iuris* in each census class yields the number of *assidui* who actually paid. The result is multiplied by the minimum census (in post-211 asses) to determine the total wealth of each census class. Those totals were added together to represent the denominator in the equation that establishes the *tributum* rate. The numerator, as explained above, is simply the *stipendium* for four legions, used as a proxy for the total annual military expenditure. The result is the rate at which tributum was levied, as set out in the following tables.

TABLE 3A   Tributum *rates for* Assidui sui iuris *according to Hypothesis I (asses)*

	Assidui sui iuris not serving	Min. wealth each	Total wealth	Tributum rate[a]	Tributum paid
1st Class	17,197	100,000	1,719,700,000	0.00336	336
2nd Class	17,197	75,000	1,289,775,000	0.00336	252
3rd Class	34,394	50,000	1,719,700,000	0.00336	168
4th Class	44,713	25,000	1,117,825,000	0.00336	84
5th Class	63,974	11,000	703,714,000	0.00336	37
Totals	177,475		6,550,714,000		

a $\dfrac{21,981,600}{6,550,714,000} = .00336$

---

23   Nicolet (1980) 61–62, 69–73; note e.g. the privilege of wearing mail armor, the *lorica*, awarded to legionaries who were members of the first census class: Polyb. 6.23.15.

TRIBUTUM IN THE MIDDLE REPUBLIC

TABLE 3B   Tributum *rates for* Assidui sui iuris *according to Hypothesis II*

	Assidui sui iuris not serving	Min. wealth each	Total wealth	Tributum rate[a]	Tributum paid
1st Class	51,595	100,000	5,159,521,879	0.0020	203
2nd Class	39,212	75,000	2,940,927,471	0.0020	152
3rd Class	35,085	50,000	1,754,237,439	0.0020	102
4th Class	27,861	25,000	696,535,454	0.0020	51
5th Class	23,734	11,000	261,071,807	0.0020	22
Totals	177,488		10,812,294,049		

a $\dfrac{21,681,600}{10,812,294,049} = .0020$

Finally, we know little about how *tributum* was actually collected save that at this date it seems to have been the responsibility of the *tribuni aerarii*.[24] How successful they would have been in obtaining what *assidui* in the lower census grades owed must remain unknown. We may suspect that the *tribuni aerarii* did not bother with them since—at least under Hypothesis I—their numbers were large and the sums they owed were fairly small. For the sake of illustration, the following tables indicate the *tributum* paid if its collection was restricted to the top three census classes.

It bears repeating that none of the foregoing can claim to represent the real distribution of wealth among Roman citizens around the last quarter of the third century, who among them paid *tributum*, or the rates at which they did so.

TABLE 4A   Tributum *paid by* Assidui *not exempted by legionary service if only those in census classes 1–3 paid according to Hypothesis I*

	Assidui sui iuris not serving	Min. wealth each	Total wealth	Tributum rate	Tributum paid
1st Class	17,197	100,000	1,719,700,000	0.0046	465
2nd Class	17,197	75,000	1,289,775,000	0.0046	349
3rd Class	34,394	50,000	1,719,700,000	0.0046	232
Totals	68,789		4,729,236,219		

---

24   Nicolet (1966) 598–604; Nicolet (1976) 46–55.

TABLE 4B   Tributum *paid by* Assidui *not exempted by legionary service if only those in census classes 1–3 paid according to Hypothesis II*

	Assidui sui iuris not serving	Min. wealth each	Total wealth	Tributum rate	Tributum paid
1st Class	51,595	100,000	5,159,521,879	0.0022	223
2nd Class	39,212	75,000	2,940,927,471	0.0022	167
3rd Class	35,085	50,000	1,754,237,439	0.0022	112
Totals	125,892		9,854,686,789		

These tables are intended only to suggest the boundaries within which those realities probably lie. They are constructed on the basis of a number of assumptions, any of which could be questioned. The critical issue, however, is whether altering any of them would change significantly the parameters to *tributum* that the tables suggest. Obviously the use of the legionaries' *stipendium* to stand for the annual cost of military operations to some degree means that the numerator in the *tributum* equation is too small. One must ask therefore whether true values for *tributum* would be significantly greater than those suggested in the tables if we knew the actual cost of a year's military operations. If that cost were double the value of *stipendium tributum* naturally would increase dramatically, but probably not much above the maximum tributum values in table 4A. These are based on an unrealistically low number of *assidui* in the first three census classes. Assuming not unreasonably that the numbers of men in these census classes were twice as large as those given in Table 4A would offset any increase in the *tributum* that resulted from doubling the annual cost of military operations.

On the other hand, a number of changes in the assumptions behind the tables would lower the values for *tributum* that they present—most importantly the number of *assidui* they assume. As noted above, the tables are based on Brunt's analysis of the population figures Polybius' reports for 225, and that analysis has recently been challenged. Elio Lo Cascio and Saskia Hin have both published important critiques of Brunt's interpretation of the figures and offered alternate analyses that put the number of Roman citizens in 225 significantly above Brunt's figure.[25] If either of these scholars is correct, the result

---

25   Lo Cascio (1994) 22–40; Lo Cascio (1999) 161–171; Lo Cascio (2001) 111–137; Hin (2008) 187–238 cf. Hin (2013). However de Ligt (2012) argues strongly for a "low count" of the population much closer to that championed by Brunt.

TRIBUTUM IN THE MIDDLE REPUBLIC

would in all likelihood necessitate a corresponding increase in the numbers of *assidui* assumed in the tables. More *assidui* would naturally spread the burden of tributum more widely and so each *assiduus* would pay less. Even if Brunt's population figures prove to be correct, however, a number of other changes to the assumptions behind the tables would likely result in lower *tributum* values. The tables assume that each *assiduus' tributum* rate was based on the minimum amount of wealth needed to qualify for a particular census class. This may not have been the case. If the personal wealth many *assidui* declared was significantly above the minimums for their census classes then the total for all citizens' wealth would naturally have been higher than the tables assume. That in turn would lower the rates at which *tributum* was levied, although for many individual *assidui* this decrease in the *tributum* rate would be offset by the higher amount of their personal wealth on which the *tributum* they paid was based. One might also suppose that most legionaries were recruited from the lower census classes, increasing the burden on those *assidui* in the higher ones. However Tables 4A and 4B already assume that no *assidui* in the 4th and 5th classes paid *tributum*. More relevant is the assumption that *tributum* would have paid for the whole of any year's military expenses. Yet some third century victories may have produced a surplus of spoils beyond the cost of winning them, and that surplus may have paid the cost of subsequent years' military operations, as was frequently the case in the early second century. When that happened naturally *tributum* was reduced or even omitted altogether. I have argued elsewhere that this was probably a rare occurrence in these years, but the possibility that it happened from time to time should not be discounted.[26] Still, only a radical revision of the citizen population upwards would decrease the lowest values for *tributum* offered in the tables.

If the tables do, in fact, represent the range within which the actual amounts of *tributum* Roman *assidui* paid fall, then the critical question is what the sums in the tables really represent: did the burdens they imposed on those who paid *tributum* amount to a lot or a little? Unfortunately no data permits us to establish a basis for comparing the amounts of *tributum* given in the tables with typical prices in this or any other period. Historians of the Roman economy therefore fall back on a hypothetical average cost of three or four sesterces for a *modius* (about 6.65 kg.) of wheat in order to estimate the cost of food and other items during the late Republic and early Empire.[27] At that date, a silver sestertius was worth four asses weighing one ounce each, one quarter of a denarius

---

26   Rosenstein (2016) 122.
27   Rickman (1980) 153–154.

TABLE 5 Tributum *as a percentage of the annual cost of wheat for a family of five*

	Table 3A	Table 3B	Table 4A	Table 4B
1st Class	0.25–0.19	0.15–0.11	0.34–0.26	0.16–0.12
2nd Class	0.19–0.14	0.11–0.08	0.26–0.19	0.12–0.09
3rd Class	0.12–0.09	0.08–0.06	0.17–0.13	0.08–0.06
4th Class	0.06–0.05	0.04–0.03		
5th Class	0.03–0.02	0.02–0.01		

tariffed at sixteen asses. The amounts of military pay that Polybius preserves however are generally assumed to be denominated in asses of more or less two ounces each, ten of which equaled a denarius in the early second century. Using these two ounce asses and again assuming that values of items in the later third century would not have differed greatly from those in the early second despite being denominated in much heavier, ten ounce asses, we can suggest what the various values of *tributum* presented in the tables might have represented in terms of the cost of the wheat necessary to meet the total caloric requirements of a family of five based on the hypothetical late republican price. Again, the results cannot be confused with reality; wheat prices often fluctuated dramatically over the course of a year and could vary greatly from place to place. The following table is offered only in order to put the *tributum* in Tables 3A and B and 4A and B in some perspective. A hypothetical family of five—two adults and three children—might have required around 1500 kg. of wheat per year (assuming they ate nothing else), about 225.5 *modii*.[28] If a *modius* of wheat cost between six and eight two-ounce asses (that is, the equivalent of twelve to sixteen late republican/early imperial one-ounce asses or three to four sesterces), the family's annual expense for wheat was 1,353–1,804 asses. Table 5 therefore presents the percentage of the cost of this hypothetical family's annual food needs as a percentage of the *tributum* that *assidui* in each class might have paid under the assumptions that inform Tables 3A and B and 4A and B.

By that measure an *assiduus* in the first census class might have paid in annual *tributum* as much as the equivalent of a third of the cost of feeding a family of five while one in the fifth class could have paid as little as one percent. The latter does not seem especially burdensome, representing only about 15 kg.

---

28 Rosenstein (2004) 67–68.

TRIBUTUM IN THE MIDDLE REPUBLIC

TABLE 6   *Slave oarsmen furnished by citizens in 214*
LIVY 24.11.7–8

Census rating in 220	Number of slaves	Pay
50,000 asses	1	6 months
100,000 asses	3	1 year
300,000 asses	5	1 year
1,000,000 asses	7	1 year
senators	8	1 year

of wheat (assuming that *tributum* was collected from citizens in this class at all). At the top end of the scale however the burden of *tributum* might seems much more onerous, particularly if it had to be paid every year. However it is possible to form some impression of the financial resources of the citizens in these classes on the basis of the incident in 214, referred to above, when the senate found a dearth of *proletarii* to man the warships being launched in that year and so elected to conscript slaves. The consuls issued an edict requiring citizens to furnish slaves and pay for them according to the criteria in Table 6.[29]

The census rating of the second group is that of the 1st census class. Men in a position to supply three slaves plus a year's pay for each of them are obviously not impoverished peasants scratching out a living on tiny farms. Their ownership implies a considerable productive capacity, and supplying the monetary equivalent of about 500 kg. of wheat, equal to the yield of perhaps two or fewer hectares, every year would not seem to have imposed a particularly heavy burden of *tributum* on *assidui* in this category. Even the owner of a single slave was clearly employing that additional labor to produce more on his farm than mere subsistence required.

The relatively modest size of the *tributum* that *assidui* paid by the later third century was a function of the number of citizens among whom the burden of financing the Republic's military efforts was shared. The size of the pool of *tributum* payers in turn resulted from the steady enlargement of the citizen body over the course of the preceding century and a quarter. One may suspect that this expansion had been a major reason for the senate's generosity in extending the franchise following the end of the Latin revolt in 338. That policy is often assumed to have stemmed from a desire to increase the pool of Rome's

---

29   On this incident see further Rosenstein (2008) 5–7, 24–26.

military manpower, and unquestionably the *patres* must have been aware that this would be a highly desirable result. But between the later fourth and mid-third century, when the bulk of the enlargement took place, the senate rarely authorized more than four legions to take the field in any one year. Few if any senators will have anticipated a mass mobilization of the citizenry on the scale that took place during the Hannibalic War. A more immediate goal was undoubtedly lightening the burden of *tributum* on individual *assidui* by spreading the burden among more payers. The result however was also the creation of the deep financial as well as manpower resources that enabled to the Republic to meet the challenge that combatting Hannibal posed.

One might go further still and wonder to what extent Roman warfare catalyzed the economic development that produced those financial resources. Once the Romans made the decision to begin paying their soldiers (whether in the later fifth or mid-fourth century), every *assiduus* had to begin producing a surplus (if he had not already been doing so before). Production of a surplus, particularly when the *tributum* began to be paid in money rather than in kind (probably around the early third century, when the *stipendium* also began to be paid in cash) meant the need to exchange surplus crops for coin. That in turn will have led to the creation or expansion of the markets where these transactions could take place and consequently an increase in overall commercial activity as well. An increase in opportunities to market agricultural products in turn is very likely to have stimulated *assidui* to additional production in order to take advantage of new or increased chances to make money. One might even surmise that some producers, perhaps many, will have elected not simply to consume their profits but instead to invest them in ways that would enable them to increase the size of the surpluses they produced and so the money they made from selling them. The slaves that *assidui* in the three top census classes owned in all likelihood represent precisely this sort of investment. And a similar decision to invest rather than consume could even have extended to the spoils some soldiers brought home from their wars. The increase in commerce that this process generated might then have opened other, additional opportunities for production, sale and profits because the *assidui* now had money to buy more of a range of other sorts of items to consume or use. One might wonder therefore whether we have gotten the whole story of the relationship between war and the Italian economy backwards. It has become increasingly clear in recent years that Rome's wars in the late third and second centuries were not ruining Italy's small farmers.[30] It could be instead that by creating a

---

30   Rosenstein (2004) 26–62.

need for *assidui* to generate agricultural profits in order to pay the *tributum*, war became the catalyst that led to the creation of a significant degree of prosperity among Romans of the middle Republic.

Military Cohesion

∵

CHAPTER 6

The Ties that Bind: Military Cohesion in Archaic Rome

Jeremy Armstrong

Since World War II, the subject of battlefield cohesion has been a major area of research in modern military studies. In the aftermath of the war, a number of scholars argued, based on evidence from both Axis and Allied forces, that unit cohesion was one of the most important factors contributing to overall military effectiveness and, ultimately, success.[1] In particular, the importance of a soldier's immediate group (squad, platoon, etc.) was emphasized in these studies,[2] with Shils and Janowitz commenting that "it appears that a soldier's ability to resist is a function of the capacity of his immediate primary group (his squad or section) to avoid social disintegration."[3] However, despite its importance, scholars struggled to define the exact nature of this cohesion. As MacCoun and Hix noted in 2010, for most of the 20th century discussions of the issue were stymied by the relatively unsophisticated and "monolithic" approaches to military cohesion prevalent in the sociological and anthropological models of the time, and that it has only been in recent years that a sufficiently nuanced framework and vocabulary have been available to describe this multifaceted phenomenon.[4] In particular, recent scholarship has argued for two broad types of cohesion, 'vertical' and 'horizontal', with vertical cohesion involving leaders and followers and horizontal cohesion referring to the bonds between the members of the primary group. These bonds in turn have two subordinate

---

1   Marshall (1947) and Stoufer et al. (1949).

2   This sociological concept was initially defined by Cooley in the early 20th century as "intimate association" and "the fusion of individualities in a common whole." Cooley (1909) 23.

3   Shils and Janowitz (1948) 281. Some of this post-war evidence has since been called in to question, however, as Shils and Janowitz used prisoners of war in their study and it has been suggested that many of the German prisoners downplayed the role of nationalistic and Nazi ideologies in their testimonies. See Siebold (2006) 186 for discussion. With regards to the ancient world, see MacMullen (1984) in support of this argument and Lendon (2004) against.

4   MacCoun and Hix (2010) 139–140. For a more on this subject in modern armies see Wong (2003); King (2006); and Kolditz (2006).

© KONINKLIJKE BRILL NV, LEIDEN, 2016 | DOI: 10.1163/9789004284852_007

aspects: task-based cohesion and social cohesion.[5] With this more sophisticated model in place, scholars have been able to delve deeper into the overall nature of military cohesion and have, particularly through a number of recent (and controversial) studies, come up with some intriguing conclusions.[6] Starting in the 1990s, scholars increasingly suggested that social considerations like racial-ethnic identification, marital status, and housing arrangements had little to no impact on unity and cohesion.[7] More recent work has confirmed the importance of cohesion for military success but,[8] perhaps surprisingly, the data has suggested that commitment to task-related goals, and not pre-existing social bonds (i.e. social cohesion) as expected, was the most important factor in determining both horizontal and overall cohesiveness, which in turn hints that this often overlooked aspect was in fact the key to effectiveness in modern military units.[9] The research and subsequent debate has also raised some significant issues, and created a new scholarly vocabulary and academic framework, for the study of morale and military cohesion more generally.

While much of the scholarship on military cohesion has emerged out of immediate concerns regarding the effectiveness of modern armies, the general principles and debates have also been increasingly adopted by those interested in ancient military studies. As Lee noted in his 1996 chapter on 'Morale and the Roman Experience of Battle', "a soldier's sense of identification with and commitment to the army in general, and especially to the particular unit of which he is a part ... plays a crucial role in the ability of soldiers to withstand the stresses of battle and to fight effectively."[10] This aligns closely with Hanson's stance, in his seminal volume *The Western Way of War*, where he noted that "the unique cohesiveness that existed among individuals within a phalanx accounts for much of the success achieved by Greek hoplites, especially in contrast to foreign troops ... the key must have been the camaraderie in the

---

5   MacCoun and Hix (2010) 139–140.

6   Many, although not all, of these studies have related to or been prompted by the United States military's 'Don't Ask, Don't Tell' policy with regards to homosexual soldiers and the various efforts to repeal it.

7   Siebold (2006) 196.

8   Oliver (1999) estimated the correlation between cohesion and performance to be r = 0.40.

9   MacCoun and Hix (2010) 156–157. These conclusions were obviously politically charged as they challenged the usual defence of 'Don't Ask, Don't Tell': that a change in the social makeup of the armed forces would damage military cohesiveness and, ultimately, effectiveness.

10   Lee (1996) 207.

THE TIES THAT BIND: MILITARY COHESION IN ARCHAIC ROME   103

Greek ranks, the confidence which grew out of the bonds among the hoplites in the phalanx."[11] Scholars investigating morale in ancient armies have, however, like scholars focused on modern armies, generally struggled to define the bonds which kept an army together and so have often reverted to eighteenth and nineteenth century models which identified concepts like pseudo-nationalism, professionalism, and *esprit de'corp* as the underlying principles.[12] The resultant models have, therefore, led to speculation that many armies in the classical world relied on a shared identity or 'social cohesion', often focused on a 'shared communal ethos' (variously conceived), as the core of their morale.[13] Indeed, as Crowley argued in his recent study of the Athenian hoplite "... the Athenian hoplite ... [fought] for his socio-political system ... and that social-political system, as an indefatigable agent of secondary socialisation, ensured that he accepted a martial set of norms and values profoundly congruent with the demands of close combat ..."[14]

This socio-political backing to military cohesion has effectively become the norm in ancient military studies, and is often used to explain a wide range of phenomena.[15] For instance, Hanson argued that when the "Athenian general Alcibiades was unable to forge two separate phalanxes into one army at Lampsakos in 409 ... [this was because] men who did not have any shared blood ties and who had no common experience under fire were hardly willing to form up together into the dense ranks of the phalanx."[16] Supporting this communal ethos is also the well-established principle of a shame-culture in the classical world, present as far back as the Homeric epics in Greek society and a regular feature of both Greek and Roman texts, which, it is argued, bolstered these existing social and political bonds via the very real penalty of social ostracism if one displayed cowardice in the battle line.[17] Added to this are often a collection of religious obligations,[18] experiential bonds (either shared battle experiences

---

11   Hanson (1989) 117–118. In contrast, however, see Lendon (2004) 445 ("The case for the decisive importance of cohesion in the Roman context is, moreover, weak").

12   See, for instance, Goldsworthy (1996) 10–11 and Sabin (2000) 13.

13   This can take a range of relatively nuanced forms, however. See, for instance, MacMullen's 1984 seminal work 'The Legion as Society,' which presented the Roman army of the early empire as a distinct social unit, or subunit, within Roman society.

14   Crowley (2012) 128.

15   See, for example, Lendon (2004).

16   Hanson (1989) 118.

17   See van Wees (2004) 162–164 for a discussion of shame in the Greek context, see Goldsworthy (1996) 251–282 for Roman examples.

18   Crowley (2012) 96–100.

or training—see particularly Veg. *Mil.* 3.9),[19] kinship-based connections,[20] and shared social constructs regarding manhood and duty, all creating a seductive argument for a social-political structure binding the soldiers together before they ever stepped onto the battlefield.

While this social structure is naturally thought to have had a number of facets, the civic aspect is typically argued to have been the most important, as it gave the overall structure to the army, provided the unifying ethos which linked the soldiers together, and typically was responsible for both leadership and direction of the force; stepping into the position previously held by bonds of kinship and family hierarchies, binding the various subgroups together into a larger cohesive army.[21] Indeed, in many classical armies, membership in the state seems to have been the only overarching, unifying bond which all the soldiers initially shared and this membership (or the desire for membership) is often understood as the basis for their recruitment. As a result, up until the present day, ancient military studies have largely focused on the social, and particularly communal/civic, bonds within ancient armies which, it was thought, formed the basis of their morale, cohesion, and identity.[22] However, as noted above, recent sociological research has demonstrated that while social and civic bonds obviously play a role in military morale and cohesion, they are often of secondary importance to task-based objectives in modern armies, and the implications of this research has yet to be applied to ancient military forces. Additionally, the models for morale which presuppose a strong communal identity or ethos as the basis for military cohesiveness have struggled with situations where the evidence suggests a successful military existed, but where evidence for a strong community ethos was lacking, often arguing that the military cohesiveness and effectiveness should be taken as proof of social and political cohesiveness.[23] The present chapter will endeavour to explore some of the implications of the modern scholarly push towards task-based cohesion in military forces as a way to explain ancient morale, with a particular focus on an ancient society which seems to have exhibited a reasonably successful

---

19   See Gilliver (1999) 92–93 for discussion.

20   Ludwig (2007) 211.

21   See van Wees (2005) 95–97 and his discussion of warfare and the archaic state. See Crowley (2012) for a Greek example of this argument. The Roman army, particularly of the early Empire, presents a slightly different model and is often seen as an associated subculture within the larger Roman socio-political construct—see MacMullen (1984).

22   See, for instance, Crowley's (2012) analysis of the Athenian hoplite.

23   Cornell (1995) 325.

THE TIES THAT BIND: MILITARY COHESION IN ARCHAIC ROME

military, but where the concepts of citizenship and community identity were evidently still developing and flexible—archaic Rome.

Problems with Identity in Early Rome

It may be appropriate here to address one of the fundamental premises behind the choice of archaic Rome as a case study—that early Roman armies lacked a strong overriding state ethos or identity which served to bind their military forces together—as this, of course, flies in the face of the literary tradition. The problems with the literary sources for early Rome are so numerous and so well examined that they virtually go without saying. The cryptic and incomplete nature of 'the sources for our sources', the numerous evident anachronisms present in the extant narrative, and the natural elaboration which characterizes ancient historiography, has led even the most positive of early Roman scholars to question the most basic aspects of the traditional narrative. T.J. Cornell for instance, who represents one of the more optimistic voices with regards to the sources for early Roman history, has noted that "the nature of the surviving evidence seems ... to warrant a cautious approach ... and [even so] it is difficult and sometimes impossible to identify precisely which elements [of the narrative] are true 'structural facts' [and which are late republican elaboration]".[24] Even this cautious position may understate the point. Livy, in the opening lines to Book 6 (6.1) of his *Ab Urbe Condita*, famously urged caution when considering his first pentad, noting that,

> ... I have set forth in five books, dealing with matters which are obscure not only by reason of their great antiquity—like far-off objects which can hardly be descried—but also because in those days there was but slightly and scanty use of writing, the sole trustworthy guardian of the memory of past events, and because even such records as existed in the commentaries of the pontiffs and in other public and private documents, nearly all perished in the conflagration of the City.
>
> trans. FOSTER

We can then, arguably with the permission of Livy himself, treat his (and Dionysius of Halicarnassus') assertions regarding a cohesive Roman state with a certain degree of skepticism. Indeed, when the entirety of the available evidence

---

24   Cornell (2005) 61.

for Rome is critically analyzed, it becomes increasingly clear that Rome did not have the same strong, cohesive, community-based identity binding it together in the sixth and fifth centuries as it did in later centuries, particularly amongst the elite who arguably represented the most important figures in the narrative, and particularly in the military realm.[25] The evident mobility of the aristocratic *gentes*, until at least the so-called 'closing of the patriciate' (*la serrata*) sometime in the late fifth century, and the flexible and changing nature of Rome's socio-political groupings and institutions down into the fourth century both suggest that the ties which bound Rome together were still forming during the early Republic, let alone during the Regal period.[26] As a result, while archaeology increasingly suggests that Rome was a thriving community and a center for elite activity from at least the start of the sixth century, with key evidence coming from regions like the S'Omobono area, the relationship between the community and the elite seems to have been a fluid and contested one.[27] Although it is likely that the region's elite were beginning to focus their attention on the community of Rome, evidence for elite habitation is surprisingly minimal and, apart from the problematic *regia* in the forum, generally seems to date to the late sixth century at the earliest.[28] So, although the evidence is far from conclusive, scholars have increasingly begun to suggest that archaic Roman society may have been a bit more disparate than previously thought, and that the very notion of a distinct Roman identity before the fourth century, at least among the elite, may well have been a mirage.

This more flexible character of the elite Roman identity, and arguably the Roman state, can also be seen in the armies of the period, where a range of models are evident. There is increasingly solid evidence for *gens*-based armies, like that of Brutus and Collatinus in 509, Appius Herdonius in the 460s, along with state sanctioned *gens*-based activity like that of the Fabii in the 470s, and finally what seem to be fully state-based armies of the 440s and later, all suggesting that a number of different types of military models were present in Rome's during the fifth century, and likely much earlier.[29] While the presence of these various types of forces is increasingly accepted by modern scholarship, how they interacted with each other and the community of Rome are still hotly debated issues.[30] One of the more significant issues to emerge from

---

25   Rich (2007) 15–16. All dates are BC unless otherwise stated.

26   See Armstrong (forthcoming) for more detailed discussion.

27   Bracato and Terrenato (2012).

28   Terrenato (2001).

29   Armstrong (2008) 47–66.

30   See Rawlings (1998) for discussion.

this new, multifaceted, view of Rome's military forces is that, given the varied social composition of the forces, the traditional model for military cohesion has proved largely inadequate when attempting to explain them. Specifically, with the relationship between the region's elite and the community of Rome increasingly under scrutiny, there is no longer a single, strong, state-based structure or ethos to explain military cohesion. In the middle and late Republic it is clear that each *gens* (and arguably each individual) was still largely concerned with personal glory and status. However, as Gelzer and more recently Hölkeskamp have argued, from the fourth century onward there was increasingly a realization and acknowledgement that they were all competing within the same 'game' focused on the politics of the community of Rome.[31] As a result, although their interests within Rome were obviously incongruent and indeed competing, they were unified in their desire to expand Rome's power and so increase the scope and scale of their achievements in that context.[32] But in the fifth century, Rome's position at the centre of this elite competition or 'game' had yet to be firmly established, although her size and prime location had already drawn quite a few powerful families into her orbit. Rome's aristocracy likely lacked the strong sense of community identity which helped unify their efforts in later centuries and instead seem to have conceived of themselves in a far more independent manner. As a result, when looking at Rome's early armies one is faced with a range of different social and political relationships and identities, many overlapping and perhaps contradictory, all existing alongside each other in Rome's military sphere.

This situation is obviously problematic as, if the sources are to be believed at all, Rome fielded reasonably effective armies in virtually all of her engagements throughout the period. Naturally it is likely that part of this reported success reflects an aversion by Roman historians to record defeats, with the confusing conclusion to the attack by Lars Porsenna in the late sixth century and Camillus' retribution for the sack of Rome by the Gauls in 390/387 representing obvious examples of this tendency. However, apart from a few setbacks, both the archaeological record and the literary narrative agree that during the fifth and early fourth centuries Rome slowly grew in power and was ultimately able to militarily dominate the entirety of central Italy by the 340s. Therefore, if the basic premise behind modern military effectiveness can be transferred to antiquity—that effectiveness is directly linked to cohesion—there must have been something unifying the heterogeneous elements within these armies.

---

31 Gelzer (1912); and Holkeskamp (1987).

32 See Harris (1985) and Rosenstein (1990) for more detailed discussion.

The answer to this conundrum lies in combining the work done on military cohesion in modern armies, and specifically the importance of task-related goals, with a more critical approach to archaic Roman warfare which attempts to remove the late republican obsession with state-based conquest and examine it for what it really represented. While social, and particularly civic, bonds may have played a role in binding together early Roman military forces, they were likely not the only, or the most important, bonds present. Instead, the cohesion and effectiveness of Rome's early armies was likely the result of a range of other connections, with short-term, task-based cohesion key amongst them.

Nature of Warfare in Archaic Rome

The traditional narrative holds that Rome was one of the most dominant military powers in century Italy during the sixth and fifth centuries, winning victories against a range of opponents and slowly building up a territorial empire in the process. Many, like Cornell, have taken the claims of the sources at something resembling face value and have interpreted the strength of Rome as that of a powerful state, with a strong communal army, controlling a territorial empire.[33] Indeed Rome, if the sources are to be believed at all, was very successful militarily during the regal and early republican periods and seems to have exerted some degree of control over a fairly large area.[34] However, the exact nature of this military power is difficult to unravel.

Despite the vivid descriptions of battles contained in Livy and Dionysius, warfare in archaic Rome and Latium seems to have been a relatively low-level and small-scale affair focused more on raiding than on conquest.[35] Rome rarely seems to have maintained dominance over communities defeated in battle, usually returning year after year to 'reconquer' her 'defeated' foes. This can be seen in examples like Rome's 'conquest' of Fidenae in the 430s and 420s (Livy 4.17–33), or Bolae in 415 and 414 (Livy 4.49–51; 4.59; 5.8; 5.13; 5.16), and countless others. Furthermore, the community of Rome's control of territory outside of the *pomerium* is itself problematic. As Roselaar has argued, apart from a few key incidents, like the taking of land from Aricia and Ardea in 446

---

33 Cornell (1995) 208–214, 322.

34 Rome's prominence in various treaties, and most notably the first treaty with the Carthaginians—traditionally dated to 509—along with the early 5th century *Foedus Cassianum*, is commonly used to support this hegemony over Latium.

35 Rich (2007) 15–20.

or Bolae in 414, the Romans did not conquer much land at all during the archaic period.[36] Indeed, the first substantial amount of land added to Rome seems to have been in 396 with the capture of Veii. Instead of land, the Romans seem to have been far more concerned with acquiring portable wealth in their warfare. This preference is explicitly stated in narrative for the capture of Bolae in 414, when land was taken only because there was no portable wealth available, as the city had been sacked the previous year (Livy 4.59). Additionally, the territory outside of the urban area of Rome was likely under the control of various *gentes*—a point which is both supported by the sources and by the names of the early rural tribes.[37] As noted previously, the relationship between the early *gentes* and Rome's urban community is ambiguous to say the least, but it is likely that this rural territory represented at best a liminal zone for Rome, which should only be considered part of Roman territory as long as the *gentes* in charge of it felt that maintaining a connection to Rome was in their best interests. Indeed, Rome (as a unified community) only seems to have become interested in controlling territory itself in the late fifth and fourth centuries, initially in reaction to the changing economy of the central Italy and later as a means to increase her military manpower in the face of a perceived Gallic threat after 390/387.

Warfare in archaic central Italy then, and particularly that practiced by the forces of Rome, seems to have been driven by motivations quite apart from state-based expansion and control during the sixth and fifth centuries, and by a range of military units which often did not require the state to form or function. By far the most active group in warfare seems to have been the *gentes*, often led by so-called *condottieri*, who raided and defended against raids and are epitomized by figures like Macstarna, the Vibenna brothers, Brutus, Collatinus, and many others. Apart from these clan-based forces, there were the forces of the various urban communities which generally seem to have been largely defensive in character, resembling simple militias, defending their territory from incursions by foreign *gentes* and the tribal entities from the interior, but not particularly interested in territorial conquest (at least during the early fifth century). This model for warfare accords well with the more flexible understanding of Roman identity and civic cohesion during the archaic period, which is particularly visible amongst the region's elite in the fifth century, as it all points toward a heterogeneous community with *ad hoc* goals driven by short-term desires—and as such stands in direct opposition to

---

36   Roselaar (2008) 33.

37   Taylor (1960) 35.

the traditional narrative presented by Rome's late republican historians. Given this position and the bias of our sources, we unfortunately lack the resources necessary to fully explain the exact nature of the military cohesion within the multifaceted collection of forces present in Rome during the archaic period. However, it may be possible to develop some structures which help to define it in broad terms. Specifically, there is evidence for both 'vertical' and, more importantly, 'horizontal' cohesive factors which may have served to unify these groups outside of the traditional socio-political and/or civic bonds.

Vertical Cohesion

The command structure, which facilitated vertical cohesion in Rome's archaic armies, needed to be able to accommodate a range of different military models and possibly internal units, including both *gentes* and more community-based forces. Due to the nature of military actions and a general desire for decisiveness, almost all military command structures are roughly hierarchical in nature.[38] However, within this there naturally exists a range of possibilities.[39] In Rome, the *gens* and the community each offered a possible model for control with discrete forms or types of legitimacy for a leader. From what we can tell about the curiate system of Rome, itself an enigmatic thing, the political model of the archaic community represented a roughly egalitarian system, with a series of magistrates (*curiones*) selected from the larger community and each vote within the *curiae* counting equally in decision-making (Gell. *NA* 15.27; Livy 1.43; Dion. Hal. *Ant. Rom.*2.14, 4.20, 84, 5.6). This system was, of course, usually headed by a single *rex*, and later a collection of *praetores/consules*, and therefore retained the hierarchy needed for military leadership, but the core structures seem to have been generally egalitarian in nature, as indeed the *reges* and *praetores/consules* were technically elected officials. The legitimacy conferred by this system likely reflected this model, with the power of the leader representing the consciously given will of the group for a limited amount of time. Conversely, the *gens* seems to have represented a much more hierarchical and personal structure, where the *potestas* of the *paterfamilias*, whose position was innate and irrevocable, held sway. In this model, the legitimacy would have been more circumscribed in nature, as it was dictated by the family relationship, but also far more potent. In the Greek world, the

---

38   Smith (1965).

39   See Westbrook (1980) 261–264 and Siebold (2006) 190.

democratically-leaning *poleis* of the Classical period generally seem to have adopted the former model, with Athens and her ten tribal generals offering the best example.[40] In mid-to-late republican Rome, the military command structure seems to have sat roughly in the middle of the spectrum, with a pair of elected consuls sharing power at the top of a more hierarchical set-up, which seems to have combined *auctoritas*, *dignitas*, and increasingly prorogated (and therefore long-lived) commands with the community granted and circumscribed power of *imperium*. In archaic Rome, however, the command system seems to have been even more extreme. Perhaps as a result of their prevailing dominance in warfare during the archaic period, the Romans seem to have initially based their command structure almost entirely on the clan-based power dynamic and only democratized it gradually over the course of the Republic. The reasons for this are varied, and may relate in part to the strongly gentilicial character of Rome's early community-based forces, which were likely formed around the core of the *gens* of the *rex*. However, it may also relate to the legitimacy offered by the two models as well. While the curiate structure which governed the community of Rome was evidently effective, it was also severely limited by the flexible nature of Roman identity and citizenship during this period. Curiate law, including the problematic *leges regiae*, seems to have held sway within the community of Rome itself, but outside the *pomerium*, and indeed on the battlefield itself, these types of civic constructs may have lacked the strength to enforce certain modes of behaviour on their own.[41] As a result, early Roman armies, whether they were gentilicial or state-based, seem to have been controlled via an intensely hierarchical combination of *patria potestas* and *imperium*—which intriguingly seem to have represented effectively the same thing in the field of battle.

It is clear that, even in later periods, both *patria potestas* and early *imperium* existed in the same sphere, as indeed was suggested by Mommsen in his work on Roman statues, and, as Nisbet and others have noted, there seems to have been a core conflict between these two concepts, particularly in the military sphere, as they were often felt to overlap.[42] This is perhaps best illustrated in

---

40   Ober (1989) 91–93. See also Crowley (2012) 121–125 for a detailed discussion of legitimacy of command in Athenian armies.

41   As Westbrook (1980, 263) has argued, "soldiers' relationship to their immediate leaders is generally considered to be a more important factor in eliciting compliance than their relationship with higher military commanders or political leaders" and very possibly the overarching political construct. See also Drogula (2007) for a full discussion of the legal restrictions inside and outside of the *pomerium* in the regal period.

42   Nisbet (1968) 218.

a famous incident from 213, when the sitting consul, Quintus Fabius Maximus, met his father, Fabius Maximus Cunctator, while riding through a camp (Gell. *NA* 2.2.13; Val Max. 2.2.4; Livy 24.44; Plut. *Fab.* 24).[43] Here, the *patria potestas* of Fabius Maximus Canctator seems to have come into direct conflict with the *imperium* of his son Quintus, and it was initially unclear who should show deference. In the end, the elder Fabius dismounted and praised his son, indicating that by this time the *imperium* of the *populus* trumped the more limited power of the *paterfamilias*, although it is arguable whether this would have always been the case. But either way this, admittedly somewhat legalistic, anecdote has often been taken as evidence that these two powers seem to have existed in roughly the same sphere of influence.[44]

Despite this point of union and the various possibilities which it raises, both *patria potestas* and *imperium* remain incredibly enigmatic for the archaic period. While the nuances of *patria poetstas* in the archaic period are likely beyond us because of the nature of the evidence, all the evidence points toward it being the ultimate source of authority in most matters within the core kinship group.[45] This can be seen in the archaic phrase used by the Pontifex Maximus in the ceremony of *adrogatio* in the *comitia curiata*, which invoked the *potestas* of the *pater, vitae necisque* (Gell. *NA.* 5.19.9).[46] Indeed, although Smith has argued convincingly that the archaic *gens* defies a simple kinship-based description, it did seem to function based on a roughly hierarchical power structure which featured *patria potestas* as a core dynamic.[47] Thus, it is possible to suggest that although it is unlikely to be the only dynamic in play, a hierarchy based on an extended form of *patria potestas* within the *gens* likely formed part of the vertical cohesion which held together gentilicial units in a military context.

What *imperium* actually meant in archaic Rome, let alone later periods, is of course highly problematic as well, but it does seem to represent yet another type of bond and an aspect of vertical cohesion. Granted, or in some way governed, by the *comitia curiata* during the regal period and early Republic, *imperium* seems to have given a *rex* (and later the praetor or consul) the formal ability to command military forces in the community's name. It is clear, however, that the nature of this bond was multifaceted. Looking at the power of

---

43 See Beck (2012) 87.

44 See Masi Doria (2000) 240–241 for discussion.

45 Westbrook (1999).

46 *velitis, iubeatis, uti L. Valerius L. Titio tam iure legeque filius siet, quam si ex eo patre matrequefamilias eius natus esset, utique ei vitae necisque in eum potestas siet, uti patri endo filio est.* (Gell. *NA.* 5.19.9).

47 Smith (2006).

generals imbued with *imperium* in later periods, we can suppose that the army of the *rex* was unified on the battlefield in part through military discipline, as *imperium* gave the general the right of life or death (*vitae necisque*) over those under his command.[48] There was also likely a religious component to the bond, although what this might have been and the connection to the auspices has been debated going to back at least as far as Mommsen, and we are no closer to a consensus.[49] Returning to the connection to *patria potestas*, however, we are offered some clues as to the nature of *imperium*. Fundamentally it seems that this bond was a hierarchical one which resembled, or may have been derived from, *patria potestas* and existed between the *curiae* of Rome and the *rex* (and later the praetors and consuls). Thus, the members of the *curiae* could be interpreted as existing, albeit temporarily and in a limited and ritually circumscribed fashion, within the extended *gens* of the *rex*, praetor, or consul.

Imperium thus provided a valuable bond which connected previously disparate entities within Rome—the *curiae* and the *rex*/praetor/consul.[50] Although membership in the *curiae* was traditionally associated with gentilicial links (a point which ultimately goes back to Laelius Felix's assertion that the *curiae* were organized according to *genera hominum*, Gell. *NA* 15.27),[51] the evidence for *gentes* being part of the *curiae* is ambiguous at best,[52] and may have represented an anachronistic invention.[53] However, if the *curiae* did contain patrician *gentes*, the granting of *imperium* to a *rex*, praetor, or consul, would have linked together a number of previously unrelated *gentes* into a single structure. The presence of *gentes* in the *curiae* is by no means certain, and

48   Drogula (2007) 451.

49   Brennan (2000) 54–57.

50   This is perhaps problematic, as it arguably introduces the concept of adhesion ("The situation where relatively unlike groups coalesce for a period of time to accomplish a common purpose") versus cohesion (where "largely similar individuals [hold] together to accomplish [a goal]") (Siebold (2006) 188). However, given that the key point of difference in these concepts is their relative 'likeness' and the presence of a preexisting connection, whether *imperium* represents a supporting bond of cohesion or an agent of adhesion is impossible to tell conclusively from the evidence.

51   See, for instance, Carandini (1997) and Smith (2006).

52   See for instance Palmer (1970) 69–71.

53   It is clear that the ancient sources believed that the *curiae* contained *gentes*, as seen by Dionysius of Halicarnassus' discussion of the formation of the Roman senate (Dion. Hal. *Ant. Rom.* 2.47) and Livy's association between the *curiae* and the entirety of the Roman population, seen in the phrase *populus Romanus Quiritum* (Livy 1.24, 1.32–33, 2.23, 3.20, 3.41–45, 6.40, 8.9, 9.10, 10.28), although he is unique is using this formulation, see Smith (2006) 200.

indeed given the strong agricultural associations of the *curiae*,[54] the association between the *comitia curiata* and the *transitio ad plebem* during the Republic,[55] and indeed the fact that the *comitia curiata* elected the first plebeian tribunes, there are hints that the *curiae* were initially associated primarily with the plebeian or proto-plebeian population in Rome.[56] But whether they were composed of patricians, plebeians, or a combination of the two, it seems clear that the *curiae* contained a heterogeneous, and likely ever-changing, collection of individuals which represented the urban community of Rome. It is, therefore, possible to suggest that *imperium* may have represented a vertical bond between the settled, urban population of Rome and a military leader that closely mirrored the bond between a *paterfamilias* and his family members.

Horizontal Cohesion

These aspects of vertical cohesion represent only part of the story, and indeed arguably the least important aspect for military effectiveness based on the research into modern armies, which suggests that vertical cohesion follows horizontal cohesion in a predictable way.[57] Although the vertical hierarchy gave direction to the force and provided one part of the overarching bond, the horizontal cohesion between the men in battle line represented the vital connection within the group.[58] As noted previously, this bond was traditionally assumed to have been based almost entirely on socio-political and civic factors in ancient armies. While this is increasingly unlikely to have been the case, it *is* probable that various types of social bonds did play a role. There are a number of different theoretical approaches to dealing with horizontal cohesion in military units, most notably looking for social capital, social function, natural development of social groups, and collective goals.[59] Although it is hard to determine anything with certainty based on the sources for early Rome, one thing which does seem clear is the importance of family for early central Italian society. As in Greece, and elsewhere in the Mediterranean, kinship formed a key bond which linked people together in both communities and clans, giving them an identity within the context of the larger socio-political and economic

---

54 Smith (2006) 206–207.
55 Smith (2006) 212–213.
56 See Magdelain (1979) for a similar argument.
57 Siebold (2006) 196.
58 MacCoun and Hix (2010) 139–140.
59 Siebold (2006) 192–193.

THE TIES THAT BIND: MILITARY COHESION IN ARCHAIC ROME 115

environment. As a result, it is only natural that the bond of kinship was a key bond between soldiers on the battlefield as well, and this is something which is generally supported by the sources.[60]

The most obvious example of the familial bond on the battlefield was within the *gentes*. It is clear that the Roman *gentes* contained a strong martial component and, while they may not have been the only military entities in central Italy, they were clearly active down into the fourth century with gentilicial units seen at Clusium before the Gallic sack of Rome and countless other occasions.[61] But family units, and indeed extended families (which may have included *clientes*), were also important outside of the *gentes* during the archaic period and, while they may not have had the size or complexity of the archaic clans, they created a social framework which was likely reflected on the battlefield as well. Looking at parallels from Classical Greece, the family bond was vital on the battlefield. Although the state was in charge of initiating conflict, armies were organized by tribes with the result that fathers, sons, uncles, etc. all ended up fighting alongside one another. Consequently, although an abstract concept of *polis* and community was obviously a factor, the ultimate motivation for battlefield unity and cohesion is usually argued to have been the relationships which existed between the friends and family members fighting alongside one another. Indeed, the Theban Sacred Band and the Spartan practice of stationing lovers side by side in the phalanx are often seen as an extension of this.[62]

Moving a bit further away from the foundational bond of kinship, it is likely that archaic central Italian armies had another, sometimes overlapping, bond focused on religion. This aspect of archaic central Italian warfare has particularly been emphasized in the work done on the famous Lapis Satricanus, which records an inscription by the *suodales* of a certain Populos Valesios to the god Marmars (usually interpreted as Mars). The title *sodales* is tricky and Roman etymologies of *sodales* (see particularly Festus 383L) do not seem to recognize a connection to *suus* as one would expect, but rather suggest a relationship which might best be described as one between 'sword brothers'. Livy used the word *sodales* on several occasions in his history of the early Republic, and in every case it carried the meaning of 'warrior-follower'. He used the term when discussing the followers of Tarquinius Superbus in Rome (Livy 2.3), again when discussing the followers of the Fabian gens who went along on their war against Veii in 479 (Livy 2.49), and when discussing Caeso

---

60 Armstrong (2013) 53–69.
61 Smith (2006) 281–298.
62 Ludwig (2007) 211.

Fabius' followers following the aristocrat's exile in 461 (Livy 3.14). Dionysius of Halicarnassus provided still further corroboration for this interpretation of *sodales* as he translated the word as *hetairoi*, a title synonymous with Alexander the Great's companions and Companion Cavalry,[63] in his description of the Fabii in 479 (Dion. Hal. *Ant. Rom.* 9.14). *Sodalis* also obviously had religious connotations, as the term was often applied to priesthoods in the late Republic and Empire.[64] While the origins and duties of these priesthoods varied, *sodales* were exclusively male and often had martial connotations.[65] In particular, one of the most famous *sodalitates*, the *sodales Salii*, wore archaic Italian armor, were the keepers of the 12 sacred *ancilia* and held ritual processions through Rome associated with the start of the campaigning season.[66] Indeed, Torelli has argued that '*suodalitas*' was an archaic social institution, possibly cemented by a special religious connection, which may have been similar to the aristocratic Germanic warrior companies described by Tacitus.[67]

These social factors would have likely played a role in creating a series of bonds between various members of the Roman community, and particularly within the elite *gentes* where the vast majority of them were located. As a result, it should come as no surprise that individual *gentes*, like the Fabii in the 470s, were able to function as effective military units whose main limiting factor was likely their size. There is very little, however, beyond the vertical bonds of *imperium*, linking the various *gentes* together, or indeed the more heterogeneous population of the *curiae*. Without doubt the political and civic bonds in Rome should not be completely discounted in this regard. Although it is clear that Roman citizenship was incredibly flexible during the archaic period and likely depended heavily on physical location and was not an inherent, portable, political identity as it was in later centuries, the existence of the *comitia curiata* and the granting of *imperium* and the auspices by that body does suggest a corporate political and military identity of some type from the regal period onward. But while membership in a *curia* did probably play a role, it is unlikely

---

63   The term was used of the companions of the Macedonian king, his noble cavalry (*hetairike hippos*) and the infantry formations forming the backbone of the army (*pezetairioi*). Most specifically it could be used of the king's inner circle of friends.

64   Scullard (1981) 30.

65   This social exclusivity became problematic during the Augustan period. While many of the archaic *sodales*, including the *Salii*, were restricted to the patricians into the Empire, the *Luperci* were reserved for the *equites* by the Augustan period and the membership of the *Augustales* was made up principally of freedmen, see Scullard (1981) 64.

66   See Beard et al. (1998) 126–128 for a brief overview of the *Salii*.

67   Torelli (1999) 17.

to have carried with it the same strong sense of obligation and civic duty associated with Roman citizenship in the middle and late Republic, at least amongst the more mobile elite who would have supplied the best equipped soldiers. It is likely that another factor was in play.

It is here that the modern research into military cohesion and morale, and particularly the importance of task-based 'horizontal cohesion', may offer an answer. Archaic Rome seems to have contained a myriad of different social groups, as even her origin myth as an asylum suggests, but, despite this heterogeneous composition, the style of warfare practiced by the Romans during the archaic period was surprisingly uniform and generally offered a very clear set of goals with a set of common interests for the soldiers involved—factors which form the crux of modern theoretical frameworks of cohesion.[68] On the defensive side, the mutual defence of territory against raiding parties was likely part of the impetus behind the creation of the first communities, and so the common defence of property would have always provided a powerful task-based bond linking soldiers.[69] Offensively, as noted above, once the anachronistic desire of Rome's late republican historians to cast everything in the light of territorial conquest has been removed, it is evident that the vast majority of Roman warfare down through the fifth century can be quickly categorized as raiding for portable wealth. In contrast to the conquest of territory, where the benefit was often ambiguous and indeed often enjoyed by those who did not engage in the conflict directly (usually via the creation and use of *ager Romanus*), the acquisition of portable wealth through warfare would have formed a direct and tangible goal which could bring together even the most disparate entities.[70] Indeed, the desire for, acquisition of, and distribution of portable wealth remained a key unifying factor in Roman armies even after the evolution of Roman citizenship and civic identity.[71] Consequently, it can be argued that the bond which unified many of Rome's early armies was at least partially economic in nature, and that this immediate, task-based bond should be seen as one of the most important horizontal bonds in early Roman armies, linking together the disparate, and possibly only loosely politically unified, groups.

Apart from the comparative principle suggested by research into modern armies, there are also hints of this multifaceted, but fundamentally economic, bond unifying archaic Roman armies in the ancient evidence. Specifically,

---

68   Siebold (2006) 192–195; Westbrook (1980) 244–278.
69   This interaction is commonly cited as the core of the 'in-group/out-group' bias. See Murphy (1957) 1018–1035.
70   *Contra*, Crowley (2012) 107–109 with regards to Greece.
71   Shatzman (1972) 177–205.

all of these bonds are visible in the institution most closely associated with Roman success in warfare: the Roman triumph. Beginning with the vertical bonds, the triumph was clearly associated with *imperium*—as only a magistrate with *imperium* could celebrate one. It also contained a number of religious elements, including the culminating sacrifice to Jupiter Optimus Maximus and the ritual feast. Within the triumph there was also an acknowledgement of the community as a whole, as the crossing of the *pomerium* and the return of the army to the city remained key aspects (and points of contention) well into the late republican period. But underpinning the entirety of the ritual is a focus on wealth—from an initial distribution of spoils outside of the *pomerium*, to the carrying of wealth through the community, and ultimately the sacrifice on the Capitoline—all suggesting that the very tangible benefits of warfare represented a key facet in this relationship, and arguably one which served to bind the soldiers together more tightly than the amorphous concept which was 'Rome' at this time.

Conclusions

All of the evidence for archaic Rome points toward a rich but diverse community. Rome's origin myths suggest that the city's first inhabitants were a ragtag mix of asylum seekers and opportunistic warriors, while the archaeology for the archaic period shows evidence for a range of cultures moving through this key crossing point on the Tiber. Frustratingly, however, the melding of these various groups and entities in Rome seems to have taken a significant amount of time. Although there is evidence for some sort of political structure in the community from an early date, Rome's elite only seemed to have established themselves as a distinct group sometime in the late fifth century and of course the 'Struggle of the Orders' suggests deep internal divides still existed well into the fourth century. The success of Rome's armies, however, has often seemed to belie this political disunity. Although Rome did suffer some setbacks during the regal period and early Republic, the narrative is generally one of repeated success. In the past, this has often been taken as implied proof of Rome's underlying social unity and that, despite the evident social and political issues within the city, the Roman people would generally unify politically for military matters.[72] Recent work into the morale and cohesion of modern armies has cast doubt on this assumption. While a civic bond may have been present in Rome's

---

72   Cornell (1995) 181–197.

armies, this does not seem to represent the most important aspect of over-all cohesion. Instead, individuals within the archaic Roman army were likely bound by a number of different, and often overlapping, bonds—including civic, kinship, religion, and most importantly economic aspects—only some of which were relevant to the state. Clearly, as Crowley noted, "the reasons why soldiers fight is as complex as human behaviour itself", and military cohesion will, therefore, always elude an attempt at a single explanation.[73] There are simply too many variables involved. However, the importance of task-based cohesion (which, in the case of early Rome, likely meant raiding for portable wealth), alongside other, socio-political factors, should not be discounted—particularly in periods like early Rome. An effective army does not mean a cohesive state, and vice versa.

---

73   Crowley (2012) 21.

CHAPTER 7

Sacramentum Militiae: Empty Words in an Age of Chaos

Mark Hebblewhite

The *sacramentum militiae*, or military oath of loyalty, has long been recognized as a core component of the relationship between the emperor and the Roman army during the early imperial period.[1] Far less, however, is known about its role in the relationship between the emperor and Roman army during the politically volatile period that stretched from the onset of the third century crisis in AD 235 to the death of Theodosius in 395.[2] This chapter is an attempt to at least partially rectify this state of affairs. After briefly examining the origins of the *sacramentum* in the republican epoch this chapter will trace its development through the early imperial period in order to establish a point of contrast for its role from 235–395. It will then establish the basic operation of the *sacramentum* during this later period, including discerning the content of the oath, the occasions on which it was sworn, and the manner in which it was sworn. These findings will then be used to explain the role the *sacramentum* played in the ongoing relationship between the emperor and the Roman army. Here two perspectives will be assessed. The first is that of the emperor. Did he consider the *sacramentum* to be a tool that could protect him against betrayal

---

1 Hereafter referred to simply as *sacramentum*. The terminology used in sources to describe the military oath is problematic. In many literary accounts of oath-taking by the army the term *sacramentum* is not used. Instead more generic language associated with oath-taking is utilized. For example, Amm. Marc. 21.5.9, 26.7.8 merely uses the verb *iuro* (to swear or take an oath) in his description. Elsewhere (26.7.17) he uses *testor* (to invoke or testify). SHA, *Max.* 24.3 does the same and simply utilizes *iuro*. Some sources use terminology interchangeably. Tacitus uses both *sacramentum* and *ius iurandum*—see Campbell (1984) 26, n. 23. The Greek texts provide similar issues. Herodian (8.7.4) uses the general term ὅρκος (oath) and then calls the oath σεμνὸν μυστήριον (the sacred secret) of empire. Zos. 2.44.3 also utilizes the generic ὅρκος. However, Veg. *Mil.* 2.5 specifically uses the term with reference to the army and it does appear with semi-regularity in other literary histories, see Campbell (1984) 26, n. 23. As will be shown the term is explicitly used in epigraphic and legal documentary evidence This strongly suggests that the military oath was indeed 'officially' referred to as the *sacramentum*. This term will therefore be utilized in the course of this paper.

2 All dates AD unless indicated otherwise.

© KONINKLIJKE BRILL NV, LEIDEN, 2016 | DOI: 10.1163/9789004284852_008

by his army? The second is that of the troops themselves. Did they believe that swearing the *sacramentum* created an actual burden of responsibility to remain loyal to the emperor? And if so, what was the basis of this responsibility and how seriously did they take it?

Origins of the *sacramentum*

The *sacramentum* remains a difficult area of research. Phang noted that, for the Republic and early empire, it "has been much studied, but is still poorly understood".[3] As for the late antique period the subject has received even less attention.[4] Much of this reticence is explained by the limited state of the relevant ancient evidence, which is heavily weighted towards the Republic and early Empire and largely consists of passing references in narrative histories. Most problematic is the evidence for the content of the *sacramentum*. While examples of its civilian counterpart survive, no extant text of the *sacramentum* is preserved.[5]

The scholarly debate regarding the development of the *sacramentum* during the Republic and into the early Empire does not need to be repeated here in detail.[6] Instead it is enough to note a number of broadly accepted conclusions. Evidence found in Livy (22.38), Dionysius of Halicarnassus, (*Ant. Rom.* 10.18.2) Isidore of Seville (*Etym.* 9.3.53) and Servius (*ad Aen.* 8.614) suggest that during the republican age soldiers swore an oath to follow the orders of the consuls, fight bravely, and not desert the field of battle. Importantly, although the *sacramentum* demanded that Roman soldiers accept the authority of the consuls while in the field, it reinforced the principle that they were servants of the Republic (Servius *ad Aen.* 8.1).[7] From its beginnings, the *sacramentum* was a public expression of fidelity to the state and its representatives. Polybius describes the ritual that would take place:

---

3   Phang (2007) 117.

4   Recently Lee (2007) 52–53 has worked to correct this.

5   The one source that offers something akin to an extant version of the *sacramentum* is found in the pages of Vegetius' *De re militari*, a work problematic in a number of respects, one of which is its portrayal of the military oath. Vegetius' portrayal of the *sacramentum* will be treated at length in the course of this study.

6   The most useful, and succinct, treatment remains Watson (1969) 44–50. See also Campbell (1984) 19–32; Phang (2008) 117–120; and Stäcker (2003) 293–308.

7   See also Dion. Hal. *Ant. Rom.* 10.18.2 and 11.43.2; cf. Brice (2011) 37–39.

The roll having been completed in this manner, the tribunes belonging to the several legions muster their men; and selecting one of the whole body that they think most suitable for the purpose, they cause him to take an oath that he will obey his officers and do their orders to the best of his ability. And all the others come up and take the oath separately, merely affirming that they will do the same as the first man.[8]

POLYB. 6.21.1–4, trans. PATON

In the late Republic the *sacramentum* became part of the new political reality of civil conflict. When ambitious magistrates sought to use the army to advance their own power, they used the *sacramentum* to try and ensure that the army was bound to them before anyone or anything else, including the needs of the Roman state. Sulla ushered in one of the most important steps in this development. Plutarch records that upon his return to Italy in 83 BC Sulla had his troops swear an oath that demanded loyalty to his person.

When he [Sulla] was about to transport his soldiers, and was in fear lest, when they had reached Italy, they should disperse to their several cities, in the first place, they took an oath of their own accord to stand by him, and to do no damage to Italy without his orders ...

PLUT. *Sull.* 27.3, trans. PERRIN

This represented a distinct development in the *sacramentum*. The traditional *sacramentum* bound Roman soldiers to the will of their commander as a practical measure more than anything else, as the consul needed a guarantee of his troops' obedience if he was to fulfill the role of battlefield commander.[9] He was, of course, still viewed as an instrument of the state. Sulla's *sacramentum*, however, emphasized the primacy of an individual commander, who was in direct conflict with the political will of the Republic.[10]

---

8   See also Festus, *Gloss. Lat.* 250L, who reports that the first full version was known as the *praeiurtio*—each soldier then simply said *idem in me*.

9   Brice (2011) 37–39.

10 It should be noted that Plutarch's evidence contains a potential ambiguity. He states that the troops "took an oath of their own accord". Does this mean that they decided to swear their oath to Sulla in an informal and spontaneous manner as opposed to Sulla himself organizing for the oath to be taken in a formal ceremony? Although Smith (1958) 33 seems to believe this to be the case, it is far more politically likely that Sulla formally orchestrated the oath-taking which was held in a formal manner.

The *sacramentum* in the Imperial Age

The *sacramentum* survived into the imperial age where it was further adapted to meet the new political reality of the emperor as supreme military commander. Now the *sacramentum* was sworn only to the emperor and no one else. Beyond the claim made by the stoic philosopher Epictetus (*Dis.* 1.14) that soldiers who swore the *sacramentum* agreed to place the safety of the emperor above all else, little is known about the content of the oath in the early imperial period. But while no extant *sacramentum* exists, copies of the civilian loyalty oath (*ius iurandum*), introduced by Augustus, and sworn annually by all civilians, have survived.[11] This evidence raises two questions. First, was the oath sworn by the army the same as that sworn by the rest of the population? If this was the case, do the preserved civilian oaths represent extant texts of the *sacramentum*? Second, if this was not the case, was the civilian *ius iurandum* in any way similar to its military counterpart?

The civilian *ius iurandum* emphasized loyalty to the *princeps* and his family. One example calls on the civilian population of the town of Conobaria to vow their concern for the safety, honour and victory of Augustus, his adopted sons Gaius and Lucius, and his grandson Marcus Agrippa. The citizens also swore that they would carry arms for the emperor if necessary, as well as actively oppose anyone the emperor considered an enemy (*AE* (1988) 723).

Despite this civilian *ius iurandum* containing a 'call to arms', it is likely that the army continued to swear a distinctly separate oath.[12] No emperor could hope to win power without the military and the singular role this institution

---

11   As with the military *sacramentum* terminology regarding the civilian loyalty oath is not straightforward as they are variously described in the source material including as *sacramenta* by Suetonius (*Calig.* 15.3) For example the so-called *Coniuratio Italiae* is not actually labeled as a *coniuratio*. Instead in the *Res Gestae* of Augustus (*RG* 25) the generic verb for taking an oath or vow (*iuro*) is used. Ando (2000) 359–360 notes, oaths to the emperor fell into two categories—"prayers for the health of the reigning emperor (*vota pro valetudine* or *pro salute principis*) and oaths to obey his commands (*iurare in verba principis*)". In this study however the civilian oath of loyalty will be referred to by the broader term *ius iurandum*, a term utilized in *CIL* 11 172 and in Pliny *Ep.* 10.52.

12   This does not preclude soldiers from also taking the general oath of allegiance in addition to their *sacramentum*. Nor does it preclude the *sacramentum* and civilian *ius iurandum* containing at least some similar language. Campbell (1984) 23–27 argues that the *sacramentum* was amended under Augustus to incorporate the general oath of allegiance sworn by the civilian population—an oath which itself was probably based on the *sacramentum* of the Republic.

played in the maintenance of imperial power must have been reflected in its *sacramentum*. That an emperor would be content to bind a soldier to his cause with the oath sworn by an ordinary civilian is, in political terms, unlikely. Von Premerstein argued for a compromise whereby Roman soldiers swore a military *sacramentum* on joining the army and then on subsequent occasions swore the same loyalty oath as civilians.[13] This position has been refuted in detail by Campbell and Stäcker, who argued that while the civilian oath may have been similar to the military oath, it cannot have been identical.[14] Thus, while extant civilian oaths provide clues as to the content of the *sacramentum*, in particular that it was designed to promote the primacy of the emperor over all others in the mind of the swearer, the two oaths were unlikely to have been identical.

Finally, there is no evidence to suggest that the *sacramentum* sworn to Augustus was copied verbatim by every emperor who succeeded him. No doubt the basic tenets of faithful military service and exclusive loyalty remained intact, but it is possible that each emperor modified the content of the *sacramentum* to fulfill his particular political needs. This is shown by the reign of Gaius, where oaths were adapted to reflect contemporary political realities. The *ius iurandum* sworn by the people of Aritium in 37 made no mention of any member of the ruling house other than Gaius himself (*CIL* II, 172). However, Suetonius (*Calig.*15.3) claims that, at some point (presumably later than the Aritium oath), Gaius included his sisters in the terms of all loyalty oaths (*omnia sacramenta*) sworn to him as Augustus.[15] Presumably one of these oaths was the *sacramentum* sworn by the army.

The occasions on which the *sacramentum* must have been sworn are also instructive. Between the reigns of Augustus and Maximinus Thrax the *sacramentum* can be linked to three distinct occasions, all of which speak to its role in the constant conversation of fidelity that existed between emperor and army. First, all new recruits swore the oath immediately upon joining the army

---

13   Von Premerstein (1937) 75.

14   Campbell (1984) 27; and Stäcker (2003) 300–301. As Campbell (1984) 28 notes, it is of course possible that soldiers also swore the civilian oath in addition to the *sacramentum*. Pliny, *Ep.* 10.52, provides interesting evidence on this point stating that both soldiers and civilians swore an oath (*ius iurandum*) on the same day. This does not necessarily mean that they swore the exact same oath in terms of language—indeed Pliny insinuates the oaths were taken separately.

15   Interesting here is the use of the plural *sacramenta* to describe a range of oaths. No doubt Suetonius is using the term in its broad sense of oaths sanctified by religious censure if broken.

(Epict. *Dis.* 1.14.15, Veg. *Mil.* 2.5, Pliny *Ep.*10.29).[16] However, unlike its republican counterpart, which covered only the campaign in which the soldier fought, the *sacramentum* covered the soldier's entire period of service.[17] Second, the *sacramentum* was renewed when a new emperor came to power.[18] This practice shows that the *sacramentum* was a pledge of loyalty to the emperor and not the state. The final occasion on which the *sacramentum* was sworn speaks to its role in the ongoing maintenance of fidelity. Although the *sacramentum* sworn by new recruits was theoretically 'permanent', emperors demanded that it be renewed annually. There remains some confusion regarding when this annual renewal occurred. Tacitus (*Hist.* 1.55) states that the annual renewal of the *sacramentum* took place on January 1st. However, this seems to have changed by the second century. Surviving letters from Pliny to the emperor Trajan (*Ep.* 10.52, 10.100–101) state that soldiers and provincials alike swore fidelity to the emperor on both January 3rd, the date of the *nuncupatio votorum*, and January 28th, the *dies imperii* of Trajan. The *Feriale Duranum* also lists the 3rd January as the day on which vows (*vota*) were taken, again strongly suggesting the recitation of the *sacramentum* (*P.Dura 54.* 2–6).[19] It is, of course, possible that the annual date was shifted to suit the needs of individual regimes, but it is more likely that the *sacramentum* was renewed more than once a year— probably on the 3rd of January and on the *dies imperii* of the reigning emperor.[20] Whatever the case, this annual renewal represents an important development in the *sacramentum's* significance. It was no longer a static tool used to ensure loyalty for a set amount of time. The emperor now used it as part of his ongoing efforts to maintain military loyalty. He recognized that the army's loyalty was not necessarily fixed and required constant attention.

---

16   See also Tac. *Hist* 4.72, As Campbell (1984) 26 notes, here Tacitus points out that the defeated soldiers would have to are told to renew their allegiance as if it was their first day of service—insinuating that the *sacramentum* was sworn by troops at the beginning of their service.

17   The bonds of the *sacramentum* had to be formally released at the end of a soldier's period of service.

18   Cass. Dio 57.3.2; Tac. *Ann* 1.7, *Hist* 1.36, 1.76, 2.6 (Otho), 2.79 (Vespasian).

19   Watson (1969) 49 argues that this was indeed the case for the early first century but that, following either Vespasian's defeat of Vitellius or Domitian's defeat of Saturninus, the date was changed to 3rd January. While it is possible that this was the case and that the date was subsequently changed in later years, the date of 3rd January given in the *Feriale Duranum*, an official document, suggests that Tacitus was in error or the date had changed. Whatever the case it is the annual renewal of the oath rather than the exact date on which it was renewed that is most important.

20   Helgeland (1978b) 1479.

Little is known regarding the physical manner in which the *sacramentum* was sworn prior to 235. This said, it is clear that it was sworn in a group setting. As previously noted, Polybius, writing of the republican *sacramentum*, claims that one soldier was chosen to recite the entire oath, after which his comrades simply affirmed their individual commitment to the entirety of that oath. Even if he is incorrect and the *sacramentum* was recited in unison *en masse*, it remained a public affirmation of loyalty to the emperor on behalf of every individual soldier and a powerful tool to link the group identity of the unit with the reigning emperor.[21] The *sacramentum* placed an added weight on the shoulders of each soldier. Breaking it was not only an act of disloyalty against the emperor, it was arguably a betrayal of the other members of the unit.

The very survival of the *sacramentum* from the Republic to the imperial age suggests that Augustus and his successors believed it had a practical use in the maintenance of fidelity. Perhaps part of the imperial confidence in the *sacramentum* can be ascribed to the strong cultural role oaths and oath-taking played in army life. The Roman army as a community had long fostered a culture where oaths were used as a way to maintain a sense of duty and discipline amongst the troops. This broader culture of oath-taking played an important day-to-day role in aspects of camp life. One oath taken by the troops forbade the theft of property within a camp (Polyb. 6.33.1).[22] The *Feriale Duranum* records another oath that was sworn every morning by the troops of the *cohors xx Palmyrenorum*, "We will do what has been ordered and will be ready for all orders," (*P.Dura*. 47).[23] The *sacramentum* then, a far more significant oath, must have been an important part of this culture.

The traditional culture of discipline was also intertwined with the ingrained religious practices of the army.[24] Campbell noted in relation to the *sacramentum* that the "bond of personal loyalty to the emperor ... was secured by ties of *religio* and sanctified by Roman military tradition."[25] Thanks to this tradition, the *sacramentum* had a strong religious hold on the psyche of the troops.

---

21 Tacitus' description of the reluctance of the German legions to swear loyalty to Galba is instructive on this point. Tacitus reports that when they were called on to recite the oath only a few voices from the forward ranks did so, with the rest maintaining silence. This suggests that the oath could have been recited en masse.

22 Brand (1968) 91–98.

23 Helgeland (1978b) 1480, n. 39. *P. Dura* can be found in Welles et al. (1959).

24 For discussion of the religious obligations and sanctions inherent in the term itself see Campbell (1984) 29–31, and Phang (2008) 117–119.

25 Campbell (1984) 25.

Nowhere is this more clearly demonstrated than in the existence of the cult of the *Sacramenti Genii*. Although the cult of the *Genii* was not itself an official part of Roman army religion, it remained integral to the expression of the imperial cult and helped foster discipline and cohesiveness in military units.[26] The cult of the *Sacramenti Genii* confirms the role of the *sacramentum* as "an active (and) judging spirit" in the lives of Roman soldiers.[27] It must have formed a constant reminder of their duty to remain loyal to the emperor. Speidel noted that the *Genius Sacramenti* functioned as "an externalized conscience, a role the less personifiable oath itself could have taken only with difficulty."[28] Literary sources also claim that the *sacramentum* placed religious obligations on individual Roman soldiers. Of particular note is Seneca's claim (*Ep.* 95.35) that anyone who deserted from the army risked breaking divine law.[29] One can also add accounts where Roman troops refused to take the *sacramentum*, such as the refusal of the legions of lower Germany to swear allegiance to Galba (Tac. *Hist.* 1.55).[30] Although the exact motivation of the soldiers in such cases remains unknown, the act of refusal suggests that the troops themselves did not consider the *sacramentum* to be an empty gesture.

Despite the range of evidence suggesting that the *sacramentum* remained important during the early imperial period, its impact must not be overestimated. Every civil conflict within the empire represented an occasion on which the *sacramentum* was broken by a sizeable segment of the army. By this measure the events of 68 and 193 alone prove that the *sacramentum* was no sure protection against military revolt.[31] Considering this, one must now ask whether the *sacramentum* retained any significance in the far more tumultuous period from 235–395. In short: did the *sacramentum* remain relevant in an epoch of severe political instability and uncertain military loyalty?

---

26 Stoll (2007) 453. Two specific pieces of evidence confirm the existence of this cult. The first, *AE* (1953) 10, comes from a Pannonian legionary fort at Intercisa reads IUDICO SACRAMENTI CULTORES. The second, *AE* (1924) 135, comes from modern day Rayak (Riyaq) in Lebanon, and reads GENIO SACRAMENTI VETERANI.

27 Von Petrikovits (1983) 193.

28 Speidel (1984) 358–359.

29 See Campbell (1984) 29–30 and Stäcker (2003) 302–303.

30 See also the reluctance of troops to swear allegiance to Vespasian (Tac *Hist.* 4.31); Stäcker (2003) 304–305.

31 Campbell (1984) 30–31.

Empty Words in an Age of Chaos?

Before answering the question posed above, it is necessary to set out what is known about the *sacramentum* from 235–395. In particular, it is necessary to establish whether the pattern of use and expression established in the earlier imperial period continued or whether significant changes took place. The first point to note is that the *sacramentum* continued to be sworn throughout the period. Literary sources explicitly confirm that the *sacramentum* was sworn during the reigns of Maximinus Thrax (Hdn. 7.9.2), Pupenius and Balbinus (SHA, *Max et Balb* 24.3, Herodian. 8.7.4–7), Diocletian and Maximian (*Acta S. Marcelli, Rec* M 2.19–21; 5.1–5), Constantine and Constantius II (Zos. 2.44.3), Julian (Amm. Marc. 21.5.9–12, Lib. 18.109–110), Valentinian and Valens (Amm. Marc. 26.7.4, 26.7.8–10, 27.7.15–17), Gratian (Oros. 7.34), and Theodosius (Zos. 4.56.2).[32] Even without explicit evidence from each and every reign during the period, this chronological spread of instances demonstrates that the *sacramentum* survived as a regular element in the relationship between emperor and army. As in earlier centuries, the *sacramentum* was connected to three occasions. First, Vegetius (*Mil.* 2.5) notes that all recruits continued to swear the *sacramentum* on their admission into the army. Second, the *sacramentum* was sworn as part of the accession process for a new emperor. When Julian seized power in 361 one he called upon the assembled soldiers to swear the *sacramentum* to him (Amm. Marc. 21.5.10). Here, Ammianus uses the adverb 'sollemniter' ('customary' or 'traditional') to describe the act of swearing the *sacramentum* suggesting that it remained a routine part of imperial accession. This could explain why few ancient authors bother to describe it in their work. Finally, the *sacramentum* continued to be renewed annually. As in the earlier imperial period there are a number of possible occasions on which this occurred. The *Feriale Duranum's* reference to the *nuncupatio votorum* held on January 3rd remains relevant as this document dates to the reign of Severus Alexander. Another possible occasion was a reoccurring anniversary of the reigning emperor. Such an occasion is suggested by the *Acta Marcelli*, in which the Centurion Marcellus implicitly renounces the *sacramentum* he had previously sworn to the reigning emperors Diocletian and Maximian,

> Marcellus replied, 'On 21st July, while you were celebrating the emperor's festival, I declared clearly and publicly before the standards of this legion

---

32   *Acta S. Marcelli* can be found in Martindale (1971–1992).

that I was a Christian, and said that I could not serve under this military oath, but only for Christ Jesus, the son of God the Father almighty.'

Acta S. Marcelli, Rec M, 2.19–21, tr. MUSURILLO

As Marcellus was neither a new recruit nor being asked to swear the *sacramentum* at the beginning of an imperial reign, the episode described is definitely an annual renewal. As to the identity of the imperial festival mentioned, it most likely the *dies imperii*. Although there is debate on the issue, there is evidence to suggest that Maximian considered 21st July as his *dies imperii*, as it was on that date he was appointed Caesar by Diocletian. This could suggest that the annual renewal of the *sacramentum* did indeed occur on the emperor's *dies imperii*.[33]

The *sacramentum* not only continued to be sworn regularly, it was also sworn widely amongst an increasingly diverse Roman army. Vegetius confirms that it was sworn by troops serving both in the legions and in the auxiliary.[34] Additionally Zosimus (4.56.1–2) in his discussion on dissension between the tribal leaders of Theodosius' barbarian *foederati* reveals that these troops also swore the *sacramentum*. Officers were also required to swear the *sacramentum*. Ammianus (21.5.10) explicitly notes that the officers present at Julian's gathering in 361 followed their men by swearing the *sacramentum*. Even officers of the highest rank swore the *sacramentum*. Orosius (7.34) confirms this in his somewhat muted criticism of the usurper Magnus Maximus, saying that the *Comes Brittaniorum* was in many ways worthy of the purple except for the fact that he had broken his oath to the emperor Gratian. Further confirmation is given by Ammianus (26.7.4), who notes that upon returning to active military service for the usurper Procopius both Gomoarius and Agilo, two senior officers who had served the House of Constantine, were "recalled to the military oath" (*revocatis in sacramentum*).

The vast majority of Roman soldiers, especially those serving on the frontiers, must have regularly sworn the *sacramentum* without the physical presence of the emperor himself. It is likely that the imperial *imagines* substituted for the emperor on these occasions. This is confirmed by the *Historia Augusta* (*Max* 24.2), which notes that in 238 the defenders of Aquileia forced the defeated army of Maximinus Thrax to worship the *imagines* of Pupienus,

---

33 See Barnes (1982) 26. However, earlier in the text (1.1–2) the event is connected to the emperor's birthday (*dies natalis*) as opposed to the day he assumed power (*dies imperii*). Either occasion is, therefore, a possibility.

34 Legions: *Veg. Mil. 2.5*. Auxilia: *Veg. Mil. 2.3*. Presumably Vegetius is here referring to the traditional *auxilia* and the *auxilia palatina*.

Balbinus and Gordian III and then swear the *sacramentum* to Pupienus and Balbinus alone. No doubt the imperial *imagines* substituted for the presence of the new emperors during the formal swearing of the *sacramentum* as well.[35]

Reconstructing the ceremonial aspect of the *sacramentum* is difficult as there is only one explicit description of such a ceremony—that of Ammianus for Julian's usurpation in 361. Ammianus (21.5.10) says that upon being asked to swear the *sacramentum* Julian's troops swore the oath *en masse* with each man pointing his sword at his throat as he spoke. Although there is no reason to suggest that this scene is invented, it is important not to assume that it forms an example of a standardized ritual practiced across the army. There is no additional evidence to suggest that Roman soldiers ritualistically pointed their swords at their throats while taking the *sacramentum*. Instead this form of oath-taking is much more likely to reflect the barbarian origins of Julian's troops.[36] However, Ammianus' account remains valuable. Leaving aside the ceremonial trappings described above, his description of the troops taking the oath in unison, followed by the officers, is likely accurate. The significance of the ritual aspect of the *sacramentum* will be discussed below.

Finally, there is the matter of content of the *sacramentum* during this period. As has been previously noted, Vegetius, who wrote in either the late fourth or early fifth century, provides the most complete version of the *sacramentum* available any period of Roman history.

> The soldiers are marked with tattoos in the skin which will last and swear an oath when they are enlisted on the rolls. That is why (the oaths) are called the "sacraments" of military service. They swear by God, Christ and the Holy Spirit, and by the majesty of the Emperor which second to God

---

35 Herodian 8.7.4 claims that Pupienus himself came to Aquileia and delivered an *adlocutio* following the defeat of Maximinus Thrax. While this is likely to be true, the army seems to have sworn the *sacramentum* to the new emperors prior to his arrival. In the *adlocutio* Herodian ascribes to Pupenius, the emperor notes that the army was now being true to the oath they had sworn. Here he must be referring to the new *sacramentum* that the army swore to the new emperors—before Pupenius himself had arrived in Aquileia. In any case, at no time was Balbinus present and it is clear that the troops only swore the *sacramentum* on his *imago*.

36 As noted previously, Ammianus uses the adverb *sollemniter* to describe the process of swearing the *sacramentum*. Although it is possible he is referring to the physical manner in which the troops swore the oath—by holding their swords to their throats—it is more likely that he is simply referring to the fact that Roman soldiers regularly swore the *sacramentum*. On the Germanic composition of the Western army, see Hoffmann (1969) 131–199.

is to be loved and worshipped by the human race. For since the emperor has received the name of the 'August', faithful devotion should be given, unceasing homage paid him as if to a present and corporeal deity. For it is God whom a private citizen or a soldier serves, when he faithfully loves him who reigns by God's authority. The soldiers swear that they will strenuously do all that the Emperor may command, will never desert the service, nor refuse to die for the Roman state.

VEG. *Mil.* 2.5, tr. MILNER

Leaving to one side the Christian language of the Vegetian *sacramentum*, the practical duties it asked of the oath taker remain similar to those asked of the swearer prior to 235. Like the republican oath it demanded that the individual soldier not desert their military duties. It also retains the later adaptations of Augustus and his successors whereby the troops would have to acknowledge the emperor as the supreme power on earth. This remained the most important part of the oath and the reason for its existence. The *sacramentum* demanded that each individual soldier faithfully serve the emperor, and only the emperor, by doing all that was commanded of him.[37] Ammianus confirms this basic tenet in his descriptions of the *sacramentum* sworn to Julian and the usurper Procopius. For the former the troops "swore in set terms under pain of dire execrations, that they would endure all hazards for him, to the extent of pouring out their life-blood, if necessity required" (Amm. Marc. 21.5.10, trans. Rolfe). For Procopius the troops "swore allegiance ... with dire penalties for disloyalty, promising to stand by him and protect him with their lives" (Amm. Marc. 26.7.9, trans. Rolfe). These two examples also highlight a crucial element missing from the Vegetian *sacramentum*: the penalties for breaking the oath. In both cases Ammianus notes that part of the *sacramentum's* formula was an explicit acknowledgment by the troops that there were penalties for breaking it. Curiously, Vegetius' version of the oath also retains a reference to 'the Roman Republic', a claim that finds some support in Herodian (8.7.4), who has the emperor Pupienus ask the troops to maintain their pledges to the senate, Roman people and emperors. This means that a permanent reference to the Roman state cannot be ruled out. As for the significance of this reference, it was at best 'ornamental'. Considering that the troops offered complete obedi-

---

37   Helgeland (1978b) 1480 makes the claim that in the fourth century the army also swore the *sacramentum* to the commander of the *scholae palatinae*. This argument is based on a much later Byzantine practice described in Procop. *Vand.* 2.18.6 and cannot be supported for the earlier fourth century.

ence to an emperor ruling by divine right, who in himself embodied the state, the reference to the *Res Publica* must be viewed as a formulaic relic.[38]

As with the earlier imperial period it cannot be assumed that the oath's content did not change. Just as Gaius included members of his family in the language of imperial oaths taken during his reign, later emperors did the same. When Constantius II addressed the rebellious troops of Vetranio he exhorted them to remember that they had sworn to remain loyal to the children of Constantine (Zos. 2.44.1–2). This suggests that Constantine had adapted the *sacramentum* to include his heirs alongside himself. The *sacramentum* then became a pledge of dynastic loyalty. One would assume that other emperors of the period could also adapt either the terms or recipients of the *sacramentum*.[39]

The real innovation of the Vegetian *sacramentum* is the introduction of the language of monotheistic Christianity whereby loyalty to the emperor is cast almost as a duty to God himself. Dating the introduction of this language is problematic as Vegetius gives no indication of when it was introduced. The Christian content of the oath provides a *terminus post quem* of the reign of Constantine, but what evidence exists suggests that it was not introduced until much later, most probably the latter stages of the reign of Theodosius. The most explicit evidence for a later date comes from Ammianus. In describing the desertion of the *iovii* and *victores* from Valens to Procopius in 365, he notes that after they deserted to his cause the soldiers "escorted him back to the camp, swearing, in the soldiers' manner, by Jupiter that Procopius would be invincible." (Amm. Marc. 26.7.17 trans. Rolfe).[40] This statement indicates that as late as 365 Roman troops swore the *sacramentum* in a pagan form (Amm. Marc. 26.7.10). A later reference from the Libanius provides further evidence that at least some military figures continued to swear the *sacramentum* in its pagan form. In his *Pro Templis*, which can be dated sometime between 380–390, Libanius claims that close companions of the emperor Theodosius

---

38   Campbell (1984) 24 sees the inclusion of the reference as either a way to "emphasise loyalty to Rome, when perhaps many of the soldiers employed had only a nominal connection with the Roman State" or a relic of republican sentiment.

39   The Tetrarchic period of course represents an interesting proposition. Would the *sacramentum* be sworn in the name of all four Tetrarchs? Or in Diocletian's name only? Furthermore would the content of the oath itself change with the shifting power relations and alliances between the Tetrarchs, especially after the death of Diocletian? Unfortunately our sources do not provide any information on these issues.

40   The language here is expansive—particularly with the use of the verb *testor*. However, considering these troops had just gone over to the cause of Procopius it would be natural that they would be called upon to immediately swear the *sacramentum*.

still took their oath to the emperor by "the gods" (Lib. 30.53). Unfortunately Libanius does not name the individuals but simply notes that they were often invited to share the emperor's table, suggesting that the individuals could be highly ranked army officers. Considering that a number of Theodosius' leading officers, including the influential Arbogastes, were avowed pagans, it is reasonable to assume that they took the *sacramentum* in its traditional pagan form. This, of course, may not have been the case for Christian officers or for that matter the rank and file troops. But at the very least it demonstrates that Theodosius had not imposed the Vegetian *sacramentum* in a universal fashion. Finally, it is also worth noting Eusebius' failure to mention any changes to the *sacramentum* in his description of the reforms Constantine introduced in order to inoculate the Christian faith into his troops.[41] If Constantine had indeed introduced a new Christian *sacramentum*, it is likely that Eusebius would have proclaimed the emperor's role in striking down the idolatrous pagan religion of the army.

Despite its late date, the Vegetian *sacramentum* remains of use in trying to reconstruct the content of the *sacramentum* sworn for the vast majority of the period in question. Importantly, it confirms that the bonds of religious duty were used to bind the oath swearer with the emperor. The language of the oath sanctifies the emperor as possessing a divine right to rule, one which makes pledging loyalty to that emperor a divine obligation on the part of the troops. This strongly suggests that the pre-Christian *sacramentum* was also couched in such terms as to make loyalty to the emperor a religious duty. Leaving aside Ammianus' statement that the *sacramentum* was sworn by Jupiter (26.7.17), one must remember that Herodian called the *sacramentum* the empire's "sacred secret" (8.7.4).[42] In a sense the Vegetian *sacramentum* shows a measure of continuity between the pagan and Christian versions of the *sacramentum*, once again illustrating the conservatism inherent in the practices of the Roman army.

---

41 Also it should be noted that the reference in Euseb. *Vit.Const.* 4.17–21 to a prayer recited by the troops is clearly not the *sacramentum*. Eusebius presents the practice as an innovation and it was supposedly recited on a weekly basis. However, his further observation here that this weekly ceremony was held outside the city (i.e.: Constantinople) suggests that only a small segment of the overall army undertook it.

42 Le Bohec (1994) 239 contends that the *sacramentum* underwent a 'secularization' before regaining its religious character in the third century. However he offers no evidence to support this argument.

Imperial Perspective

Having established the *sacramentum's* survival, the next question to ask is whether the emperor continued to view it as an effective way of enhancing his bond of loyalty with the army. The survival of the *sacramentum* and its continuing connection with key imperial events strongly suggests that the emperor viewed its impact favorably. Specific examples reinforce this view. Herodian recounts that, in his 'last ditch' pleas to the army, Severus Alexander invoked the fact that the rebellious new recruits had sworn the *sacramentum* to him. He was reportedly incredulous that they should break their *sacramentum* and believed that it offered him some protection against rebellion (Hdn. 6.9.2). Following the death of the Maximini in 238, the *sacramentum* was an important part of the emperor Pupienus' attempts to bind the army of Maximinus Thrax to the new regime. In a formal address to these troops Herodian has the new emperor reinforce the religious obligations and sanctions of the *sacramentum*. He tells them that their decision to follow the new regime meant that they were now at peace with the gods by whom they had sworn their military oath. As has been noted above he calls the *sacramentum* the "sacred secret" of the empire. Similarly, Zosimus has Constantius II invoke the *sacramentum* in his efforts to return the troops of the usurper Vetranio to his cause (2.44.1–2).

The emperor's positive attitude towards the *sacramentum* is also evident in a number of martyrologies. In the *Acta Marcelli*, the Centurion Marcellus was not condemned for publicly proclaiming his Christian faith or for symbolically rejecting his duties by throwing down his belt and centurion's staff. Instead he is executed for his rejection of the *sacramentum*,

> Agricolanus said: 'What Marcellus has done merits punishment according to military rules. And so, whereas Marcellus, who held the rank of centurion, first class, has confessed that he has disgraced himself by publicly renouncing his military oath, and has further used expressions completely lacking in control as are recorded in the report of the prefect, I hereby sentence him to death by the sword.'[43]
>
> *Acta S. Marcelli* 5.8–10, trans. MUSURILLO

---

43   It is interesting to note that the Theodosian Code, in a *lex* dating from 385, directly links the *sacramentum* with the *cingulum*. See *CTh* 7.2.2. This suggests that there could be no doubt what Marcellus was rejecting when he cast aside his *cingulum*.

Agricolanus could have charged Marcellus with any number of offences but chose to instead cite the breaking of the *sacramentum*. This implies not only that the commitments made in the *sacramentum* covered Marcellus' many crimes, but that it was the rejection of the *sacramentum* that was most worrying to the military hierarchy.[44]

The *sacramentum* sworn by Julian's troops in 361 provides further evidence to the imperial mindset. Seeking to quickly march east and confront Constantius, Julian thought it necessary to seek a firm assurance of loyalty from the Gallic army, perhaps due to the nature of his upcoming endeavor. If the troops were to swear a formal oath to Julian, their actions in the upcoming civil war, against the emperor they had previously sworn fidelity to, would not be those of oath breakers. Could the *sacramentum* help a usurper justify his endeavor in the eyes of his troops? This would almost certainly seem to be the case for Procopius, who used the *sacramentum* to help bind the first substantial force of troops who joined his cause (Amm. Marc. 26.7.8–10). These units, who Ammianus says were newly raised for Valens' Thracian campaign, would have only recently sworn the *sacramentum* to Valens. Having gained their support with the promise of riches, Procopius was able to legitimize their betrayal by asking them to swear the *sacramentum*. It is possible that the usurper felt that troops who had deserted to him out of financial desire could not be trusted but that the *sacramentum* provided him with at least some assurance. Finally, it is worth again noting the decision of Theodosius to require his barbarian *foederati* to swear the *sacramentum*. This was presumably because he believed it would be of some help in maintaining their tenuous loyalty. Indeed, Zosimus' description of the debate that raged between Eriulphus and Fravitta regarding the oaths they had sworn to Theodosius forms a partial vindication of this belief (4.56.1–3).

Other forms of evidence also illuminate imperial thinking in relation to the benefits of the *sacramentum*. One of the *signa militaria* carried by the Roman army featured a raised open hand.[45] Von Petrikovits linked this open hand *signum* with the oath gesture taken by Roman soldiers. He persuasively argued that since the taking of the oath was done in the presence of the military standards that this particular standard "constantly reminded the soldiers to

---

44 Helgeland (1978a) 781–782. A similar case can be found in the *Acta S. Maximiliani* 3.6–10, where a young recruit is executed for refusing to take the *sacramentum*.

45 Helgeland (1978b) 1479. For Domitian see *RIC* II (Rome) n. 260, 288 a and b, 306; 320–321, 346. For Trajan see *RIC* II: (Rome) n. 439. Images of this 'open hand' can be found in Von Petrikovits (1983) 180.

136 HEBBLEWHITE

fulfill their oath conscientiously".[46] That numismatic representations of this type of standard are often paired with reverse legends relating to the loyalty of the army strengthens this interpretation.[47]

Surviving imperial edicts also demonstrate that the emperor considered the *sacramentum* to be a serious and binding agreement with his troops.[48] Imperial authorities used the *sacramentum* to distinguish between civilian and military life. The *sacramentum* was an important legal right of passage that "had the effect of dividing a soldier's life into two periods; his former civilian existence and his present career".[49] A *lex* issued by Constans talks of the transition from military to civilian life in terms of being discharged from the *sacramentum* (*CTh* 7.1.4). When a soldier was to return to civilian life at the end of his service he would undergo a ceremony in which he would be formally relieved of the burdens of his loyalty oath. The soldier who had undertaken this process was now considered to have returned to civilian life (*PSI.* 1026 l. 5–11. *Dig.* 49.18.2).

An Oath Honoured?

While it is clear that the emperor continued to see value in the *sacramentum*, it is more difficult to assess the attitude of the troops. Did swearing the *sacramentum* cause Roman soldiers to identify more closely with the emperor? The lack of explicit evidence regarding the attitudes of the troops themselves makes it tempting to instead judge their view of the *sacramentum* via the level of military discord in the period. This measuring stick would, of course, suggest that the army did not take the *sacramentum* seriously. However, such a conclusion is premature and simplistic. Indeed, it is possible to argue that

---

46 Von Petrikovits (1983) 191.

47 Von Petrikovits (1983) 182.

48 It should be noted that the term *sacramentum* does not seem to have been used in a completely uniform manner within preserved legal codes. At times, such as in the examples cited below the term specifically refers to the military oath and the status and obligations it placed on the troops. At others times, such as in *CTh* 7.1.6–7, the term seems to be used far more broadly to refer to a soldier's period of service. This is not problematic and again reinforces the important symbolic role of the *sacramentum*. A detailed study of how the term *sacramentum* is used in legal sources would be welcome and would perhaps assist in shining more light on how the oath was perceived by both soldier and emperor. Sadly, such a task falls outside the scope of the current paper.

49 Helgeland (1978b) 1501.

SACRAMENTUM MILITIAE: EMPTY WORDS IN AN AGE OF CHAOS 137

the *sacramentum* continued to be viewed by the troops with a great degree of respect and reverence.

Previously noted factors remain relevant here. Foremost among these is the religious force of the *sacramentum*, which remained potent for troops who feared the spiritual consequences of being considered as oath breakers. The Christian literary tradition, which places the *sacramentum* as central to the idolatrous, pagan religion of the Roman army, provides a unique perspective on this issue.[50] This is most clearly demonstrated by Tertullian who specifically discussed the incompatible nature of the *sacramentum* with the oath sworn by a Christian to his god, stating, *Non conuenit sacramento diuino et humano, signo Christi et signo diaboli, castris lucis et castris tenebrarum non potest una anima duobus deberi, deo et Caesari* ("There is no agreement between the divine and human sacrament, the standard of Christ and the standard of the devil, the camp of light and the camp of darkness. One soul cannot serve two masters—God and Caesar," Tert. *de Idol.* 19.2),[51]

To Tertullian the *sacramentum* was no empty ritual. It represented a serious pledge of total allegiance to the Roman emperor, one that meant a Christian soldier could not fulfill his responsibilities to God. One could not be a soldier of God and of the emperor. Tertullian's outrage at the *sacramentum* is only really explicable if the oath was generally considered to place real religious obligations on an individual soldier. Indeed, the pious Christian heroes of the martyr genre are presented as being well aware of the threat that the *sacramentum* represented to their Christian faith. Either they would refuse to

---

50 This source material must be handled carefully as there is a possibility that Christian authors deliberately over-emphasized the significance of the *sacramentum* in order to make their argument that the secular militia and the militia Christi were totally incompatible. Nonetheless despite this proviso there is no reason to completely discard this evidence as Phang (2008) 120 has done by labeling it "rhetorical and polemical ... unreliable as evidence for the significance of the *sacramentum*." Instead, while exaggeration on the part of these writers is possible, it does not completely divorce their accounts from reality. One must also remember that Tertullian's decision to use the term *sacramentum* to describe the Christian rite of baptism is significant. He at least believed the *sacramentum* was a very important part of military life—and viewed it as having sacred meaning to the troops who swore it. By describing the baptism process with this term he was emphasizing the importance of the rite to those who underwent it—and the fact that it made total allegiance to the emperor incompatible with their declared allegiance to god.

51 See also *De Corona* 11.1. For Tertullian's contrast between the human origin of the *sacramentum* and the spiritual contract between a Christian and God, see Michaélidés (1970) 56–57.

swear the oath, or they would be presented, like St Martin, as being "forced" to swear the *sacramentum* (Sulp. Sev. *V. Mart.* 2.5). Whether pagan soldiers serving under the Christian *sacramentum* felt similarly constrained, or indeed valued the oath less because of their own religious proclivities, remains unknown.

The force of the *sacramentum* continued to be strongly buttressed by the long tradition of habit and military discipline within the institution of the army that instilled in the troops a reverence towards oath-taking.[52] A letter written by an unnamed new recruit serving in Egypt sometime in the 380s confirms this attitude. In the letter, the recruit writes to a certain Herminus to tell him he has sent his wife and mother to him in order to receive a cash payment he is owed, as the oath he has sworn constrains him from leaving his post (*P.London* 3.982). This otherwise mundane correspondence reveals the day-to-day hold an oath like the *sacramentum* may have held over individual troops. A *lex* preserved in the *Codex Theodosianus* expresses the bond of identity troops believed the *sacramentum* could create with their emperor. It tells of a group of veterans who approached Constantine and shouted, *Auguste Constantine, dii te nobis servent: vestra salus nostra salus: vere dicimus, iurati dicimus* ("Augustus Constantine! The gods preserve you for us! Your salvation is our salvation. In truth we speak, on our oath we speak," *CTh* 7.20.2, tr. Campbell). Even though this oath must have lapsed when the veterans left Constantine's service, they considered the bonds it had created with the emperor so strong that they invoked it to demonstrate the importance and the sincerity with which they were making their request. Even if this scene has been invented, it is highly plausible that the idea of a Roman soldier invoking their *sacramentum* to make a request of the emperor was an accepted if not a common practice.

Additional clues can be gleaned from the literary sources. In 238 the Governor Capellianus counted on this bond in his attempts to ensure that the soldiers of *Legio III Augusta* did not join the revolt of Gordian. Before marching against the usurper, Capellianus successfully retained the soldiers for the cause of Maximinus Thrax by urging them to maintain the *sacramentum* they had sworn to their emperor (Hdn. 7.9.3). Another particularly potent example of this bond is recounted by Ammianus. While on campaign in 359, Constantius II personally came under attack by the Limigantes. Ammianus noted that the Roman army considered that their greatest duty was to aid him (19.11.13). Consequently Roman soldiers attacked with great bravery and saved Constantius' life (19.11.14). The actions of these troops in rushing to Constantius' protection

---

52   Campbell (1984) 29–32.

mirror the main requirement of the *sacramentum* as recounted by writers as diverse as Vegetius, Epictetus and Ammianus: that is to value the safety of the emperor above all else.

The ritualized nature of the act of swearing the *sacramentum* was also important as it helped enforce a group identity linked to displaying loyalty to the emperor. Libanius describes the manner in which the *sacramentum* helped bind the army together in loyalty to Julian's cause:

> By now he had assembled a force remarkable not so much in numbers as in morale. The men bound each other by compacts and agreements to go to every length and endure every hardship to ensure victory, and to fear only the disgrace that would be the consequence of negligence.
>
> LIB. 18.109, trans. NORMAN

Important here is that in swearing loyalty to the emperor each soldier was also reinforcing his duty towards his comrades. Unless an entire unit acted together, individual soldiers who renounced their oath were not only rebelling against the emperor but also against their own comrades. The passion with which Julian's army took the *sacramentum* on this occasion is discussed in more detail by Ammianus:

> And when all had been bidden to take the usual oath of allegiance, aiming their swords at their throats they swore in set terms under pain of dire execrations, that they would endure all hazards for him, to the extent of pouring out their life-blood, if necessity required; their officers and all the emperor's closest advisors followed their example, and pledged loyalty with like ceremony.
>
> AMM. MARC. 21.5.10, trans. ROLFE

Ammianus' description is especially significant as he was himself a soldier and therefore must have taken the *sacramentum* to both Constantius II and Julian during his career. Ammianus at least, through this portrayal, believed that the troops who took the *sacramentum* to Julian did so with great vigour and sincerity. This fervour is particularly evident in the physical manner by which they took the *sacramentum*, with each soldier pointing his own sword at his neck as he recited the oath. This behavior brings to mind another scene described by Ammianus whereby high-ranking leaders of the Quadi swore an oath to Rome begging that swords be placed to their throats to symbolize the seriousness of their actions (17.12.16). On this occasion Ammianus further noted that the Quadi venerated their swords as gods, thus implying that they planned to remain true to their oath (17.12.21). As has already been suggested, the bulk of the troops gathered for Julian's accession ceremony were barbarian recruits and it is reasonable to assume that they, like the Quadi, took their *sacramentum* in this fashion to publically express the fact they considered it to be a binding statement of fidelity to their new emperor.

Julian's *sacramentum* ceremony offers one final example of how the practice could strengthen the bond between emperor and army. Ammianus noted that after the rank and file soldiers had sworn the *sacramentum* to Julian, the officers and Constantius' closest advisors followed. However, one among them, the Praetorian Prefect Nebridius, refused to follow, "Alone among all the prefect Nebridius, with a loyalty that was firm rather than prudent, opposed him, declaring that he could by no means be bound by an oath against Constantius, to whom he was indebted for many and repeated acts of kindness," (Amm. Marc. 21.5.11, trans. Rolfe). This refusal infuriated the troops and they rushed to kill the Praetorian Prefect, who was saved only by the personal intervention of Julian. This anger is instructive. First, the reaction of the troops shows that they believed that Nebridius' decision not to swear the *sacramentum* was an explicit demonstration that he would maintain faith with Constantius II. His rejection of the *sacramentum* was a rejection of their decision to proclaim Julian as a legitimate emperor. Second, their actions demonstrate the group identity that the swearing ceremony helped to create. Nebridius' actions in rejecting the group decision to support Julian were deeply resented and the gathered troops felt that had been betrayed by a fellow soldier. (Amm. Marc. 14.2.20).[53] Both Nebridius and the troops who supported Julian clearly believed that the act of swearing the *sacramentum* was no empty gesture. They saw it as a living, breathing commitment of fidelity that needed to be honoured.

As noted earlier the tradition of oath-taking had a powerful hold on the Roman army as an institution. This hold was not merely confined to the rank and file, and permeated as far as the emperor himself. During his Persian campaign Julian quelled a potential mutiny by swearing an oath on the great military achievements he planned to make in the coming months. According to Ammianus, Julian's preparedness to couple his intentions with oaths gave the troops courage and belief in the emperor's claims (24.3.8–9).[54] This reaction

---

53   Despite his rank as Praetorian Prefect, Nebridius was no career civil servant and had undertaken military ventures in the past. As *Comes Orientis* in 354 he led a force that relieved Castricius in Isuaria. See Martindale (1971–1992) Vol. 1, 169 (Nebridius 1).

54   In doing this, Ammianus notes that Julian was following in the footsteps of the great military emperor Trajan who emphasized his statements by making oaths.

from the troops once again suggests that they must have also viewed the *sacramentum* with great reverence. A final reference shows the continuing hold the *sacramentum* could have on troops who otherwise had no great affection for their emperor. In his lengthy account of the usurpation of Procopius, Ammianus describes a risky gambit the usurper took in the opening moments of the Battle of Mygdus. Just as battle was being joined the usurper rushed into the open space between the armies and called out to the troops of Valens,

> So this is the old loyalty of Roman armies and their oaths bound by firm religious rites. Is this your pleasure, my brave men. All this mass of Roman swords uplifted for strangers! That a base Pannonian should shake and trample upon the world, to gain a throne which he never so much dared to pray for, we groan for your wounds and ours.
>
> AMM. MARC. 26.7.16, trans. ROLFE

The implication of the words Ammianus ascribed to Procopius is clear: the troops' fidelity to an unworthy and unpopular emperor was based more on their deep respect for the *sacramentum*, both in terms of its long tradition within the Roman army and the religious sanction it threatened, rather than any real affection for Valens.

Conclusions

As this chapter has shown, despite the turbulent political nature of the period 235–395 the *sacramentum* survived to remain a constant feature of the relationship between the emperor and the Roman army. It was sworn to the emperor alone and asked the individual solider to express absolute fidelity to his cause. As in the earlier imperial period, it remained an important marker of military service and ongoing fidelity to the emperor. Each soldier not only swore the *sacramentum* upon joining the army but also at the start of a new emperor's reign, and then again at least once yearly for his entire reign. The continued existence of the *sacramentum* in such an active manner suggests that the emperor continued to consider it of some use in maintaining the loyalty of the army. This conclusion is further strengthened by evidence suggesting that the troops themselves were loath to break the *sacramentum* because they feared the religious and practical consequences of such an act.

However, despite this, the power of the *sacramentum* to maintain the ongoing loyalty of the army to a particular emperor must not be overestimated. Usurpations were common from 235–395, and each time a part of the army rebelled against the reigning emperor they were breaking the *sacramentum* they had sworn to him. Instead, the actual effectiveness of the *sacramentum* ultimately depended on the existing attitude the army had towards the emperor. The *sacramentum* could strengthen the army's fidelity to an emperor but it could not in and of itself create loyalty to his cause. A clear example from the later fourth century confirms this point. As has been noted, Julian's troops took the *sacramentum* with great fervour when he made his final break from Constantius II. However, this fervour did not exist in a vacuum. It was the product of Julian's string of military successes and the faith the troops had in him as *imperator*. As Ammianus noted, they believed he was a "great and exalted leader and (as they knew from experience) a fortunate victor over nations and kings" (21.5.9, trans Rolfe). Here the *sacramentum* strengthened an existing bond of loyalty that existed between Julian and his troops. When this event is contrasted with the usurpation of Procopius, the limitations of the *sacramentum* in and of itself become apparent. Although Procopius received the *sacramentum* from a range of units that had joined his nascent cause, many of these same units deserted him during their first major confrontation with Valens' forces (Amm. Marc. 26.9.7). With no record of military success or existing goodwill from his troops there was no foundation of loyalty for the *sacramentum* to strengthen.

Military Authority

∴

CHAPTER 8

Circumscribing *Imperium*: Power and Regulation in the Republican Province

Ralph Covino

It is all too easy to think of the power of a Roman magistrate in the field as being nearly absolute within his three main areas of authority, namely in maintaining the civil order, conducting judicial affairs, and leading the military. Cicero makes this quite clear to his brother at *Q. Frat.* 1.1.22 while dispensing his fraternal advice, highlighting the absence of any true appeal or complaints procedure in the provinces as well as their distance from the senate's watchful eyes. Despite the level of limitation placed on Roman magistrates at home that emerged at or near the time of the creation of the magistracies themselves, it was not the case that similar restrictions were imposed on those serving abroad from the very start. Imbued with *imperium*, the fullness of Roman power which allowed a magistrate both to command and to compel, a magistrate's power beyond the *pomerium* could be truly awesome, leading to some of the classic tales of extreme behaviors from across the republican period. A prime example of this is the story told of T. Manlius Torquatus from c. 340. Torquatus acquired the sobriquet 'Imperiosus' for having ordered his son to be put to death for having broken with established rules for military discipline. His son had abandoned his post to valiantly fight in single combat, in emulation of his father's own deeds though against his orders. That he defeated his enemy was insignificant compared to the power of his father's *imperium* (Livy 8.6–7).[1]

Opportunities for abuse of one's *imperium* clearly abounded, especially once Rome acquired dominion over lands beyond the Italian peninsula. This naturally led to a growth in the number of rules and regulations issued which were intended to prevent provincial governors from acting, particularly from acting rapaciously.[2] These came principally in the form of *repetundarum* legislation,

---

1 On this example see Oakley (1985) 392–410 and Hölkeskamp (2010) 117–118. For a late republican example that does not confuse matters by being a family squabble, see the case of the T. Turpilius Silanus who was executed by Q. Caecilius Metellus, the later Numidicus, in 108 at Sall. *Iug.* 69.4.

2 See now González Romanillos (2003) 136–156.

© KONINKLIJKE BRILL NV, LEIDEN, 2016 | DOI: 10.1163/9789004284852_009

beginning with the *lex Calpurnia* of 149 and its successors such as the *lex Acilia* in c. 122, the *lex Servilia* of 100, and ultimately the *lex Cornelia* of 81 and the *lex Iulia* of 59, which have been much studied.[3] Violations of such laws occasioned a number of the sensational trials of the Republic, such as that of G. Verres, the notoriously corrupt governor of Sicily during the 70s.[4] Owing to the existence of epigraphically attested law in addition to the known show-piece trials, disproportionate stress is put on the threat of a *repetundae* charge as incentive to good gubernatorial behavior. This detracts from the attention which might be paid to senatorial *consulta*, pieces of legislation, and customary rules that developed whose intent was also to rein in the governors. Such have received only cursory treatment in the past.

Lintott, in his *Imperium Romanum*, devoted nearly the whole of his sixth chapter, "Restrictions on Magistrates and the Punishment of Delinquents", to a discussion of the activities of the *quaestio de repetundis* with only scant mention of the other enactments which bound a magistrate.[5] This has been the normal method of dealing with such matters for over one hundred years. In his early attempt to deal with the Romans' system of provincial administration, for example, Arnold stated his belief that a governor's authority was "essentially absolute" in his province and, indeed, able to be equated to monarchical power. To his mind, the seat of the governor of Sicily in King Hieron's palace in Syracuse was sufficient to illustrate that point succinctly. Nevertheless, he discussed his perception of such checks, as he saw them to exist, most notably via *repetundae* laws.[6] Beyond that, though, he also commented on the means by which the senate could exert its control over the provincial governors during the republican period. He noted that the senate possessed the right to determine the number

---

3 On *repetundae* in general, see Lintott (1981a); Ferrary (1983); and Mossakowski (1993). On the *lex Calpurnia*, see Cic. *Brut.* 106 with Richardson (1987); Forsythe (1988); and Venturini (1997). See also Sherwin-White (1982). On the *lex Acilia* see Badian (1954); Hands (1965); and *CIL* I² 583 with Mattingly (1970, 1979); Lintott (1992) and Crawford (1996) 39–112. On the *lex Servilia*, Mattingly (1983). The provisions of the *lex Cornelia* are not known to the same degree as those of the epigraphic law; whether Sulla altered much beyond the composition of the jury cannot be known. Similarly, the extent to which it is tralatician cannot be assessed. On the *lex Iulia*, Oost (1956).

4 Alexander (1990) notes over thirty trials *de repetundis* from the establishment of the *quaestio* under the *lex Calpurnia* to the time of Verres. See also Riggsby (1999) who suggests that the defense in all such cases could indicate that the threat of potential prosecution following their term in office could fatally weaken a governor's authority.

5 Lintott (1993) 97–107.

6 Arnold (1914) 64–65.

CIRCUMSCRIBING IMPERIUM

of troops in a province (Cic. *Pis.* 16), their pay (Plut. *Pomp.* 20), conduct with allied states and princes more generally (Caes. *B Gall.* 1.35; Cic. *Fam.* 15.2.4), as well as, ultimately, the distribution of honors to returning governors; such gave the senate a power which was, for Arnold, sufficient to ensure cooperation and curb outrage. Arnold's assessment, and indeed those of similar ilk, note especially the fact that the magistrate was to present his accounts on completion of his period of service following his return and, if necessary, have *acta* ratified prior to their having any validity.[7]

Stevenson's account of the governors, while still discussing their near omnipotence, also noted that they were "subject to a certain amount of control from the senate" and he echoed Arnold's list of senate-style checks. His presentation of the system built on that of Arnold by focusing on the threat of prosecution and highlighting the importance of avoiding offending those with powerful *patroni*.[8] These types of account provide what is perhaps too heavy a focus on controls stemming from Rome and, particularly, from *repetundae* legislation. A noted exception is Drogula's recent treatment of the *lex Porcia*, a piece of legislation which attempted to ban governors' movement outside of their allotted provinces without the express permission of the senate; however, while excellent in its analysis of the law and its context and rationale, his account presents as largely new for c. 100 any sort of interest in limiting governors' behavior beyond extortion legislation prior to that time.[9] As this chapter seeks to show, circumscribing *imperium* was not a novelty of the late Republic.

Livy and other authors divide Roman history into two distinct spheres, *domi militiaeque*, that which happened at home in Rome and that outside of it, the latter designated *militiae* to emphasize the martial character of that which

---

7 Arnold (1914) 65–70, citing Cic. *Pis.* 25 and *Att.* 6.7, on the account books and, of course, Mommsen for acts' ratification.

8 Stevenson (1939) 66–72. It should be noted that both Arnold and Stevenson's accounts also cover, at some length, the role which a province's so-called *lex provinciae* played in curtailing governors' activities, on which see below.

9 Drogula (2011) esp. 93. The account at 121 of magistrates' fearing the high penalties associated with violating the *lex Porcia* is highly reminiscent of previous works that make the same case for fear of *repetundae* legislation. It is worth noting that Alexander (1990) lists no known charges for violations of the *lex Porcia* following its passage down to the time of Sulla, by which time the law's provisions had undoubtedly become incorporated into Sulla's *lex de maiestate* per Lintott (1993) 23. Lintott (1981a) 196 notes that the *lex Porcia* opened a new range of offenses to be sure; however, he points out at (1999) 102 that the *rei publicae causa* loophole in the attested law (δημοσίων χάριν πραγμάτων) was likely a wide one indeed.

148 COVINO

transpired beyond the *pomerium*.[10] This civil-military divide colors the way in which scholars have addressed both abstract concepts such as *imperium* as well as the magistracies themselves.[11] Such is only natural given the way in which the origins and development of the offices and any restrictions put on them are presented in the sources. The experience of the governors of Sicily, Rome's first territorial possession outside of peninsular Italy and arguably the best documented owing to the existence of Cicero's *Verrines*, however, can provide a better insight into the reality of magistrates' operations in the field by demonstrating that a split between civil/administrative affairs and military ones is artificial when it comes to thinking about the wider rule set which apply to magistrates in the field. Drawing examples from several provinces but particularly on the experience of the governors of Sicily, then, this chapter will argue that restrictions passed at Rome which touch upon military action in the provinces, including the *lex Porcia* of c. 100 which so interested Drogula, were but a part of a wider series of restrictive efforts which ought to be considered together. In doing so, it seeks to better understand such laws' place within the development of the provincial system and its concomitant impact on the provinces.

Background

Truly unrestricted exercise of power and authority was apparently not a part of the Romans' experience of government. If Cicero's account of the content of the *libri pontificii* is to be believed, checks existed even during the period of the monarchy. In his *Republic*, Scipio stated that the pontifical books reveal that the kings' decisions were subject to an appeal process, a process that he claimed was later echoed in the content of the Twelve Tables (Cic. *Rep.* 2.31.54).[12] The traditional Roman narrative surrounding the end of their monarchy, though, centers on the seventh king, L. Tarquinius Superbus, and his tyrannical abuse

---

10   See Gargola (1995) 26 on the *pomerium* as the dividing line between *domi* and *militiae*. For the traditional view, see Mommsen (1887–1888) 1.61–75.

11   Armstrong (2013), for example, approaches a grant of *imperium* as a military contract with an ultimate culmination in the ritual setting of the triumphal procession. Pina Polo (2011) preserves this dichotomy by focusing solely on that which happens to the consuls at home. See also Brennan (2004) 34.

12   For a dim view on priestly records of this sort, see Cato quoted in Gell. *NA* 2.28.6. See also Poucet (1985) 33–71.

CIRCUMSCRIBING IMPERIUM

of power. It was said to have been among the causes of the overthrow of his government, though not the proximate one. As a result of their experience of Tarquin and his family, according to Livy, following the dissolution of the monarchy, the Romans sought to place limitations on their new leaders' ability to exercise their *imperium* which was vested in them by the people following election (1.60.4).[13] To his mind, this was done in order that the citizenry might be preserved from future abuse of governmental power at the hands of these newly created magistrates or, worse, from a reversion to monarchy and the tyranny associated with it as they had experienced under the reign of the Tarquins. From the very start, then, the stories told of the monarchy and subsequent birth of the republican system of government at Rome highlight the checks which existed on those who would wield the power that would become known as *imperium*.[14]

Cicero believed the *imperium* of the consuls to be regal in its nature: *atque uti consules potestatem haberent tempore dumtaxat annuam, genere ipso ac iure regiam* (*Rep.* 2.32.56).[15] Livy reported that the earliest consuls commanded the full power of the kings and used their *insignia* including the *fasces*; however, even he recognized that the authority of the kings was not transferred completely. In his reconstruction, the consuls held the *imperium* whereas certain other dimensions of the old regal power were farmed out to other individuals or bodies, such as with the monarch's religious authority's being transferred to the *Rex Sacrificulus* (2.2.1).[16] The dual consulship which had certainly emerged following the time of the Licinian-Sextian rogations was limited by factors embed-

---

13   See also Ogilvie (1965) 230–231 and Livy 2.1.6–7; sadly there is a *lacuna* of around five pages preceding Cic. *Rep.* 2.31.53 which may have provided a Ciceronian account of the Romans' actions at this point of the establishment of the consulship in Scipio's narrative. Cic. *Rep.* 2.25, 2.31, 2.33, and 2.38 indicate that the *comitia curiata* granted new monarchs their *imperium*, though whether this was the practice for magistrates during the Republic is subject to debate as per Brennan (2000) 19 and Stasse (2005). See also Caspar (2011) 165–167.

14   For previous studies on the nature of *imperium*, see Heuss (1944); Magdelain (1968); Awerbruch (1981); Beck (2011); Simón (2011); and Drogula (2007) with its review of prior bibliography. On the word *imperium* itself, see Richardson (1991); (2003).

15   The regal power connection is echoed in "*regio imperio*" of Cic. *Leg.* 3.3.8 as well as in Polybius' account (6.12) of the Romans' government. It is curious to note that Cicero at *Rep.* 2.32.56 conceived of the *imperium* of the dictator as 'new' despite its similarity to regal *imperium*: *novumque id genus imperii visum est et proximum similitudini regiae.*

16   The *Rex Sacrificulus* was also known as the *Rex Sacrorum*, on whom see Cornell (1995) 239–241. See also Ogilvie (1965) 237–238.

ded within its very nature as well.[17] An individual consular magistrate's ability to act without restraint was prevented by the presence of a colleague possessed of the negative power of the veto and the cycle of annual elections ensured that no man could dominate the state for an indefinite period.

Geographic restrictions on the exercise of *imperium* existed depending on the location of the consul either at home or in the field, *domi militiaeque*. The consul lacked power over life and death to a certain boundary whereas outside of it said power did exist; it was necessary in order to lead armies and enforce discipline.[18] Such was symbolized by the presence of the axes embedded in the *fasces* born by the lictors (Livy 2.1.7–2.2).[19] Limitations on the free exercise of authority and power were incorporated at the point of the creation of subsequent offices and magistracies as well; such limits were thus not tied explicitly to the possession of *imperium*. The case of the Tribunate of the Plebeians, for example, traditionally brought into existence following the events of the first *secessio plebis* in 494 as a counterbalance to the consuls, shows this to be true.[20] The tribunes could exercise their right to bring aid and to veto only within the *pomerium*. Beyond it, after one mile, their word normally carried no authority.[21]

---

17 On the complicated nature of the relationship between consular and praetorian *imperium* prior to the establishment of the *praetor peregrinus*, see Bunse (2002) 29–43 with Cornell (1995) 226–230 and Brennan (2000) 58–78. See also Holloway (2009) 71–75.

18 The traditional view of Mommsen extended a consul's power at home to the tenth milestone; however, Giovannini (1983) 7–30 has challenged this. For a revision of the traditional understanding of *imperium domi*, see, again, Drogula (2007).

19 See also Ogilvie (1965) 235–236. On the axes and the *pomerium*, see Livy 29.9.2; maintaining public order within the bounds of the *pomerium* fell to minor magistrates. On this see Lintott (1968) 89–106 with Nippel (1995) 85–112. On the lictors more generally, see Gladigow (1972); Kolb (1977); and Marshall (1984).

20 Cic. *Rep.* 2.53.58: *ut contra consulare imperium tribuni plebis sic illi contra vim regiam constituti*. On the origin of the office as one brought about "specifically against the almost unlimited power of the magistrates and its abuse in support of patrician interests", see Raaflaub (2005b) 199.

21 The requirement that a tribune be present within the city to provide *auxilium* if necessary normally prevented them from leaving the city's bounds for more than a single night as per Gell. *NA* 13.12.9: *ac proptera ius abnoctandi ademptum quoniam, ut vim fieri vetarent, absiduitate eorum et praesentium oculis opus erat*; the tribune's station was located near the *tabula Valeria* as per Coarelli (1985) 53–62. Exceptions could be made when tribunes' powers were needed beyond the city such as those described at Livy 9.36.14, which contains the unusual case of two tribunes having been sent out on a mission by the Senate, and 29.20.9–11 where two tribunes of the plebs, again on official business, depart Rome for the provinces. See also Mitchell (2005) 154–155.

CIRCUMSCRIBING IMPERIUM   151

Even with such restrictions, Livy reported that in the first half of the fifth century debates raged among the Roman people as to the nature of governmental power, particularly *imperium*, and how it might be circumscribed. The Tribune G. Terentilius Harsa is reported, for example, to have inveighed in several speeches before the people at the unregulated nature of consular power in particular,

> For [he said that] it was only in name that it [the word consul] was less hateful than king; it was almost crueler. In fact, for the one [king] they had accepted two [consuls], possessed of an unregulated and boundless power, who free and unbridled themselves turned against the plebs all of the fears of the law and all of its punishments.[22]
>
> LIVY 3.9.3–4

He sought to have a commission of five men created to limit and define the extent of consular *imperium* in 462 but was defeated (Livy 3.9–5, 13). Livy reported that attempts to have Terentilius' plan enacted by law continued sporadically during the subsequent decade with little success down to the time of the decemvirs and the creation of the regularized quaestorship (Livy 3.9–31).[23] I have argued elsewhere that the creation of the elected quaestorship formed part of this on-going process of establishing checks on consular power beyond those stemming from the creation of the office.[24]

These checks were sufficient at the dawn of the republic owing to the relatively small size of the nascent community and the Romans' limited territory. They were also very much 'place-based' in that they chiefly applied to limiting actions specifically within the bounds of the city rather than outside of it. The impact of this lack of restriction placed on the use of governmental authority and power abroad, however, only became clear during Rome's transmarine expansionary period and thereafter.

---

22   See Ogilvie (1965) 411–413 who states "There are no strong grounds for doubting the historicity of Terentilius' motion" at 411. It is important to note, however, that Dionysius of Halicarnassus does not contain an account of it.

23   Forsythe (2005) 202 argues his belief that the laws of the Twelve Tables did not contain any of the restrictions or definitions called for by Terentilius Harsa owing to the lack of citations thereof in subsequent discussions of *imperium* in the Republic and owing to the Twelve Tables' having been compiled under the authority of the consuls, *imperio consulari legibus scribundis.*

24   Covino (2011).

152   COVINO

Lands beyond Italy

After the Romans had subdued, by various means, the Italic peoples of their peninsula and brought the Greek cities of southern Italy under their aegis, their attentions were drawn to neighboring Sicily. Their previous conquests in Italy were solidified via a pastiche of treaties and other arrangements made with the conquered so as to ensure Roman dominance. They attempted to replicate this system in Sicily. Indeed, before Sicily became a Roman province, Rome had established treaties with Messana, Centuripae, Halaesa, Segesta, and Halicyae. However, following the First Punic War, the Romans encountered a new situation, that of dominion. They came to acquire control over not just Sicily's cities, but also its peoples and their affairs, at least on the Carthaginian portion of the island; following the war, only Messana retained its *foedus*-relationship with Rome.[25] The cities which had sided with Rome during the war were declared to be *civitates immunes ac liberae* (Cic. *Verr.* 2.3.13).[26] The remaining cities, classed as *decumanae* or *censoriae*, became liable for the first time to Roman taxation.[27] Such was the state of Sicily with regard to the Romans in the lead-up to the Second Punic War, a series of cities linked to Rome by a number of devices and arrangements, but with no clear guiding trend to their arrangement beyond that of convenience and reciprocity for services rendered.

It was during this period that the Romans came to see the need to have an *imperium*-holding magistrate present within the province.[28] From 227 and the

---

25 Cic. *Verr.* 2.5.50 discusses the old-style *foedus* relationships of Messana and Tauromenium which persisted into his era; Tauromenium and Netum likely received their treaties during the Second Punic War. On the latter, see Cic. *Verr.* 2.3.13 and Polyb. 1.10. Finley (1968) 117 discusses the series of *ad hoc* measures employed by the Romans in Sicily instead of the old manner of alliances and half-citizenships employed in Italy.

26 At least four cities received this designation during or closely following the First Punic War, Halaesa, Halicyae, Centuripae and Segesta; Panhormus received it later as per Scramuzza (1959) 235. Diodorus Siculus relates that many more had held the status, but had had it revoked following their actions during the Second Punic War (23.4.1); see also Livy 25.40.4; 26.40.14 and Badian (1958) 37.

27 On the basis of Cic. *Verr.* 2.2.137 it can be surmised that there were around 60 cities liable to the tithe in Sicily at the time of Verres, given that during his administration 120 censors were appointed, two each per city, for the collection of municipal taxes; on this see Finley (1968) 124 *contra* Scrumuzza (1959) 328–329 n. 2. Cicero describes the number of cities of the *censoriae* class as being *"perpaucae"* ("very few") at *Verr.* 2.3.13; on this see Pritchard (1975).

28 *Contra* Appian who, at *Sic.* 2.2, stated that Sicily received a "στρατηγός" ("praetor") annually following the declaration of peace in 241.

allotment of G. Flaminius as praetor for Sicily, the province acquired magistrates who were to serve as regular governors, most often praetors, save for times such as during emergencies such as the later Servile Wars when a consular governor, such as P. Rupilius in 132, was deemed a necessity. These governors would come to act in many capacities in addition to quelling slave insurrections; they would defend Sicily's harbors from pirates, sell the grain contracts in accordance with Sicily's unique *lex Hieronica*, and adjudicate cases in some trials and nominate others to do so under the terms of the aforementioned Rupilius' edict, the so-called *lex Rupilia*. The governors were not alone in their task, of course, as they had their own personal staff and junior magistrates, Sicily's unique dual quaestorship, to assist them.[29] The governor of Sicily's chief duties focused on the maintenance of public order in order to ensure that the grain-tithe flowed to Rome. With no colleague to counter his actions in a manner similar to a consul, at the dawn of the regularized governorship his authority in the province was absolute and not subject to any check.

Some of the evidence certainly gives the distinct impression that a magistrate's word outside of Rome remained law and was indisputable even into the late republican period—and this tends to color perceptions of gubernatorial activity. Complaints against such activity emerged starting in the third century. A tribunal was set up to deal with the complaints of the Locrians in 204 (Livy 29.8.1–9.12, 16.4–22.12). However, long before the *lex Porcia* and even before the first of the *leges de repetundis*, actions were being taken by the senate so as to curtail magisterial power in what would become Rome's more regularized provinces. A relatively obvious illustration of this in action regards the erection of fixed borders for *provinciae* around this time. Sicily did not seem to have had magistrates specifically assigned to draw up its borders as had, for example, Spain in 197 (Livy 32.27.11); however, its status as a province with easily understandable geographic limitations undoubtedly precluded such an act.[30] Nevertheless, the establishing of borders anywhere within their world, of even starting a process which could in some way limit the area in which a magistrate could exercise his *imperium*, does rather suggest that even at the

---

29   On Sicily's dual quaestorship as a unique oddity, see Harris (1976) esp. 104. In addition to the quaestors there were, of course, others in the governors' retinue such as *scribae* like Verres' *scriba* Maevius named at Cic. *Verr.*, 2.3.171. On the *scribae*, see Badian (1989). On legates, such as G. Cato, see Badian (1993).

30   On the erection of borders in the Spains, see Richardson (1986) 6–8, 55–57 and Curchin (1991) 28–29. Of course, concerning the province of Sicily, it must be recalled that the Sicilian *provincia* encompassed more than just the island. The province contained not only Sicily proper, but also its outlying islands. See Hansen (1991) 184.

head of the second century the Romans were thinking about the transition of their *provinciae* from being distinctly military zones to more administrative ones. Such is similarly reflected in other steps which were taken dealing with what Drogula called "minor regulations".[31] He argued that the restrictions on the number of slaves allowed in a governor's train as well as the inscriptional records from Colophon (*SEG* 39.1243, 1244), regardless of their dating, which urged magistrates to respect the authority of free states, are of little import.[32] While this is, perhaps, true, it is impossible to deny that both of these, as with the erection of borders, represent additional incremental steps towards limitation.

To these might be added instructions issued by the senate which were identified by Lintott and which started to emerge from them following the Macedonian wars (Polyb. 18.44).[33] Several of these rules are best understood as restricting the actions of those travelling abroad on official service. Cicero in the *Verrines* stated explicitly that a magistrate's purchasing a slave in his province was forbidden, save for replacing one that has died (Cic. *Verr.* 2.4.9); Crawford believed such a regulation dated from the time of the *lex Calpurnia* in 149.[34] When one of Scipio Aemilianus' slaves died on a journey he was undertaking on behalf of the senate, he replaced them with one from home rather than one acquired in the province; Jacoby understood the fragment of Poseidonius preserved in Athenaeus (6.273a–275b) to suggest that Scipio was overreacting to the regulation prohibiting slave purchase.[35] Cicero stated: *Videte maiorum diligentiam, qui nihildum etiam istius modi suspicabantur, verum tamen ea quae parvis in rebus accidere poterant providebant* ("Behold the attentiveness of our forefathers, who as yet suspected no such conduct as this, yet nevertheless made provisions for things which might happen in affairs of little import" Cic. *Verr.* 2.4.9). The focus on the small things implies that the restriction on purchases abroad was not limited to just slaves; indeed, he generalized on this point in the subsequent passage when he highlighted the potential for any magisterial purchasing to become pillaging in short order (Cic. *Verr.* 2.4.10). The epigraphic *lex agraria* of 111 also may provide evidence of this sort of smaller restriction on provincial profiteering for at least Africa province. Crawford held

---

31   Drogula (2011) n. 12.

32   On the slaves, see Lintott (1981b) 176.

33   Lintott (1993) 44, 202 n. 5–6.

34   Crawford (1977) 51.

35   *FGH*, 87, F59 = 265.K. That the number of slaves in an official's retinue might also be limited is suggested by the examples cited in the passage. On interpretations of this passage, see again Crawford (1977) 51.

CIRCUMSCRIBING IMPERIUM

that the beginning of l. 55 of the *lex agraria* was framed to prevent men serving in the province from benefitting from the clause.[36]

Cicero's use of the term "*maiorum*" denotes a long-standing interest in curtailing magistrates' rapacity which was not just with regard to small matters. Such is confirmed by several passages from Livy. Regarding Spain in particular, he wrote of a *senatus consultum* of 171, *ne frumenti aestimationem magistratus Romanus haberet neve cogeret vicensimas vendere Hispanos, quanti ipse vellet, et ne praefecti in oppida sua ad pecunias cogendas imponerentur* ("A Roman magistrate ought not set the price of grain; neither should he force the Spaniards to sell their twentieth part at the price he wills. Officers ought not be imposed on their towns for the purpose of collecting money," Livy 43.2.12). The requisitioning of grain, therefore, represented a prime avenue for opportunistic enrichment; however, it was not the only one. A scant few years later in 167 Livy reported that the senate had issued a *consultum* discontinuing the leasing of the Macedonian mines and rural estates and that similar bars were present for Illyricum (45.17–18). The immense revenues generated from these activities were also opportunities for administrative exploitation, both for Roman and local magistrates. Around this time as well the senate issued instructions to the Greeks, via a pronouncement read out at Thebes, which informed them that no one should contribute anything to Roman officers for the war effort save for that which had been officially allocated by the senate (Livy 43.17.2).

Each of the preceding measures, stemming from Rome itself, is seemingly innocuous if considered in isolation. Viewed as individual responses to particular situations, they undoubtedly offered no serious challenge to the totality of a magistrate's *imperium*. However, taken corporately they represent a steady diminution of magisterial authority and freedom to act. Each step, like these taken in the second century, though, transitioned the magistrate's power away from that of a military governor towards that of an administrator. A magistrate's capacity to act, though, was tied implicitly his *imperium* as related to the military sphere. The *Tabula Contrebiensis*, which dates to 87, demonstrates this most ably. The tablet provides an account of the resolution of a local water dispute between two groups of indigenes by G. Valerius Flaccus, the longtime governor of Spain, who had adjudicated it; however, it is the means by which he did so that is most instructive and revelatory about the nature of his authority and the power being deployed, for they were those of a military leader. Stewart placed the praetor's power to act judicially firmly within the military sphere on the grounds of the ritual procedures involved, namely the surrender of the

---

36   Crawford (1996) 171.

foreign population to Rome's authority, or *deditio*.[37] On the tablet's l. 14, both Valerius Flaccus' name and his filiation are provided. One might expect that his current proconsular position would follow; however, the line reads *"Iudicium addeixit C. Valerius C.f. Flaccus imperator."* This led to Birks, Rodger and Richardson's conclusion that he was called *imperator*, and not by his gubernatorial title, on account of the fact that this is whence the authorization for the proceedings stemmed.[38] This conclusion would seem to be supported by Tacitus when he discussed a decision made by the *Imperator* L. Mummius in 146: *"idem regis Antigoni, idem imperatoris Mummii iudicium"* ("Just as King Antigonus, so too did the *imperator* Mummius judge," *Ann.* 4.43).[39]

Despite the near absolute authority of governors, which stemmed from the military dimension of their *imperium*, which can be seen in the example of Valerius Flaccus from the *Tabula Contrebiensis*, it is clear from the above that from the early second century the senate's instructions and legislation had begun to blur the lines between civil and military power. As noted, many of these steps were taken even before *repetundae* legislation accelerated the trend of more centralized accountability measures following the passage of the *lex Calpurnia* (*de repetundis*) in 149. With the pacification of the provinces, over time the unrestricted *imperium* of the war-leader segued into the rise of the administrative governor. This process was accelerated by the rise of customary rules and regulations set by the governors themselves as precedents, edicts, or *leges datae*. It is in regard to these sort of province-specific restrictions that deal with justice and peace, then, that the example of Sicily is truly revelatory.

Rules for Sicily

By the time of Cicero, it is clear that the individual cities and inhabitants of the province of Sicily were granted specific sets of rights and privileges which

---

37 Stewart (1998) 204–207.

38 Birks, Rodger, and Richardson (1984) 48. See also the previous Richardson (1983) 40.

39 It must be noted that *IOlympia* 52.52–55 and 63–66 may cast doubt upon this as verification, though, given that it is a contemporary inscription which recorded the events in question. There, Mummius is called ὕπατος (here "consul") and ἀνθύπατος ("proconsul") which clearly makes no reference to his status as an *imperator contra CIL* I² 626, a votive inscription of his on the temple of Hercules Victor, which states that his provincial actions were *"ductu auspicio imperioque eius"* ("conducted under his power and auspices"). On the former inscription, see Ager (1996) case nos. 150 and 159; on the latter, Palombi (1996) 23–25.

CIRCUMSCRIBING IMPERIUM

governors recognized and affirmed. As noted, some of these were stipulated by treaty. Others, particularly with regard to taxation and justice, were bound up in Sicily's so-called *lex Hieronica* and *lex Rupilia*. This is clear from Cicero's claims, for example, that Verres swept the former away with a single order even though maintaining it had been a part of his praetorian edict (*Verr.* 2.2.34, 39).[40]

The sources do not indicate that the province as a whole had ever been granted anything resembling a cohesive law or even body of law which dealt with taxation, justice, and other administrative matters for the province. Such a concept of a law for the province has been termed by scholars a *lex provinciae* and, in the past, the *locus classicus* for such an institution has traditionally been the *lex Rupilia* of the Sicilian governor of 132/1, P. Rupilius.[41] The fact that both the *lex Hieronica* and the *lex Rupilia* are reported by Cicero to have been respected by all previous governors demonstrates that their inclusion in the governors' edicts is the proper manner in which to understand each system. That these laws remained in effect was due to the fact that they worked for the Sicilians, at least until the time of Verres' governorship. Despite the fact that they are never described by the sources as being all-encompassing, Rupilius' efforts, as well as those of his predecessors, however, did establish for the province precisely the sort of province-specific regulation which could constrain magistrates' actions therein. As with the regulations and instructions stemming from Rome, Sicily's *lex Hieronica* and *lex Rupilia* are pointed and detailed. They placed on the magistrates' *imperium* yet more restrictions, though they were of a more customary type and could, thus, be abrogated as did Verres, though not without some cost.

Scramuzza and, later, Lintott both asserted that there was a general settlement concerning Sicily, which occurred after 241, and provided for the organization of the province's taxation—that stemming from the treaty of Lutatius.[42] The precise nature of taxation in Sicily prior to the addition of the Kingdom of Hieron, however, is difficult to determine with any degree of certainty.[43] Cicero certainly believed that no changes were made at this point to existing local tax

---

40 See Lintott (1977) 184–186.

41 Proponents of the *lex provinciae* concept include, among others, Carcopino (1914); Cobban (1935) 161–171; Stevenson (1939) 68–69; Marshall (1969) 255–271; Sherk (1984) 138–139. On the *lex Rupilia* in particular see especially Scramuzza (1959) 246–248; Hoyos (1973) 47–53; Lintott (1993) 28–32.

42 Scramuzza (1959) 234; Lintott (1993) 30. On the treaty of Lutatius see Polyb. 1.62.8–63.3; Zonar. 8.17.

43 For attempts to do so, Pritchard (1970b) and Serrati (2000).

158   COVINO

structures (*Verr.* 2.3.12–15). Appian went further and stated that in 241 the Sicilians were charged an agricultural tithe by the Romans (*Sic.* 2.2) which would seem to be corroborated by the testimony of Livy who reported that the Sicilians paid their taxes in kind at this point (23.48.7). This would suggest that the Carthaginians had implemented an agri-tithe covering their portion of the island.[44]

This is particularly relevant considering that the Romans regularized the province in 227 when they began to annually assign praetors to govern Sicily. Care of the grain and the supply thereof to feed Rome's legions was always of paramount importance.[45] The *foedus* for Tauromenium dating from 213 obligated the city to sell grain to the Romans (Cic. *Verr.* 2.5.56).[46] While the inclusion of a grain-requirement within their treaty does not necessarily point to a tithe-system for the rest of the province at the time, it certainly affirms the Romans' interest in procuring Sicilian grain during the period and thereafter. The consul M. Valerius Laevinus in 210, following the fall of Agrigentum, urged the Sicilians to turn their attentions back to agriculture and toured the province inspecting and assessing which lands were under cultivation and which were not (Livy 26.40, 27.8.18).

Under the Sicily-specific *lex Hieronica*, which was not a single law but rather what came to be the customary name for their tithe's taxation and collection system regardless of its implementation date, the 'tax-in-kind' required of the Sicilians, who were obligated to pay it, was one tenth of the annual harvest.[47] Verres' successor L. Caecilius Metellus wrote to the cities of Sicily urging them to plant their crops with the *lex Hieronica* in mind (Cic. *Verr.* 2.3.44). Granaries within the province may have been constructed to take it specifically into account.[48] The tithe, though, was not limited just to grain. Until 75, it was the

---

44   Rickman (1980) 37–38 suggests that the Carthaginians had perhaps taken up to one quarter of the annual grain crop. See also Carcopino (1914); Pritchard (1970a); (1970b).

45   Serrati (2000).

46   On the city and its relationship with Rome, see as well App. *Sic.* 5.

47   See especially Rickman (1980) 36–42 and Pritchard (1970b). On the use of *lex* as a method or custom see *OLD s.v. lex*, ref. 5a.

48   Sjöqvist (1960) 130 believes that the four smaller rooms and the subdivisions of the main two rooms of the east granary at Morgantina could have been used as regular compartments for the storage of grain and possibly other produce due to Rome under the *lex Hieronica*. He speculated that the main chamber could be subdivided into regular intervals according to the buttresses on the walls' interiors *contra* Bell (1988) 313–342 who discounts this on the basis of the placement of the doors. The granary was undoubtedly arranged with the *lex Hieronica* in mind regardless. For a more recent treatments, see

CIRCUMSCRIBING IMPERIUM

responsibility of the quaestors of Sicily to auction off the tithes for the minor crops such as wine, oil, beans, fruit, and other produce (Cic. *Verr.* 2.3.14–15, 18–19, 149).[49] Magistrates of several ranks were heavily involved in the administration, auctioning, and even the collection and quality control of the tithes.[50] Disputes arising were settled under different provisions to other provinces as well; instead of the normal process with regard to a *conventus*, three *recuperatores* were selected to settle actions relating to the tithe's collection.

Sicily's example of the province-specific *lex Hieronica*, a set of rules concerning taxation, was undoubtedly unique in that it was most probably a continuation of previously established custom. It is indisputable that the implementation of the *lex Hieronica* carried with it a large degree of potential for personal enrichment and that it was undoubtedly frequently exploited. Nevertheless, magistrates serving in the province were at least in principle bound to uphold its tenets, especially if such formed a part of their edict, as with Verres, and, later, Metellus. Nearly the whole of Cic. *Verr.* 2.3, the *De Frumento*, stands as a testament to the former's refusal to be so bound.

The trial of Verres demonstrates that no less binding on the magistrates of Sicily were the provisions of rules which were set down by P. Rupilius. The consul Rupilius had been dispatched to Sicily in order to bring about the closure of the first slave war which had interrupted the grain flow. That a consular governor had been sent clearly indicates the importance which Rome attached to maintaining stability and order in the province to this end. Once he had dealt with the slaves, he, along with a ten man senatorial commission, set out to reorganize the province's affairs.[51] When speaking of what came to be known in the province as the *lex Rupilia*, Cicero and the scholiasts who followed him repeatedly emphasized the role which the commission played (Cic. *Verr.* 2.2.32,

---

Deussen (1994); Bell (2007a); (2007b). For evidence of other Sicilian granaries such as the Royal Syracusan, see Livy 24.21.12 and the public granary of Tauromenium, *IG* XIV.423.

49   See also Lintott (1993) 75 and Carcopino (1914) 78–80; the addition of beans to the list of other minor fruits mentioned by Cicero follows *IG* XIV.423–425 (= *SEG* XLVI.1247).

50   On the quaestors as the bookkeepers for grain-related issues, see Cic. *Div. Caec.* 32. On the governors and their role in grain inspection, see Cic. *Verr.* 2.3.172. On Cicero himself and his two grain-related checks, see Cic. *Verr.* 2.3.182. On the role of merchants in the process, see Cic. *Verr.* 2.3.176.

51   Ten man commissions of this nature were a common feature of Senatorial policy in the second century; Flamininus had one with him in Greece in 196, as did Manlius Vulso in Asia in 189. On them see Polyb. 18.42.5; 44.1; Livy 33.30; Polyb. 21.24.9; and Livy 37.55.7. Spain saw two such commissions in around the same period, with one in 133 and the other in the mid-90s. On these, see App. *Hisp.* 94.428 and 100.434.

39, 40, 90; Ps.-Ascon. 294 Stangl; Schol. Gronov. 334 Stangl). The identities of the ten men who assisted Rupilius are unknown and, indeed, it is likely that they were unknown even in Cicero's time, as his failed attempt to ascertain the identities of a similar commission sent to Corinth in 146 demonstrates (Cic. *Att.* 13.33.3).

The *lex Hieronica* and the *lex Rupilia* appear frequently together within the *Verrines*. Cicero repeatedly draws a parallel between the two Sicilian systems and highlights that they are integral to understanding the governor's role in the province. As with the *lex Hieronica*, the *lex Rupilia* is only known through references to it, which has led to considerable and wild speculation as to its breadth.[52] Among the sole things which can be stated with absolute certainty about the *lex Rupilia* is that it detailed the stipulations for the administration of justice within the province. Cicero stated:

> These are the laws of the Sicilians. A case brought by one city's citizens against another is tried by their own city's laws. In a case between a Sicilian and another Sicilian of a different city, the praetor will appoint judges by lot according to the decree of P. Rupilius, commonly called the *lex Rupilia*, which was enacted by Rupilius following the advice of a ten man commission. When a private citizen brings suit against a community or a community against a private citizen, the senate of some other city will try the case, after each has rejected unsuitable bodies for that purpose. When a Sicilian is being sued by a Roman citizen, a Sicilian judge is appointed; a Roman judge is appointed when a Roman is being sued by a Sicilian. In other cases, it is the custom to propose judges from the local body of Roman citizens. Cases between farmers and the collectors of the grain-tithe, however, are tried in accordance with the grain law which is commonly called the *lex Hieronica*.
>
> CIC. *Verr.* 2.2.32

Thus, the inhabitants of the province know what is expected of them in terms of the grain tithe under the *lex Hieronica* and their legal rights under the *lex*

---

52 Carcopino (1914) 73 and Scramuzza (1959), for example, include within the *lex Rupilia* provisions for the alteration of civic constitutions in their assessments, owing to confusion with another of Rupilius' acts in the province, namely the constitution for Heraclea, on which see Covino (2013) 21–22. They also believe that the *lex Rupilia* changed patterns of land-holding in Sicily based on the acts of Rupilius' consular colleague Popillius in Italy as described at *CIL* I².638. Such is not attested anywhere in the ancient sources. Marshall (1967) also views the judicial piece as a part of wider Sicilian provincial charter.

CIRCUMSCRIBING IMPERIUM

Rupilia. Governors knew that for peace and order to be maintained that they, too, would have to abide by the strictures set down.

Governors' edicts which contained provisions for systems like the *lex Hieronica* and the *lex Rupilia*, like all provisions which were taken on from previous governors, required affirmation in order to be binding.[53] Many provisions would be carried over from governor to governor as a matter of course. Given this, there must have been a great deal of limitational accretion—to the extent that some provisions may even have been forgotten though still on the books. Such was certainly the case in terms of a provision of the edict of M'. Aquillus, the consular governor who brought the second slave insurrection to an end, described at Cic. *Verr.* 2.5.7.[54] Prior to his departure from the province, Aquillus issued the order which banned slaves from the carrying of arms in an effort to bring an end to the slave insurrections of the type he had just concluded. This order was maintained by at least one subsequent governor of the province, L. Domitius, and likely thereafter.[55] According to the story told by Cicero, Domitius was presented with a huge boar and asked to meet the man who had killed it. The man in question was a Sicilian *pastor* and, as such, ought not to have been legally entitled to carry a weapon under the terms of the edit Domitius had carried over. Shortly after having been brought before the governor, the *pastor* was questioned as to the manner by which he killed the boar and when he replied that he had done so with a spear, the governor ordered his immediate crucifixion.[56] While a presumably rustic and humble herdsman may be excused from knowing the intricate details of the praetors' *edicta*, such would certainly not have been the case for a provincial governor.

Conclusions

Cicero's Verrine orations reveal many things, not only about Verres himself and Sicilian governmental peculiarities, but also, and possibly more importantly, about magisterial culture and expected behaviors during the late Republic.

---

53 On the inclusion of the provisions of the *lex Rupilia* system in the edict of Verres, see Cic. *Verr.* 2.2.90; for the *lex Hieronica* in that of Metellus, see Cic. *Verr.* 2.2.63 and 2.3.123. On edicts in Sicily, see Maganzani (2007).

54 Obseq. 45 places Aquillus as governor in 100 *contra* Livy *Epit.* 69 which indicates the year 98.

55 Domitius' term in office is difficult to ascertain; Broughton (1952) 7 and 560 assigns an uncertain date of 97.

56 The cautionary tale is repeated at Val. Max. 6.3.5.

They depict Cicero's premise that, as a governor, Verres ought to have remained at all times keenly aware of the rules which ought to have impeded his actions as a magistrate and that he largely failed to abide by them. The presentation of the severity of his abrogations and the evidence pertaining to them, of course, resulted in his fleeing the trial before the verdict could be reached.[57] However, none can argue that the portrait which Cicero painted of Verres is free from bias or obfuscation; it is as problematic if not more so than the *Pro Cluentio* after which Cicero famously boasted of having occluded the jurors' vision (Quint. *Inst.* 2.17.21).[58] Steel and others have repeatedly made the case for this.[59] The bulk of the *Verrines,* as a published but not delivered work, draws the obvious comparison to the *Pro Milone.*

As a source, the *Verrines* portray Verres as an individual product of his class and his era. It is certain, though, even from Cicero's presentation, that Verres (or at the very least, those among his intimate circle of advisors) did, in fact, possess the required awareness of precisely those the rules, laws, and customs which Cicero claimed that he violated during his time in office. He could not have come up with his ingenious methods of circumventing the *lex Hieronica* without an expert's knowledge thereof.[60] He knew that his delegation of naval commands and judicial duties was legal, but only just. Indeed, Marshall even argued some time ago that "So far from promoting judicial corruption, Verres may have even been seeking to prevent it" when offering his concluding thoughts on the *lex Rupilia.*[61] Far from being ignorant of the many rules, Cicero's Verres actually reveals himself to be a master of them, a trait which he used to his own personal advantage when he thought he could get away with it.

Much later in his career, in the *De Legibus,* Cicero would state what he believed to be the ideal knowledge set for a Roman Senator. He specifically asserted that senators should know the number and disposition of Rome's troops, its financial resources, its allies, friends, and tax-paying subjects in addition to the status, condition, and rights under treaty applied to each—all from memory (*Leg.* 3.41). To these must be added knowledge of the different limitations on his *imperium* that had accrued over time. Lintott, of course,

---

57   Butler (2002) is definitive on the evidence itself.

58   Wilson (2000) 134–160 provides an alternate account based on archaeology to that which appears in the literary material.

59   Steel (2004) and (2007). See Frazel (2009).

60   See Cic. *Verr.* 2.3.38; Cicero presents Verres' innovation there as being a novelty, whereas at 2.3.39, 102, 112, 115, and 129 it seems to have been a regular part of the *lex Hieronica* system.

61   Marshall (1967) 413.

CIRCUMSCRIBING IMPERIUM

asserted that "there was no question of a Roman magistrate in the Republic being supplied with a handbook of standing orders" and this is undoubtedly correct as a statement of policy.[62] However, it is similarly unquestionable that part of being a provincial governor in the late Republic must have included a hyper-awareness of the many rules and regulations, both statutory like the *lex Porcia*, as well as customary and specific to his province such as the *lex Hieronica* and *lex Rupilia*. From the early second century, these had become a fact of gubernatorial life.

The importance of the *lex Porcia* in the development of empire and the provincial system has, thus, been overstated. The transition from active military commanders to neutered governors such as Pliny during the empire begging for constant clarifications on the right course of action to take was not achieved overnight; rather, it was a long process during which new customs and limits appeared gradually, more often than not as reactionary measures both civil and military. While public pronouncements such as the *Tabula Contrebiensis* continued to emphasize for the provincials that the military dimension of *imperium* was still a determining factor in the expression of Rome's power, the more pacified and civilized provinces came to be governed by men whose *imperium*, though still military in nature, was cordoned by myriad limitations.

---

62   Lintott (1993) 44.

CHAPTER 9

The Delian and Second Athenian Leagues: The Perspective of Collective Action*

James Kierstead

Both of the naval leagues led by Athens in the classical period—the Delian League of the fifth century and the Second Athenian League of the fourth— have been the subject of extensive and lively debate. Scholars have argued about the popularity of the Delian League; about when and if it hardened into an empire; and about whether the Second Athenian League was any different. These debates all dealt in an implicit way with the question of whether (or to what extent) Athens oppressed and exploited its allies at various points of the classical period. Recent scholarship has rightly focused in on this question, which should be considered the central normative consideration in the debate about the Athenian naval leagues.

This chapter proposes a new way of looking at the Delian and Second Athenian Leagues—from the perspective of the social scientific theory of collective action. This new perspective aims to change our view of the leagues by affecting our positive or descriptive understanding of inter-state relations in the classical Aegean and by altering our normative or moral evaluation of Athens' treatment of its allies within its naval leagues. This chapter will lay out a model of collective action, examine whether the ancient evidence supports the model, and finally consider how the exercise might increase our positive understanding of the Athenian naval leagues. The more complex question of what implications the perspective of collective action might have for our normative evaluation of the naval leagues will have to be tackled at a later stage.[1]

---

* For comments on earlier drafts, I thank Sam Asarnow, Matt Simonton, Matthew Trundle, and Simon Perris. I also thank an anonymous reviewer who provided comments at a late stage.

1 For a prospectus of what future points will be considered, see the conclusions.

© KONINKLIJKE BRILL NV, LEIDEN, 2016 | DOI: 10.1163/9789004284852_010

Two Types of Collective Action

The novel perspective offered in this chapter is that of the social scientific theory of groups, specifically the theory of collective action. This section presents the basic assumptions and parameters of the model and distinguishes two types of collective action. These are hierarchical—characteristic of large groups and reliant on a coercive hegemon—and cooperative—characteristic of smaller groups and marked by a more cooperative way of operating. Emphasis is placed on the three predictions of the theory of groups: that groups work differently depending on their size, that coercion is often necessary for collective action, and that the costs of joint activity are often borne to a disproportionate degree by the largest members of groups.[2]

The account of collective action offered here shares many of the basic assumptions of the theory of rational choice, whose central assumptions are that individual agents seek to increase their utility in a rational way.[3] In utilizing models of collective action, I am not claiming that all actual individuals always act rationally to increase their utility. Such models are useful, however, because analyzing human actors as if they acted in that way turns out to have significant explanatory force. It should be noted that 'utility' is a place-holder for our specific conception of what is good for a particular set of agents (so that it does not, by definition, rule out other-regarding preferences, such as altruism). The only two assumptions about the utility of Greek *poleis* that are made in this analysis are that it includes survival as a *polis* and that it includes material prosperity as a significant ingredient, though not necessarily the most important.

The theory of collective action describes the conditions for enterprises engaged in jointly by rational agents. In joint enterprises, groups seek to secure for themselves certain goods, and in the provision of collective or public goods, a special set of conditions apply. Public goods are 'non-excludable' and 'non-rivalrous', that is they cannot be prevented from being consumed by all and one person's consumption of them will not reduce the amount available to others. Clean air is a good example of a public good as, once provided, nobody can be

---

2 This account of collective action is drawn mainly from the classic analysis by Olson (1965) 5–65 and, for his formal model, 22–33. The conception of hierarchical collective action, in which group-members consent to being monitored by a hegemon, can be traced back to Hobbes (1651/1968). For models (and real-world examples) of cooperative joint activity, see Ostrom (1990).

3 For a good introduction to the theory of collective action (as espoused by Olson) within the context of the rational choice paradigm, see Hindmoor (2006), especially 102–128.

prevented from enjoying it and my consumption of it will have no impact on the amount of it available to you. A birthday cake, by contrast, is not a public good as, if I make one, I can easily prevent others from having access to it and my consumption of part of it will reduce the amount available to other people.

Because public goods can be consumed by anyone, the problem of free-riding arises. Free-riders are agents that do not contribute to the provision of a public good but who enjoy the good anyway. Someone who fails to pay taxes towards the provision of clean air, but consumes that air in any case, provides an example. Free-riding can be minimized by monitoring members of a group, identifying shirkers, and punishing them. Groups can ensure this happens either by ceding power to a hegemon—who will identify and punish free-riders on behalf of the group's other members—or by members taking measures to identify and punish free-riders themselves. I call the first hierarchical, and the second cooperative collective action.

Which type of collective action a group ends up employing will depend on the size of the group because, as the size of a group increases, the difficulty of identifying free-riding also increases. The main reason for this is that, in large groups, common goods may continue to be provided in spite of shirking by a few members, thus making it difficult for others to notice that free-riding is taking place. Large groups are further characterized by greater complexity, so that it is often in the interests of other members to cede the tasks of monitoring and punishing free-riders to the most powerful member or members of the group. In small groups, members can happily monitor each other and will be quickly alerted to any free-riding by the noticeable decline of the common good that they have been working to provide.

This is the first key prediction of the theory of collective action—that small and large groups will be organized differently. The second prediction is that, because of the temptations of free-riding, coercion is often necessary to ensure that collective action takes place.[4] Because coercion is necessary for the provision of certain goods sought by each member, it may be in the interests of those members either to take coercive measures themselves, or to assist a hegemon

---

4   For empirical support for this claim, see the study of trade unions in Olson (1965) 68–70. Olson finds that union membership is often compulsory, and that coercion is often employed to enforce group measures. This is not, of course, an attack of trade unions in particular but an illustration of the nature of collective action in any large group. See also his expansive conclusion: "Collective bargaining, war, and the provision of basic governmental services are alike in that the 'benefits' of all three go to everyone in the relevant group, whether or not he has supported the union, served in the military, or paid the taxes. Compulsion is involved in all three, and has to be." Olsen (1965) 90.

THE DELIAN AND SECOND ATHENIAN LEAGUES 167

in imposing them. This may be done even as it remains in the interests of individual members of the group to try to stop contributing without suffering any consequences.[5] As a result, the fact that one member of a group stops contributing need not imply dissatisfaction with the aims pursued by that group.

The third and final feature of groups predicted by the theory is that, in large groups employing hierarchical collective action, the costs of joint action are often borne disproportionately by the hegemon.[6] This comes about partly because 'smaller' members of a group tend to consume less of whatever public good is provided. They thus have less of an incentive to contribute towards the continuing provision of that good. But it is also a result of the greater capacities of 'larger' members to identify and punish free-riders. This may give 'larger' members opportunities for simple rent-seeking (that is, the extraction of resources without any provision of goods in return). But we should be open to the possibility that, in the absence of such over-reaching, it is the 'larger' members of groups who, despite their direct involvement in coercion, are being exploited by the 'smaller' members, and not vice versa.

Collective Action in the Leagues

The central argument of this chapter is that the classical Athenian naval leagues were examples of hierarchical and cooperative types of collective action. This section will set out the evidence that both leagues were formed in order to secure the ongoing provision of two linked public goods: security and a functioning maritime market (an important condition for the prosperity of island *poleis*). Both leagues employed a central fleet to secure the Aegean against foreign and Greek enemies, as well as pirates, and both paid for it with a tax levied on all members and assessed according to their economic productivity.

Though security against Persian encroachment was the main purpose for which the Delian League was founded, and security against Spartan domination the main purpose for which the Second Athenian League was founded, both these leagues were also used to fight against different external threats as they arose.[7] The goods of security and prosperity were linked, because clearing

---

5 Olson (1965) claims that in trade unions, attempts to avoid paying dues are as frequent as votes that payment of dues should be compulsory (86).

6 Olson (1965) 35 calls this the "surprising tendency for the 'exploitation' of the great by the small."

7 It may be that the Greeks involved in these alliances were wrong to worry as much as they did about particular external threats. In particular, some in recent years have downplayed

168   KIERSTEAD

the sea of foreign enemies went hand in hand with clearing the seas of pirates and this second achievement was a necessary condition for a functioning maritime market.[8]

The Delian League grew out of the alliance of 31 *poleis* that defeated the Persian invasion of 480–79.[9] Thucydides stated that the Athenians' pretext for assessing the first contributions from the allies was "to make up for the things they had suffered by laying waste the territory of the Persian king" (Thuc. 1.96.1).[10] This is probably best taken as an intrusion of the historian's

---

  the effect that a Persian conquest of mainland Greece would have had; see Osborne (1996) 342: "What then had the Greeks saved themselves from by defeating the Persian invasion? They had saved themselves from an imposed end to inter-city conflict." He later admits that "they had also saved themselves from prolonged domination by a particular political group" within their own *poleis*—those close to the Persians: Osborne (1996) 343. My view is that however serious the threats against them were in reality, many classical Greeks believed themselves to be in considerable danger (physical, financial, cultural, religious, and/or political) from various foreign invaders at various points in time. But I cannot argue the point in detail here. Those who think that the Greeks who worried about the Persians were wrong to do so are free to replace the word 'security' in this chapter with a formula such as 'freedom from what many Greeks believed to be threats, whether not they were in actuality.'

8 It goes without saying that the question of the relationship between prosperity and markets is a complex one, which continues to divide both expert and lay opinion and to serve as one of the main markers of our political allegiances. It is certainly not my aim to solve this question here, and the only assumption that I require for the purposes of my argument is a modest one: that the classical Greeks believed that there was some positive relationship between a functioning maritime market (free from the predations of pirates) and their prosperity. Again, this does require that there *was* such a relationship, although I believe that there was. On maritime trade see Bresson (2000) and Bresson (2008). On economic growth in ancient Greece see Morris (2004). My argument does not require me to subscribe to assumptions such as the superiority of unregulated markets (in fact, it assumes that functioning markets require at least minimal 'regulation' through the form of basic security). Nor does it require me to believe that maritime trade was the *only* foundation of Aegean prosperity; other factors such as agriculture no doubt played some role.

9 All dates are BC unless otherwise stated.

10 πρόσχημα γὰρ ἦν ἀμύνεσθαι ὦ ἔπαθον δῃοῦντας τὴν βασιλέως χώραν. Rhodes (1985) 7 may be right that δῃοῦντας is indicative more of wanton damage than of a plundering expedition; whether Thucydides is presenting Athenian aims as revenge or as profiteering in this sentence, though, he is certainly not presenting them as protection. It might be worth recalling here that the alliance depending on contributions was preceded by a less formal Ionian alliance established immediately after the Persian Wars; its first and only act was to recapture Sestos, which was in the hands of the Persians (Thuc. 1.89.2–3).

THE DELIAN AND SECOND ATHENIAN LEAGUES   169

familiarly cynical viewpoint into an otherwise apparently objective narrative of events. But another passage suggests that, whatever Thucydides thought was the purpose of the league, some of Athens' allies believed that it was founded in order to continue the war against the Persians which was seen, even after Plataea and Mycale, as incomplete. This passage is contained in a speech by representatives of the Mytileneans, who remind the Spartans that they joined the alliance with the Athenians "when you had abandoned the war against Persia, and they stayed on for the work that was leftover," but then insist that "we became allies not for the enslavement of the Greeks to the Athenians, but for the liberation of the Greeks from the Persians" (Thuc. 3.10.2–3).[11]

The early history of the league seems to make clear that its central purpose was to push back the Persian sphere of influence. As Thucydides reported, the first action of the league was an attack on Eion, which was occupied by the Persians (Thuc. 1.98.1).[12] Soon afterwards, at the Eurymedon River, the Athenians and their allies won a double victory on land and at sea over the Persians, destroying some 200 Phoenician ships (Thuc. 1.100).[13] Around 460 we find the Athenians taking the fight to the Persians again in Cyprus along with their allies and with some 200 ships (Thuc. 1.104.2, 1.110).[14] Around the same time, they supported an Egyptian revolt against the Persian king, in concert with the allies; the revolt ended in 454 with the destruction of nearly

---

11   Thuc. 3.10.2–3: ἡμῖν δὲ καὶ Ἀθηναίοις ξυμμαχία ἐγένετο πρῶτον ἀπολιπόντων μὲν ὑμῶν ἐκ τοῦ Μηδικοῦ πολέμου, παραμεινάντων δὲ ἐκείνων πρὸς τὰ ὑπόλοιπα τῶν ἔργων. ξύμμαχοι μέντοι ἐγενόμεθα οὐκ ἐπὶ καταδουλώσει τῶν Ἑλλήνων Ἀθηναίοις, ἀλλ᾽ ἐπ᾽ ἐλευθερώσει ἀπὸ τοῦ Μήδου τοῖς Ἕλλησιν. Though there is reason to distrust the narrative content of the speeches in Thucydides as statements of fact, they may represent a more reliable guide to the sort of arguments contemporary actors were making. Why should we believe the second part of this last sentence, but have doubts about the first? As will emerge more fully later in this chapter, free-riders have incentives to present themselves as unjustly oppressed, even when that is not the case. But, as with Thucydides himself, the Mytileneans would surely have attributed some other purpose to the league at its onset (such as simple Athenian greed) if it was not well recognized that it was founded as an anti-Persian alliance.

12   Enslaving the inhabitants was admittedly a robust course of action; it is not clear how many of the individuals affected were Greeks and how many Persians; the phrase Μήδων ἐχόντων suggests the latter were in the ascendant before the intervention. For the leadership of Cimon the son of Miltiades (instrumental at Marathon) as itself a link with the Persian Wars, see Hornblower (2002) 19.

13   The date is disputed; Sealey (1976) 250 argues for 469 on the basis of Plut. *Cim.* 8.

14   For the date, Hornblower (2002) 30.

250 ships (and the death of a large number of Athenians).[15] Between the initial intervention and the allied defeat, part of the fleet may have been involved in anti-Persian operations in Phoenicia.[16] In spite of the huge blow to alliance's forces in Egypt, the following year (453/2) it launched yet another anti-Persian expedition of 200 ships to Cyprus, and intervened against pro-Persian factions within Erythrai and Miletus (Thuc. 112.2).[17] This brings the story down to roughly the middle of the fifth century. It is at this point that one interpretive tradition has located a rupture in the nature of the alliance: ceasing to be a league, it became an empire instead, with Athens' role passing, to use Koehler's terms, from *Hegemonie* to *Herrschaft*.[18] The re-dating of many of the crucial inscriptions has undermined the orthodox narrative about a 'hardening' of Athenian actions at this point. But the issue we should remain focused on here is whether the league still had any purpose as an anti-Persian alliance after the 460s and 450s.

Some scholars believe that victory at the Eurymedon River in the early 460s was already sufficient to deprive the Delian League of its ostensible purpose, and to convert subsequent Athenian leadership into rule. This was Koehler's view, and is the view adopted by two leading modern textbooks.[19] But in view of

---

15 For the presence of the allies in the Athenian expedition even to the bitter end: Thuc. 1.110.4. For the Athenian dead, see Meiggs and Lewis (1969) 33, a list of casualties from the Erechtheid tribe; admittedly we do not know how many of the individuals listed died in Egypt as opposed to the other conflicts also mentioned in the prescript (Cyprus, Phoenicia, Haliae, Aigina, Megara).

16 I would not press this particular point, based on the mention of Phoenicia in the prescript of Meiggs and Lewis (1969) 33 with the editors' comment on 75: "No literary source mentions fighting in Phoenicia at this time, but if, as many think, the main part of the fleet returned from Egypt after winning control of the Nile, a raid on Phoenicia is easy to accept."

17 ἐς δὲ Κύπρον ἐστρατεύοντο ναυσὶ διακοσίαις αὐτῶν τε καὶ ξυμμάχων Κίμωνος στρατηγοῦντος. Note the explicit reference to ships contributed by the allies. Erythrai and Miletus: Meiggs and Lewis (1969) 40 and 43.

18 Koehler (1869) 101, quoted by Liddel (2009) 21. For a brief survey of the 'orthodoxy' on the mid-century transformation in modern scholarship see Kallet (2009) 50–54 and note her salutary skepticism on 51: the orthodox picture "is constructed heavily on undated documents and literary evidence of often-dubious reliability—a house of cards is not an entirely inaccurate formulation."

19 For Koehler (1869), see again Liddel (2009) 21. The leading modern textbooks mentioned are those of Hornblower (2002) and Morris and Powell (2010). Hornblower (2002) 35: "The end of the Persian War did not bring an end to the Athenian empire, though the existence of the confederacy was now harder to justify." Morris and Powell (2010) 281: "The battles

THE DELIAN AND SECOND ATHENIAN LEAGUES

the events subsequent to Eurymedon summarized immediately above (a catastrophic defeat by Persian forces in Egypt; two large expeditions to Cyprus; Persian interference in two *poleis*), the view that the battle in Pamphylia marked the end of the need for Greeks to fight Persians may appear somewhat naïve, from both a theoretical and empirical perspective. The theoretical consideration points to an apparently common assumption by historians that pushing the Persians out of the Aegean would have rendered the Delian League otiose. But this assumption is facile; scholars who are inclined to accept it might consider ceasing to pay contributions to their local police force on the grounds that levels of crime in their area are currently low. It is arguable that the only reason that the Persians were not more of a threat in the Aegean after the 450s is precisely because of the continuing existence of the Delian League. This brings us to the empirical evidence which suggests that Persia continued to be a threat (or a perceived threat) to Greek city-states well into the Peloponnesian War, and that the Delian League continued to function as an anti-Persian alliance into that period. In 440 the Persian satrap Pissouthnes was involved in an attempted oligarchic takeover of Samos (Thuc. 1.15).[20] Near the beginning of the war, in 428/7, the Athenian general Paches intervened to defeat pro-Persian factions (supported by Pissouthnes) that had taken over Colophon and part of Notion (Thuc. 3.34).[21] According to some Ionian and Lesbian exiles (probably oligarchs) in the Peloponnesian fleet, Pissouthnes stood ready to support anti-democratic subversion in Ionia (Thuc. 3.31). Based on his reading of the Greek and Lycian inscriptions on the funerary monument at Xanthos, Peter Thonemann has argued that the Athenian siege of Kaunos in the early 420s can be explained by the likely presence in the city of a Persian overseer from around 428.[22] Thonemann also suggested that, had Thucydides given us more information, "it might well be that the 'cold war' between Athens and Persia, from the early 440s to the peace of Epilykos in 423 ... would start to seem significantly warmer."[23]

---

at the Eurymedon River made many Greeks wonder whether they any longer needed the naval alliance." Although theirs is not presented as a scholarly work, it is a pity that they cite no contemporary sources in support of this.

20 For the activities of Pissouthnes, see Thonemann (2009) 173.

21 Brock (2009) 161 suggests that the slowness of the Athenian response demonstrates a relative lack of interest in anti-Persian operations in this period. But considering the situation Athens found itself in at the time (fighting another major enemy and ravaged by plague), that it responded at all might seem the most remarkable fact about the episode.

22 Thonemann (2009).

23 Thonemann (2009) 179.

Critics might object that, whatever the evidence for warm—or lukewarm—conflict between Athens and Persia during the Peloponnesian War, evidence of hot conflict (actual fighting) between the two polities is nonexistent. And surely, these critics would insist, this should lead us to conclude that by the Peloponnesian War, at least, Athens and its allies had little to fear from Persia. Instead, Athens clearly used its allies' tribute in the period of the Peloponnesian War primarily to fight Sparta and its allies, and not Persia. Of course, it could be argued that in defending its allies against Sparta, Athens was still simply fulfilling its role as hegemon within a large group, and helping ensure the provision of security to that group's members. But here there would be a strong objection, which is that Athens' fifth century allies consented to a league that would defend them against Persia, not Sparta. This objection has a number of normative implications, and not all of them can be pursued here.

But one important consideration is that the Spartans and Persians colluded, especially in the second half of the war, to return many of the Ionian Greek *poleis* to the Persian Empire (Thuc 8.18, 37, 58). The Peloponnesian War was a struggle between Sparta and Athens for strategic control of the Aegean and mainland Greece. It also involved an ideological rupture between democracy and oligarchy. But, from a longer-term perspective, it can be seen as a phase in the struggle of the Ionian city-states for independence from Persian overlordship. In this particular phase of the struggle, the Athenians defended the Aegean and the coastal *poleis* of Asia Minor from the Persians, and the Persians sought to regain control of them through their Spartan proxies. If we view the war partly in this light, the implications for our reading of the evidence is clear. We should not expect to find much evidence for direct conflict between Athens and Persia, since Persia's attempts to undermine Athenian strategic goals were, in this period, conducted through Sparta.

Before turning to the provision of security by the Second Athenian League, three further pieces of evidence that league became empire sometime around the middle of the fifth century need to be briefly explored. Specifically, these concern the moving of the league treasury to Athens, the tribute lists, and the so-called Peace of Kallias. These incidents should be considered in light of the empirical evidence already discussed. If the evidence that Persia continued to threaten Greek interests in the Aegean after the middle of the century is sound, then there is an argument that the League's operation continued to be justified. The moving of the league treasury to Athens from Delos in 454 would then simply be an expedient prompted by the destruction of the allied fleet in Egypt in the same year and the threat to allied control of the Aegean

THE DELIAN AND SECOND ATHENIAN LEAGUES 173

(and of Delos itself) that this disaster presented.[24] The evidence of the tribute lists—that there was a precipitous drop in payments between 449/8 and 447/6—may reflect defection by a number of allies, but does not necessarily change the nature of the relationship between Athens and its allies.[25] Against the background of a continuing Persian threat, a decline in attested tribute contributions, however sharp, does nothing in itself to mark out the end of a period of collective action and the beginning of a period of imperial exploitation. The Peace of Kallias has of course been extensively debated but there are some important points to stress. Firstly, the peace is not mentioned in any contemporary source, and indeed much of the corroborating 'evidence' for the peace is taken from Plutarch (*Cim.* 13.4–6, *Per.* 12.3–4).[26] Secondly, making peace does not necessarily entail that military alliances are no longer necessary. As Greek history regularly demonstrates, treaties are often broken and, if the Peace of Kallias had any enduring existence, this may have been because the Delian League underwrote it.

In the case of the Second Athenian League, it is easier to show that the league in question was founded for the purpose of security. This is because it is clearly stated on the Aristoteles Decree that the League is being founded in order to ensure that its members remain free and autonomous from Spartan control.[27] The threat that sparked the formation of the Second Athenian League was not Persia but Sparta. It might be thought that this removes the possibility that the league provided security from pirates in the Aegean as well as from Sparta. But, it should be remembered that, in this period, Sparta was able to project its power across the Aegean, and the Second Athenian League was, like its predecessor, focused primarily in that arena. In the case of the Second Athenian League, the enemy that the league was formed to combat quickly ceased to be a threat. In 371, only 7 years after the formation of the Second Athenian League to contain Spartan aggression, Sparta was defeated by the Thebans at the battle of Leuktra. Because of the loss of Messenia that followed, this defeat on land would have a major effect on Sparta's ability to project power into the Aegean arena. But states on the Aegean seaboard were at no loss for other strategic threats in the 360s and 350s, mainly from the north. Alexander of

---

24 The moving of the treasury is often linked to the beginning of our series of tribute lists. But this may be simple chronological coincidence; and even the chronological coincidence is not firmly established. See Pritchett (1969).

25 Kallet (2009) 51, "In one year, either in 449/8 or 448/7, cities did not pay their dedication, and therefore, by inference, paid no tribute." See further Meritt (1925).

26 See Kallet (2009) 51; Lewis in *CAH*[2] 5, 121–127.

27 Rhodes and Osborne (2003) 22, ll. 9–10.

Pherai captured Tenos sometime before 362, and also took Peparethos, defeating an Athenian relieving force. Philip of Macedon took Amphipolis in 357, Pydna sometime soon afterward, Methone around 354, and finally Maroneia. He also destroyed Olynthos in 348, enslaving most of its inhabitants.[28] If this erases any doubt that members of the Second Athenian League were in need of protection in the middle decades of the fourth century, it may also make us wonder whether the League ever acted robustly to deal with the threats that existed. But the evidence that Athens sent relieving expeditions is strong. Relief missions are attested on behalf of the Euboians (377/6), Thebans (376), Zakynthians (374), Akarnanians (373/2), Peparethians (361/0), the Euboians again (357), and the Neapolitans (355–4).[29]

So much for the provision of the good of security by Athens' naval leagues in the classical period. What about a market? This second good is considered to be a consequence of the provision of the first. In the course of securing the Aegean against the threat of Persia (and, in the fourth century, other powers), Athens also cleared the sea of pirates and made low-risk trade between island and coastal *poleis* possible.

But is there any more direct evidence that Athens' naval leagues had a positive impact not only upon their members' security, but also on their access to a market? A passage in Isocrates may point to a resurgence in piracy in the 390s and 80s, between the dissolution of the Delian League and the establishment of the Second Athenian League, suggesting that the leagues did indeed have a positive effect on the security of maritime travel during the periods of their operation (Isocr. 4.115, from 380). Another passage, from Demosthenes, indicates that the protection of the freedom of the seas was viewed as an Athenian prerogative, at least for a time ([Dem.] 7.14–15). An inscribed decree from the 420s provides even more direct evidence that Athens occasionally intervened to ensure that its allies could trade freely. Among other measures concerning Athens' ally Methone, it resolves that an embassy should be sent to Perdikkas of Macedon to tell him to allow the Methonians to sail the seas freely, to enter his territory for the purposes of trade, and not to injure them as they do so. A later mention in the same inscription of the Athenian soldiers stationed at nearby Potidaea drops a heavy hint that the Athenians are willing to back up this declaration with force if necessary.[30] We might add to this a number of

---

28 Most of the cities mentioned were listed as members of the Second Athenian League on the Aristoteles Decree, e.g. ll. 85–87 (Peparethos, Tenos, Maroneia).

29 Cargill (1981) 144–145.

30 Meiggs and Lewis (1969) 65, ll. 17–24; the implicit threat comes at l. 28.

THE DELIAN AND SECOND ATHENIAN LEAGUES

Athenian measures that could be interpreted as encouraging trade. For the Delian League, the most important of these is the Standards Decree of 413, which imposes the use of common standards for weights and measures, as well as encouraging the use of Athenian coins.[31] It could be argued that the decree significantly lowered transaction costs for members of the Delian League, though the move would certainly have had costs for most of the allies as well as benefits, and its imposition was undeniably heavy-handed.[32] In this last regard, it differs substantially from fourth century measures such as the Law of Nikophon, which sought to guarantee the quality of Athenian coins. While the earlier law depends upon coercion, the latter seeks only to increase trust in Athens' now widely-used currency.[33] Finally, there is evidence in the sources that contemporary Athenians thought of their provision of security to the Aegean in terms of collective action. A passage in Isocrates may indicate that fifth century Athenians occasionally sought to deny freedom of passage to non-tribute payers (Isocr. 8.36).[34] This makes perfect sense in terms of collective action: those who do not contribute to the provision of a good do not deserve to benefit from it (and must be punished to ensure that others are dissuaded from free-riding). It makes little sense if Athens is modeled as a pure rent-seeker, since excluding traders from the sea is costly in itself and would presumably have had an injurious effect on Athenian trade.

---

31   Meiggs and Lewis (1969) 45. The dating of this decree to 413 (or at least to the 420s or 410s, rather than to 450–446, as suggested by Meiggs and Lewis) should now be regarded as firm. See Papazarkadas (2009) 72 (mid-420s at the earliest, probably closer to 414); Kroll (2009) 201–203 (413); Rhodes (2008) 51 (c. 425).

32   See Bradeen (1960) 269. Note the observation of Kroll (2009) 199–200 (on the basis of the figures in Figueira 1998) that "of the approximately 282 allied poleis, over two thirds were small communities that had never coined." For these *poleis*, it is arguable that the introduction of Athenian currency was an unmitigated boon. For the decline of 'local' currencies in *poleis* that already had them see Kroll (2009) 200; Figueira (1998) 49–179.

33   Rhodes and Osborne (2003) 25. For a reading of the law in terms of providing the good of a trustworthy currency to all its users see Ober (2008) 220–240, esp. 225: "The law's apparent goal is ensuring that silver coinage remains a reliable and low-transaction-cost exchange mechanism."

34   Isocr. 8.36. Bonner (1923) 195 also adduces Xen. [*Ath. Pol.*] 2.12 and Aristoph. *Ach.* 530–537 (the Megarian Decree). But though in both cases Athens seems to be restricting others' use of the sea, in neither case is this clearly because of failure to pay tribute.

176   KIERSTEAD

Some Predictions: Size, Coercion, Exploitation

In this final section, we return to the three predictions of the theory of collective action emphasized above to see if they are confirmed in the case of the Athenian naval leagues. This chapter argues that the two leagues did indeed operate differently partly because of their size; that they depended to a large extent on coercion, coercion which was often supported by league members; and that Athens, as hegemon, despite the constant influx of revenues from allies, bore considerable costs in helping ensure the successful joint activity of the leagues.

The first prediction was that large and small groups would develop different features. Though the literature provides no precise number of members after which a small group becomes large, several of the characteristics that define small groups (e.g. the ability of members to monitor other members with ease) might encourage us to think that a small group is one with tens of members, and a large group one with hundreds of members or more. The number of members in the Delian League varied throughout the fifth century as members left and joined the alliance for various reasons. The best information about the number of League members comes from after the appearance of the tribute lists. Using them we can state that in the 430s the number of states never rises above 175, and reaches its high point in the extraordinary reassessment of 425, when Athens laid claim to tribute from no fewer than 400 states.[35] The number of members in the Second Athenian League also varied (especially during the Social War, from 357 to 355), but it was never greater than at the foundation of the League in 378. We have an excellent idea of the number of members at the foundation of the league, because we can count the states that had their names inscribed on the Aristoteles Decree. There are about 60.[36]

It seems appropriate, then, to class the Delian League as a large group and the Second Athenian League as a small group, and to expect them to display some of the features predicted by the theory of collective action. After the end of the Persian Wars, members of the Delian League did indeed cede powers to monitor and punish defectors to a hegemon, Athens. That they did so because they could not tolerate the high costs of monitoring a large number of other members (and of punishing free-riders) is in full accord with the theory (Thuc. 1.99.3). The Second Athenian League, by contrast, was more cooperative. It was

---

35   See the commentary on Meiggs and Lewis (1969) 69.
36   For a fuller discussion of the exact number of states on the Aristoteles Decree (probably 59), see Cargill (1981) 45–47. Since some names have been added in between the original columns of text, we can rule out the possibility that many additional names were inscribed under the break in the inscription.

THE DELIAN AND SECOND ATHENIAN LEAGUES 177

founded on the explicit promise to maintain the freedom and autonomy of its members, a promise which Athens seems never to have broken (although the point is a controversial one).[37] The second prediction that was drawn from the theory of collective action was that successful joint action in both types of group would depend partly on coercion, which would often have the support of other group members. The first part of this statement—that Athens often coerced its allies—is so routinely emphasized in recent scholarly writing on the classical naval leagues that there would be little point in my reviewing the evidence for it here. The second part of the statement—that Athens' allies often joined in with the hegemon's coercive acts—is not a particularly controversial one but is usually passed over in silence, or at least without a great deal of attention being drawn to it.

John Ma is one of the few to look squarely at this phenomenon, and the question he asks about it is a good one, "Why did the allies ... collude actively in their own subordination?"[38] We may now be in a position to answer this question. The allies "colluded in their own subordination" because they valued the security and prosperity that were provided by the operation of the Athens-led naval leagues and because they recognized that failure to punish free-riders would result in these public goods no longer being provided. We should not be too quick then to take rebellions of individual *poleis* as evidence of the unpopularity of Athens, or of its naval leagues. Citizens in various *poleis* may well have wanted to opt out of the Athens-led project of naval domination of the Aegean. But they may also have simply been using the language of dissatisfaction and injustice to mask their disinclination to continue bearing the costs of a collective enterprise.

The third and final prediction that we must return to here is that 'larger' and more powerful members of groups would risk being disproportionately burdened by the costs of collective action. In a general sense, the pattern of contributions to the Delian League appears to bear this out, with less than a third of members paying almost 90% of the tribute, and 70% of the allies together contributing little more than a tenth of league funds.[39] If we focus in on the con-

---

37 For the explicit promise, see the Aristoteles Decree ll. 9–10. For an argument that Athens kept its promises, see Cargill (1981) 132–162.

38 Ma (2009) 230 with n. 12 for evidence of allied participation in suppressing the Samian revolt. Ma's reference to the prisoner's dilemma game may indicate that he has intuited that one answer to his question may come from the strategic analysis of Athenian and allied interactions.

39 For the precise figures (and the calculations and data behind them) see Nixon and Price (1990).

178   KIERSTEAD

tributions of Athens in particular, it is arguable that this extraordinarily large
member of the Delian League was more exploited than exploitative.

Scholars have repeatedly pointed to the amounts of silver flowing into
Athens as evidence of the hegemon's exploitation of its allies. But to establish the fact of exploitation, these figures would need to be complemented by
some estimation of the value of the goods Athens was helping to secure for its
allies.[40] A complete cost-benefit analysis would be complicated by the fact that
Athens' allies (as noted above) also occasionally contributed to the allied cause
through their actions, and by the number of separate sources of revenue and
expenditures of the League. Preliminary considerations (for example: that the
tax burden represented by tribute seems to have been less than 5% of the value
of trade in a given city) are not suggestive of gross exploitation on the part of
the Athenians.[41] But some measure of the value of the exchange entered into
by members of the Delian League may be reached if we restrict ourselves to the
cost of triremes, and the contributions that most states paid.[42]

A first approach to this problem is presented by Thucydides' account of the
early history of the Delian League. After a few years, he wrote, Athens did more
than its fair share of the League's military duties, but on the other hand it
was now increasingly easy for it to put down allied revolts (Thuc. 1.99.2). For
this fact (says Thucydides), the allies were themselves at fault: on account of a
disinclination to go on military expeditions, most of them, so that they would
not have to leave home, arranged to pay the requisite contribution by being
assessed for money, not ships (Thuc. 1.99.3).[43] If we accept the clear meaning
of Thucydides' text, we must conclude that, in the perception of most of the
allies, it was more costly for them to build and man ships than to pay the

---

40   This point is grasped by Popper (1945) 181: "In an attempt to evaluate the significance of
      these taxes, we must, of course, compare them with the volume of the trade which, in
      return, was protected by the Athenian fleet."

41   In 413, Athens levied a 5% tax on imports and exports instead of tribute (Ar. *Ran.* 363),
      apparently convinced that it could raise more money this way.

42   Rhodes (1985) 8 n. 16 cites the calculation of Ruschenbusch (1983), on the basis of the
      payments in the tribute lists that almost 60% of the members of the Delian League would
      not have been able to pay for a single warship for one campaigning season. For these states,
      being defended by a fleet of warships might have represented especially good value in
      return for their monetary contribution.

43   ὧν αὐτοὶ αἴτιοι ἐγένοντο οἱ ξύμμαχοι· διὰ γὰρ τὴν ἀπόκνησιν ταύτην τῶν στρατειῶν οἱ πλείους
      αὐτῶν, ἵνα μὴ ἀπ' οἴκου ὦσι, χρήματα ἐτάξαντο ἀντὶ τῶν νεῶν τὸ ἱκνούμενον ἀνάλωμα φέρειν
      ("The allies themselves were the cause of this: for on account of their disinclination to go
      on military expeditions, most of them, so that they would not have to leave home, arranged
      to pay the requisite contribution in the form of money rather than ships," Thuc. 1.99.3).

THE DELIAN AND SECOND ATHENIAN LEAGUES 179

Athenians to do these things for them (not only in financial terms, but also in terms of personal hardship).[44] The allies had strong incentives to reach an accurate estimation of the real costs of paying for their own triremes; but they may still have been wrong to believe that they were getting a good deal.

A standard figure for the cost of operating a trireme for a single month in the late fifth century is 1 talent of silver (Thuc. 6.8.1). This, however, looks like a minimum figure that includes only a crew's pay (*misthos*) while on operations, and does not cover the costs of building and maintaining a trireme or of training the crew. When Mytilene sent 10 ships to Athens in the spring of 428 "according to the alliance" (i.e. instead of tribute), this would have necessitated an outlay of 10 talents a month *at the very least,* and probably for the entirety of the five-month sailing season (Thuc. 3.3.4). It would appear that Mytilene spent in the region of 50 talents in 428. But we know from the Athenian Tribute Lists that by far the largest single payment from an allied state was 30 talents, and most paid much less. Even Finley has to conclude that "for the maritime states, tribute often meant a reduced financial burden, in some years a substantial reduction."[45]

We might wish to play with the figures proposed in various ways and many legitimate alterations might be suggested.[46] If Mytilene's ships only sailed for three months, the figure would be lower (although still equal to the highest single annual payment of tribute). Matthew Trundle offers different figures for the upkeep of triremes which may alter the calculations materially.[47] But it is hard to deny that the calculations are nicely balanced. In other words, it is by no means evident that Athens spent less on triremes for the allied fleet than it recouped from the allies. It is clear that profits, if there were any, can hardly be described as grossly exploitative. And it remains a possibility that the Delian

---

44   Finley (1978) 110 flatly refuses to accept the clear meaning of Thucydides text here, insisting that the allies' reluctance stemmed from the fact that they were "serving an alien, imperial state." Finley does not cite alternative sources for the view that the allies switched to contributions of coin for these reasons. In any case, the idea that a league of mostly Ionian *poleis* would have considered Athens 'alien' is problematic.

45   Finley (1978) 113.

46   Finley (1978) works on the basis of half a talent a month (for which there is some evidence); this would yield a figure of 25 talents for a full sailing season, still close to the largest recorded tribute payment.

47   Trundle's evidence, presented in this volume, suggests that sailors could be paid 3 obols rather than one drachma for a day's work. This aligns Trundle's figures more closely with those of Finley. But it should be remembered that even the lower figures still represent minimum costs for states running their own triremes.

League confirms the prediction of the theory of collective action that large members of groups will be disproportionately burdened by the costs of joint endeavours.

Conclusions

This chapter proposed a new perspective on the classical Athenian naval leagues, that of the theory of collective action, where the leagues are viewed as groups seeking to provide themselves with the related public goods of security and a market. It has focused on positive political theory and what it can contribute to our understanding of the dynamics of the naval leagues. This perspective is preferable at the explanatory level to accounts which model Athens as a simple extractor of rents. Ian Morris has recently proposed that we view the Delian League as an example of incipient (and ultimately incomplete) state-formation. To the extent that state-formation can resemble hierarchical collective action, my account is in broad sympathy with his.[48] To the extent that Morris' model is dependent upon viewing Athenian citizens as an elite seeking to expand its power and consumption, our accounts differ.[49]

The issue of whether Athens was a pure extractor of rents makes clear that even questions of positive analysis can have normative implications. There is no space here to perform the delicate work of deciding what the normative implications of viewing the Athenian naval leagues as examples of collective action might be.[50] Because of this, full discussion of such issues will need to wait for another time; although it may be helpful to highlight some salient points for future research.

A review of the literature reveals that an implication of some recent writing is that violence is sufficient for imperialism (and that imperialism is an illegitimate form of authority). But the view that all forms of coercion are illegitimate, though apparently widespread among Greek historians, is not a mainstream view among political theorists. Coercion can be legitimate or illegitimate, depending on the circumstances in which it is exercised. I suggest that coercing free-riders may represent a legitimate use of authority under certain conditions. One of these conditions is that membership of the group in which

---

48   Note, for example, that I view allied contributions as a tax.

49   Morris (2009) drawing on the model of state-formation in Tilly (1990).

50   As Olson (1965) 29 n. 47 notes: "no general moral conclusions can follow from a purely logical analysis."

THE DELIAN AND SECOND ATHENIAN LEAGUES 181

free-riding occurs must be consensual, at least initially. To support the claim that the Delian and Second Athenian Leagues were founded in a consensual manner, we must return to the primary source material.

There are a number of considerations that complicate the moral balance-sheet that need to be addressed in any such analysis. These include instances of undoubtedly exploitative and oppressive behavior by Athens, from massacres of neutral states to the imposition of cleruchies. Equally important are the varieties of drift facilitated by the nature of long-term collective action: the drift away from the league's original aims towards slightly different aims; the drift from the allies' predictably inaccurate expectations of what the league would be like to what it turned into in reality. Another form of drift, though it is not one that is a necessary part of collective action, saw the league shift from an explicitly consensual organization to one that lacked functioning collective institutions. Ironically perhaps, the descent of the Athenian naval confederacies into imperialism was brought about partly by the failure of democratic states to construct and maintain robust institutions at the international level that linked actions taken in the name of the leagues with the consent of their members.

The picture that emerges from these considerations is not one in which Athens is cleared of any wrong-doing. Nor, however, is it be one in which Athens is assumed to share in all the characteristics of more purely exploitative empires. Instead, I suggest that any accurate narrative of the rise and fall of the classical Athenian naval leagues will be informed by the tragic character of long-term collective action. The Athenians who were present at the creation of the Delian League did not go about their work with exclusively altruistic motives; nor, however, were their motives exclusively rapacious. The allies who entered into the League had a similar mix of motives, and beside those who were eager to take the fight to the Persians there seem to have been others who would rather that somebody else did it for them. The change in the nature of the Delian League was a tragedy for both sides, a tragedy produced more by the unwilled yet inexorable laws of collective action than by a contingently predatory hegemon.

Irregular Warfare

∴

CHAPTER 10

'Warlordism' and the Disintegration of the Western Roman Army[*]

Jeroen W.P. Wijnendaele

Some of the problems inherent in applying the label 'warlord' to the field of history can be shown in the following three cases. Between 976 and 989 the eastern Roman emperor Basil II's regime was shaken by the insurrection of his former generals Bardas Scleras and Bardas Phocas.[1] Both men were members of the rich military aristocracy of Anatolia and were able to muster local forces strong enough to directly challenge the central authority in Constantinople and score several victories over dynastic armies. Basil had to go to considerable lengths to suppress both rebellions, even enlisting the aid of Vladimir I of Kiev at the price of offering the hand of his younger sister Anna—a privilege previously unheard of for any 'barbarian ruler'. For most of the sixteenth century, Japan was subjugated to domestic strife and warfare during the so-called Sengoku period ('the country at war'). The traditional political order, in the form of the Shogunate and imperial court, lapsed and gave way to a fractured dominion of provincial *daimyo*. Unification of the country would only be achieved violently after the battle of Sekigahara in 1600 by Tokugawa Ieyasu and the establishment of his dynasty. During the third quarter of the fifth century, several western Roman commanders revolted against reigning emperors in Italy, detaching former areas of the Empire in the process and maintaining an independent stance. They possessed forces significant enough to achieve this, yet were unable to pursue grander military aims beyond their respective domains. They can, therefore, be seen as catalysts for the accelerating demise of western imperial rule in this century. These three episodes of world history are significantly separated from each other in terms of both space and time. Each occurred in different contexts and feature different historical processes

---

[*] An early version of this chapter was presented at the Warfare in the Ancient World symposium (University of Auckland, 10 July 2012). I am indebted to Dr. Jeremy Armstrong (University of Auckland), Dr. Mark Hebblewhite (Macquarie University), Dr. Fiona Tweedie (University of Sydney), Bradley Jordan and David Rafferty (University of Melbourne), and the anonymous peer reviewers for their feedback.

[1] All dates are AD unless otherwise stated.

and outcomes. Yet all three highlight military leaders who have been dubbed 'warlords' by modern historians analyzing these episodes.[2] As these cases illustrate, the label can and has been used in an idiosyncratic way.

Outside the field of (ancient) history, however, the study of warlords and 'warlordism' has attracted significant attention in political sciences and third world studies over the past few decades. This chapter seeks to highlight the dynamics behind the changing nature of the western Roman army in the first half of the fifth century, specifically through the concept of warlordism. The aim of this chapter is to evaluate the usefulness of the concept to a study of the western Roman army in the fifth century by providing an interdisciplinary overview of the concept, an introduction to the methodological problems for applying it to this era and domain, and ultimately a case study. Specifically, this chapter present an analysis of the career of the African field commander (*comes Africae*) Bonifatius and his relationship with his warrior retinue—the so-called *buccellarii*—as a case study for the 'semi-privatization' of the western Roman army. Bonifatius' rise to power signified a digression from traditional means to claim political and military authority through usurpation of the imperial office, whilst also setting a precedent for future commanders to foster personal control over their armed forces. This case study will argue that middle-ranking commanders, such as Bonifatius and Aëtius, broke the monopoly of violence hitherto exercised by the emperor and his court. But before we can attempt to apply the label to the western Roman army in the fifth century a few *caveats* have to be raised in regards to methodology.

'Warlords'—What's in a Name?

The modern term 'warlord' finds its most likely antecedent in the German word *Kriegsherr*, a title both used by generals during the reign of the Holy Roman Emperor Maximilian (1508–1519) and the emperors of the new German Empire (1871–1918) to denote their constitutional role as supreme commander (*oberster Kriegsherr*). However, the study of warlordism in social and political sciences originates with the rise of the Chinese *Jūnfá* during the demise of the Manchu Empire and its Qing dynasty.[3] Provincial military commanders attained more and more power, stepping into the political vacuum of the crumbling empire left after incessant insurrections and rebellions. Most of these men had been

---

2 Respectively: Blaum (1994); Elison and Smith (1987); and MacGeorge (2002).

3 Waldron (1991); McCord (1993); Waldron (2003).

former imperial generals, often driven into revolt by the forces they enlisted, and motivated by the dynasty's inability to deal with foreign threats and incursions. As the 'Warlord Era' (*Jūnfá shídài*) progressed, western commentators started naming these rival commanders 'warlords'.[4] The wars were highlighted by fierce internecine competition over economic resources to secure local autonomy. To maintain the loyalty of their personal forces, the 'warlords' were pressured into providing them with goods and money, often violently extorting the local population under their control. Yet it has to be noted that not all of these 'warlords' behaved as economic predators and some set themselves up as reformist governors, enacting policies to improve general conditions for their people. Even after the *Jūnfá* era ended, due to the alliance of the nationalist Kuomintang and Chinese communists in 1928, certain 'warlords' continued their independent stance—both within official governments, but also occasionally resorting to violence against the newly established political regimes.[5] Only with Mao's victory in 1949 and the establishment of the communist state did the phenomenon phase out in China.

Even though the warlord concept, as a research topic, briefly went out of vogue, during the cold war, the term became popular in the field of history to describe military leaders in medieval western Europe and Japan.[6] Yet in the aftermath of the USSR's collapse it gained new significance in analyzing increasing political instability in the third world. In particular, countries such as Somalia and Afghanistan have been described as 'failed states', where centralized structures have disappeared all but in name, or states fail to provide basic services and responsibilities to exercise sovereign government. Instead, both countries host a plenitude of warlords holding sway over vast areas.[7] At the same time, it has to be pointed out that not all of these warlords content themselves with being *de facto* rulers over fragmented territory. Some even push to become legitimate state rulers, the most notable case being that of Charles Taylor who established himself as president of Liberia.

There are reasons to remain cautious when applying, in a straightforward fashion, any conceptual approaches to warlordism in the modern world to the ancient world. In particular, the rapid evolution of technology and global forms of mass communication have no parallels in the pre-Industrial world, and already causes much difference between the Chinese warlords of the early twentieth century and their twenty-first century counterparts in sub-Saharan

---

4 Bloch (1938); Ch'en (1968).
5 Sheridan (1966) 14–16.
6 Smyth (1984); Simms (1987).
7 Herbst (1997); Little (2003); Giustozzi (2012).

Africa or Afghanistan. However, the most useful aspects of modern warlordism which could also be adapted by scholars of the ancient world is to avoid a top-down approach and instead look at factions behind independent military actors, the economy of violence, and patronage enabling these actors to become autonomous powers during the collapse of centralized state authority.

The field of ancient history has several authors who consistently use the label 'warlord', some even going as far as to provide a conceptual analysis for the later Roman Empire.[8] The term is usually used in regards to the (western) later Roman Empire, though the republican era has also been analyzed under this concept.[9] Both Whittaker and Liebeschuetz offer the most thorough abstract analysis and focus on the themes of declining central authority and the rise of local strong men, enabled through private patronage. Even in these cases, however, the term 'warlord' can be applied to a heterogeneous collection of actors: the *phylarchs* of Arabic tribes, sporadically providing support to both eastern Roman and Persian forces or acting as independent brigands, the leaders of barbarians clans on the outskirts of the Empire, such as the Sueves in Gallicia, the first leaders of the Merovingian Franks; and ultimately officers of both the western and eastern Roman army.

This problem is also reflected in political sciences, where much disagreement remains over what exactly constitutes 'warlordism'. Already during the *Jūnfá* era, attempts were made to define warlords. Chang Kai-Shek, who himself would later be vilified as a *Jūnfá* by Mao Tse-tung, declared in 1927 that a warlord was "a man who lacked a political principle ..., occupied an area, had an insatiable lust for money and property, loved his own skin and depended on imperialists for support."[10] The stress on self-serving motivations and external influence should not come as a surprise during an age of Chinese nationalism, the pressure on China from foreign powers, and the fact that warlords were often political rivals of one another. Sheridan, whose monograph on the *Jūnfá* era was one of the first proper academic works on warlords, proposed a very neat definition: a warlord is a person who "exercised effective governmental control over a fairly well-defined region by means of a military organization

---

8   Whittakker (1993); MacGeorge (2002); Liebeschüetz (2007). It is interesting that MacGeorge herself expressed dissatisfaction with the title and did not wish to elaborate it fully, except as a useful catch-all term to describe the military officials in her monograph. Other scholars use it consistently for the fifth century imperial West, see: Muhlberger (1990); Harries (1994); Williams and Friell (1999).

9   Ñaco del Hoyo and López-Sánchez (forthcoming). The notion of Sulla being a warlord has been considered but unsurprisingly rejected by Keaveney (2005).

10   Ch'en (1968) 575.

'WARLORDISM' AND DISTINTEGRATION OF THE WESTERN ROMAN ARMY 189

that obeyed no higher authority then himself."[11] His definition was enthusiastically adopted by Africanists and has remained prominent thanks to its elegant simplicity and applicability. Consequentially, however, in social sciences the label can be used to denote a great variety of leaders who use armed followings to enforce their power.

In order to differentiate between the diverse actors who have been labeled warlords, Giustozzi discards 'conflict entrepreneurs', traditional local rulers, and political entrepreneurs who may or may not be clan-based.[12] In his view warlords are "non-state politico military actors who have military legitimacy, but little or no political legitimacy".[13] In other words: they have the loyalty of the military class and enter politics either because they are 'orphans' of a state or as the result of an evolutionary process within its armed forces. The distinction with the former groups is important because warlords find themselves directly imbedded in the process of war and derive legitimacy from it, while their political aims take priority over the acquisition of goods. These leaders need to possess sufficient military prowess, demonstrate leadership in battle, while at the same time being able to manage the logistics and supply of their organizations. An important addition to these conceptual considerations is Vinci's analysis of the warlord-organization.[14]

Even though the term 'warlord' linguistically refers to a single individual, one of the most useful approaches to the phenomenon is to look beyond warlords as individuals and focus instead on warlord-organizations. It is the organization that he holds authority over that truly enables his power base. Vinci takes this one step further by downplaying the territorial aspect of the traditional warlord definition. Vinci's definition of a warlord, which this chapter wishes to adopt, is that of "the leader of an armed group that uses military power and economic exploitation to maintain fiefdoms which are autonomous and independent from the state and society."[15] The fiefdoms do not need to constitute territorial control as the primary factor to denote authority. Members of the warlord group do not need to be born into the group and, through the recruiting-process, members can become part of the warlord's patronage network. As a result, the warlord's fiefdom is not necessarily the territory he controls but, first

---

11   Sheridan (1966) 1.

12   Giustozzi (2005) 10–12. Conflict entrepreneurs are defined as "financiers acting behind the scenes or leaders of violent movements motivated only by greed." Traditional local rulers are "tribal leaders who mobilize support from within their own clan or tribe."

13   Giustozzi (2005) 9.

14   Vinci (2007).

15   Vinci (2007) 328.

and foremost, his military organization. This is a crucial element that can be seen in the behavior of certain defeated warlords in modern Africa, who simply abandon their territory and move on with loyal followers to a different territory.

It is equally important to distinguish warlords from those military actors based on clan-structures, as can be seen in some parts of Afghanistan and Somalia, or insurgencies, since the latter rely on civilian support. By contrast, warlord organizations have shown tendencies to turn against elements of local populations who share the same ethnicity. This is possible since members of the warlord organization identify themselves as being part of a separate political community. Warlords are completely autonomous from the state they find themselves in and the state will have 'failed' when it loses control over the population within their territory. They are able to do so because of their military power—their organization is effectively an army and should be distinguished from militias. This military power might not be sufficient to overthrow a state, yet it can be powerful enough to establish temporary control over a piece of territory. Even within the state it can be used as a 'power reservoir', allowing the warlord to maintain his authority. Ultimately, the warlord's leadership can be rooted in charismatic, patrimonial, or even legal authority, which holds his organization together.[16] His soldiers are not induced by ethnic or clan motivations to fight for the warlord's cause, and instead it is an economic incentive, the warlord's patronage, which ensures their allegiance.

These distinctions are important and can be used for the ancient world to differentiate late Roman warlords from tribal leaders such as the Arab *phylarchs*, who derived their legitimacy from clan-based structures, or so-called barbarian leaders, who could legitimize themselves through kingship over the ethnic core of their people.[17] Equally important is the matter of military

---

16 These aspects were already noted by Weber in his groundbreaking essay *Politics as a vocation* (originally given as a public lecture in 1919). Weber explains that before the arrival of modern states, leaders were able to receive the devotion of people through "purely personal 'charisma'". He elaborated that this type of charismatic leadership can be seen throughout all historical eras, in the form of magicians and prophets, or "in the elected war-lord, the gang leader and *condotierre*." Weber also coined the concept of 'monopoly of violence' in this essay, which he defines as the prime factor in a state as "a human community that (successfully) claims the monopoly of the legitimate use of physical force (*das Monopol legitimen physischen Zwanges*) within a given territory".

17 It is this notion that ultimately causes hesitation to qualify the Gothic leader Alaric as a warlord. Certainly, Alaric did have a chequered career as an imperial general (397–400 in eastern Roman service, 405–408 in western Roman service). But during the times he found himself stranded outside imperial service, he maintained authority over his

legitimacy. Even though there are several cases of military dissidence noted among fifth century western Roman officers, who either put themselves outside the political framework or were forced to do so, these men still carried the loyalty of their troops. Therefore, their soldiers deemed them to be legitimate leaders. However, in order to maintain the loyalty of these troops, successful and repeated patronage is necessary. Especially the notions of a fiefdom as relating to a group-structure that still needs to be supplied by its leader's patronage, and the warlord as a battle-hardened 'orphaned' general who continues politics outside the institutional framework through his military legitimacy, are crucial to understand the rise of warlord-commanders in the western Roman army, such as Bonifatius. Before considering the latter's career, however, it is important to survey the historiographical and military context of his era.

Tantalizing Sources

The fifth century remains a momentous era in the history of the Roman Empire. It witnessed the dissolution of the western imperial office and the replacement of its governance over Roman provinces by so-called 'barbarian' rulers.[18] This geo-political process known as the 'Fall of the (western) Roman Empire' has been the scope of endless debate and will probably remain so.[19] The simple fact is that the literary sources for any analysis of political or military matter

---

following as a king. Furthermore, even at the lowest estimates, his group ranged in the lower ten thousands, far more numerous than any other western Roman general's personal retainers. Hence this group was able to maintain its cohesion and developed a separate ethnic identity as the Visigoths after Alaric's death, while its leaders remained kings. On the debate surrounding Alaric's leadership and his following, see: Liebeschuetz (1992) 82–83; Kulikowski (2007) 161–177; Halsall (2008) 200–206.

18   Modern scholars still tend to speak of a 'Western Roman Empire' and 'Eastern Roman Empire' as if these were separate states. It remains beyond doubt that during most of the fourth and fifth century imperial territory was governed by more than one emperor who kept his own court, administration and army. Especially as the fifth century progressed, it became impossible for one emperor to impose his authority over the entire realms even when he was a legitimate candidate, as can be seen in the case of Honorius after the death of his brother Arcadius (408) and Valentinian III after the death of his uncle Theodosius II (450). Despite these practical issues of administration, the Empire constitutionally remained a single state, see Kornemann (1930); Sandberg (2008).

19   The 21st century has seen an increased reevaluation of this theme, see Wijnendaele (2011). For a full narrative of fifth century events, see Heather (2005); Halsall (2007). Among older scholarship, very useful studies remain, such as Bury (1923); Stein (1947).

192 WIJNENDAELE

in this period remain frustratingly meager. Already a century ago, one of the foremost scholars of the later Roman Empire pointed out that:

> The fifth century was one of the most critical periods in the history of Europe ... but ... there is no history of contemporary events, and the story has had to be pieced together from fragments, *jejune* chronicles, incidental references in poets, rhetoricians, and theologians ... Battles, for instance, were being fought continually, but no full account of a single battle is extant. We know much more of the Syrian campaigns of Thothmes III in the fifteenth century B.C. than we know of the campaigns of Stilicho or Aëtius or Theoderic.[20]

The problem becomes especially pervasive in regards to evaluating the role of the western Roman army.[21] The fourth and sixth centuries provide a wealth of detail for imperial forces through the histories of Ammianus Marcellinus, for the period 354–378, and Procopius of Caesarea, for the reign of Justinian I (527–565).[22] Both Ammianus, as officer in the *protectores domestici*, and Procopius, as secretary to the *magister militum* Belisarius, had military experience and provided eyewitness accounts of battles and imperial campaigns. Despite the long gap separating both authors, one scholar observed that it is of fundamental importance to note that the Roman army did not see any major reforms and remained virtually the same in its basic structure. In other words: the same army that in the past had been blamed for the loss of the western provinces in the fifth century, not only managed to retain the eastern emperor's sovereignty of nearly his entire domain in the same period, but also reestablished direct control over significant tracts of the western Mediterranean in the sixth century.[23]

Similarly, what little information we have from fifth century sources about battles fought by the army shows an altogether not undistinguished track record. In direct confrontations, against internal or external enemies, the imperial army retained the upper hand until the death of the emperor Majorian (461).[24] The armies of Stilicho managed to crush Gildo's rebellion (398), fight

---

20   Bury (1923) vii.

21   The late Roman army has attracted a lot of thorough scholarship in recent years, see Elton (1996); Nicasie (1998); Le Bohec (2006); Lee (2007).

22   Studies on Ammianus are abundant. For military matters especially, see Crump (1975); Austin (1979); Matthews (1989). On Procopius, see Cameron, Av. (1985); Kaldellis (2004).

23   Elton (2007) 309.

24   As Bury already noted, however, we possess very little detail about the actual battles being

'WARLORDISM' AND DISTINTEGRATION OF THE WESTERN ROMAN ARMY 193

Alaric's Goths twice to a standstill (402), and bring Radagaisus' huge confederation to complete surrender (406).[25] His eventual successor, Constantius, oversaw the destruction of the usurpers Constantine III, Jovinus, Attalus Priscus, Maximus, and the revolt of Heraclian, whilst bringing the Visigoths under control. Even more surprisingly, very few battles were actually lost in major confrontations against external enemies.[26] Finally, many units of the western army survived the eclipse of western imperial office. The life of Severinus depicts vivid scenes of Roman garrisons in Noricum maintaining their profession in the 460s, but gradually withering away due to lack of supplies and payment coming from Italy (Eugipp. *V. Sev.* 4.1–4). Procopius noted that in Northern Gaul, during the early sixth century, Roman soldiers continued serving as *limitanei* and still maintained the traditional military customs of their ancestors (Procop. *Goth.* 1.12.16–18). The *scholae palatinae*, crack units, raised by Constantine I and functioning as imperial bodyguards, were only abolished by Theoderic the Amal after his victory over Odoacer in 493. We should be very cautious, therefore, to explain the disintegration of the late imperial west in strictly military terms.[27]

Outside the narrative sources, however, there are two important documents dealing directly with the western Roman army in the first half of the fifth century which immediately highlight some of the constraints hampering the imperial war machine. The enigmatic document, the *Notitia Dignitatum*, preserves a snapshot of the imperial bureaucracy and military establishment in the late Roman West during the 420s.[28] At face value it portrays an impressive state

---

fought in the fifth century. The only battles that go beyond a simple statement of outcome
are Constantius victory over auxiliaries of Constantine III (Sozom. *Hist. eccl.* 9.13–15) and
the so-called battle of the Catalaunian Plains between Attila's Huns and allies, and a
motley crew of diverse barbarian allied forces together with what was left of the western
field armies under the leadership of Aëtius (Jord. *Get.* 197–213).

25   For details, see: PLRE 1 'Stilicho' 853–858; Mazzarino (1942); Cameron, Al (1970) 43–188; O'
Flynn (1983) 14–62.

26   A notorious exception are the Vandals under the successful leadership of Geiseric (PLRE
2 'Geisericus' 496–499), who managed to corner the local field army of Africa in 430 and
defeat an eastern Roman relief army in 431. Geiseric managed to capture Carthage in 439,
forestall Majorian's invasion of Africa in 460, and defeat the eastern Roman armada of
Basiliscus in 468.

27   On the fall of the Roman Empire as a military failure see: Ferrill (1988). Few scholars accept
Ferril's mono-causal interpretation, but he has valuable things to say on the deterioration
of western Roman armies in terms of discipline and cohesion.

28   The exact date of the *Notitia* remains a conundrum. The eastern half most likely dates to
Theodosius I's campaign against the western usurper Eugenius in 394 yet the western copy

apparatus that puts control over western imperial forces in the hands of its most senior general (the *magister utriusque militae*).[29] But detailed analysis of military units show the detrimental attrition the army had suffered in the preceding decades during the reign of the emperor Honorius (393–423), which witnessed unprecedented deterioration of central authority over its western provinces. These losses are best explained through the many usurpations, military unrest, barbarian incursions, and civil wars which had plagued the imperial West as much as the apex of the so-called 'crisis of the third century'. Britain was *de facto* severed from the Empire, whilst control over the outer Danubian provinces became nominal at best. Meanwhile, large parts of Italy, Spain, and Gaul were thoroughly wrecked through civil war and foreign incursions. A prime victim of this turmoil was the western Roman army whose field units (the *comitatenses*) experienced massive casualties and suffered heavily from attrition throughout these decades, as reflected in the *Notitia*. Out of 181 units of the western *comitatenses*, 76 had perished between the eve of the battle of the Frigidus, the wars against Alaric and Radagaisus, and the countless civil wars against usurpers. The loss in manpower could only be compensated in make-shift ways by transferring garrison troops into the mobile field units; from 97 newly raised regiments, 62 had been frontier troops upgraded 'on paper'.[30]

Despite these tremendous losses, the balance of power had been restored in favor of the Empire during the second decade of the fifth century by the *magister utriusque militiae* Constantius.[31] In Gaul, barbarian war-bands such as the Visigoths and Burgundians had been pacified and settled as imperial auxiliaries (the so-called *foederati*), while the Vandals and Alans had been thoroughly cowed in Spain. Contemporaries enthusiastically spoke about "restored

---

      was several times edited and amended after this date. The last datable western military unit is the newly raised *Placidi Valentiniaci Felices*—a clear reference to Valentinian III who was born in 419 and crowned emperor in 425. See Kulikowski (2000).

29   The emperor is curiously absent from this document, though that can be explained through the ideological nature of the *Notitia* where inclusion of the imperial office might put visible limits to his theoretical omnipotence. The *Notitia* is very antiquarian text clearly demonstrating ideological purposes rather than practical, see Brennan (1996). The concentration of so much western military power into the office of *magister peditum praesentalis* is probably explained due to the unique position Stilicho found himself in after the death of Theodosius I, as regent of Honorius and the most senior commander who had survived the battle of the Frigidus.

30   Jones (1964) 355.

31   *PLRE* 2 'Constantius 17', 323–325; O' Flynn (1983) 63–73; Lutkenhaus (1998).

order" (*ordo renascendi*).[32] Yet less than two decades later, the author of a military manual decried the "barbarization" of the army and nostalgically pleaded for a return to the fabled citizen armies of the Republic. The *Epitoma Rei Militaris* of Vegetius is one of the most important, yet complicated, sources for Roman military history.[33] The work is composed of four books that provide detailed instructions on all elements a commander should know on how to wage war successfully with a properly trained army. To achieve this Vegetius went through the recruiting and training of soldiers, the formation and structure of units, battle tactics and strategy, administration, fortification, siege and naval warfare. The aim of Vegetius' *Epitoma* was to offer a guide on how to reform the Roman army of his day. In his opinion, the contemporary army was lacking in discipline and effectiveness because of widespread barbarization. To counter these defects, the army had to be reformed in the ways of the armies from the Republic and early Empire, of which he held a highly idealized view.

One should be careful though, not to take these criticisms and recommendations at face value. Vegetius was not a man of military experience himself. Instead, he composed his work by compiling all sorts of antiquarian sources, such as Cato Maior, Cornelius Celsus, Frontinus, Paternus, and *constitutiones* of Augustus, Trajan, and Hadrian. The work is rife with archaic/anachronistic terminology and language to describe the contemporary army. Vegetius spoke about the Roman army in antiquarian language, even using republican terms such as *hastati* and *triarii*, which often obscures the vast differences between

---

32   Rut. Nam. *De Reditu* 1.139–140; Oros. 7.42.15–16. On this theme, see especially Matthews (1975) 329–388.

33   There still remains significant disagreement whether the *Epitoma* was written during the reign of Theodosius I or Valentinian III. For the strongest argument in support of Valentinian III, see Charles (2007). No reference can be found to any divided or shared authority over the Empire in Vegetius. Because of the nature of imperial panegyrical literature, to which Vegetius' work has strong similarities, it rarely happened that imperial colleagues—especially if they were the senior Augustus' children—were not mentioned. Theodosius I was known be very prudent about maintaining the fiction of imperial collegiality. A more important factor is Vegetius' statement that during the reign of Gratian infantry ceased to wear armor. Though most likely an exaggerated claim, this would make more sense during the reign of Valentinian III when the greater part of the western army may have been composed of *foederati* among whom helmets and body armor are rarely attested. Furthermore, Vegetius' section on naval warfare makes more sense during Valentinian's reign when the Vandals conquered North Africa and acquired a fleet that enable them to widely engage in piracy.

the army of his day and the one he idealized.[34] Despite these *caveats*, Vegetius' criticisms of the army reflect the anxiety felt on military security in the 440s. During this same period, which saw the Vandals' acquisition of the richest African provinces, the western emperor Valentinian III issued a decree stating that the imperial treasury was no longer able to properly equip and supply its veteran forces.[35] Somewhere in the quarter century amid the ideological expression of supremacy established in the *Notitia* and the enthusiastic reports of contemporary writers, and the antiquarian appeal for reforms by Vegetius and Valentinian III's admission of bankruptcy, important internal developments took place that fundamentally altered the balance of authority inside the western Roman government. This evolution can be traced in the careers of Bonifatius and Aëtius.

Bonifatius, Aëtius, and the Rise of Warlord-Commanders (422–432)

Between 422 and 432, the *comes Africae* Bonifatius was arguably one of the most influential men in the western Roman Empire.[36] We do not have any information about his life prior to his military career, except one dubious source stating that he was a Thracian. However, there is good reason to assume that Bonifatius probably was an African of Romano-Punic stock.[37] At Marseilles, in 413, he became a war hero by personally wounding the Visigothic king Athaulf (Olymp. *fr.* 22.2). Several years later he reappeared in Numidia as tribune of a unit of Gothic *foederati*. Here he began a remarkable friendship with St Augustine, the bishop of Hippo Regius, and exchanged several letters with him in

---

34   Yet even Ammianus Marcellinus, despite his extensive military service, often leaves out specific terminology when describing the institutional framework of the imperial army. Such practices owe much to the conventions of classizing historiography, as can also be noted in the history of Zosimus. Only when copying Olympiodorus in the final sections of his work is the reader suddenly faced with a wealth of military nomenclature. However, Olympiodorus was already considered an unusual historiographer by his compiler Photius for his fascination with numbers, economic material and non-Greek terminology.

35   *Nov. Val.* 16. On the impact of African tax revenue on the western imperial treasury, see Elton (1996) 125–127.

36   *PLRE* 2 'Bonifatius (3)' 237–240; De Lepper (1941); Diesner (1963); Wijnendaele (2015).

37   The name Bonifatius was typical for Christians in late Roman Africa, but is rarely attested outside this province. The only source who states that he was Thracian is the forger of the so-called Pseudo-Bonifatius. On his Thracian origin, see Pseudo-Bonif. *ep.* 10. On his African origin, see Clover (1993a).

'WARLORDISM' AND DISTINTEGRATION OF THE WESTERN ROMAN ARMY   197

which theological matters were discussed. In these years he rapidly acquired fame as a skilled soldier and officer in his battles against the local Mauri tribes of the pre-Sahara zone.[38]

His reputation had become so great that he was recalled to Italy when Constantius III died in 422. Most probably, it was then that he married his second wife, Pelagia.[39] She was the daughter of the Gothic nobleman Beremudus, providing her husband with a *Gefolgschaft* of his own.[40] The contemporary historiographer Olympiodorus noted that such soldiers were called *buccellarii*, meaning "biscuit eaters", a term which could refer to a mixed body of Roman and Gothic soldiers (Olymp. *fr.* 7.4, 12). The word appears for the first time in the title of the eastern Roman commander of the *catafractarii*, an elite cavalry unit, and generally implied a special close relationship between the retainers and their commander-in-chief.[41] Although this particular concept of a noble's household already existed in the barbarian world, the *buccellarii* nevertheless were a Roman institution, as seen in its adoption in the Visigothic lawcodes by its Latin name.[42] Officially, these were state troops who received their equipment and supplies from the imperial administration, but were remunerated by their respective officers. These retinues were not just mere bodyguards, but often contained enough manpower to form armies in their own respect. The most famous, but exceptional, case is that of Belisarius, who was claimed to have 7000 *buccellarii* drawn from his household.[43] We do not have any information on the strength of Bonifatius' *buccellarii*, but it seems to have been impressive since, as we shall see, they were to allow him to remain in power for more than a decade through extraordinary circumstances.

In 422 Bonifatius joined the campaign of Castinus, the new *magister utriusque militiae*, against the confederation of Vandals and Alans in Baetica. Boni-

---

38   On theological matters, see Aug. *ep.* 185, 189. On command in Africa, see Aug. *ep.* 220.

39   *PLRE* 2 'Pelagia 1' 856–857.

40   Gil Egea (1999) 496–502.

41   *Not.Dign.* [or.] 7.25: *Comites catafractarii Buccellarii iuniores.*

42   For the development of the *buccellarii*, see Diesner (1972); Liebeschütz (1986); Schmitt (1994).

43   Liebeschütz (1996) 232, n. 14 notes that Belisarius had no more than 5000 soldiers during the siege of Rome, which included a vast proportion of regular infantry. Even in the Justinianic era cavalry rarely represented more than a third of mobile field armies, hence Belisarius' *buccellarii* did not likely exceeded 1500 men. The earliest recorded numerical strenght of such retainers is that of the Goth Sarus, who had 200 or 300 men at his disposal in 410 (Olymp. *fr*, 6; Zos. 6.13.2). These numbers are more akin to that of regular sixth century officers in Procopius' histories.

fatius' forces, however, never actually arrived there since the two commanders found themselves quarrelling with each other from the start (Prosp. *s.a.*
422). Castinus aspired to become a new *generalissimo*, while the empress Galla
Placidia was trying to prevent that.[44] As a result, Castinus found it prudent
to remove Bonifatius from his command. He may have suspected that the latter was a partisan of Placidia, but far more alarming was Bonifatius' marriage
with Pelagia. This substantially increased Bonifatius' military power—power
that lay outside Castinus' command sphere. By dismissing Bonifatius, Castinus
short-circuited this potential liability by depriving the junior officer from essential funds needed to maintain his retainers. Bonifatius refused to stay with his
superior any further and retreated to Africa where he acquired, albeit in suspi-
cious circumstances, the supreme command of the regional forces there (Hyd.
70 (78); *Chron. Gall.* 511, 571).[45] That Bonifatius was able to get away with such
a clear act of insubordination reveals that it must have been regarded as too
dangerous to oppose. Castinus continued with his campaign, but was soundly
defeated, while Bonifatius used his new power to engage in campaigns against
the Mauri tribes whom he pacified (Olymp. *fr.* 42; Hyd. 69 (79)). The full extent
of Bonifatius' power would become most apparent after Honorius' death in 423.

Prior to this event, the empress Galla Placidia had been exiled to Constantinople with her infant son Valentinian (Olymp. *fr.* 40; Prosp. *s.a.* 423).
Initially the eastern emperor Theodosius II was not inclined to deal with the
situation in the West. Most probably he wanted to remain sole emperor, as contemporaries observed, while having Castinus, who was designated consul for
424, as his representative in Italy (Prosp. *s.a.* 423; Hyd. 73 (82)). Bonifatius, however, continued to support the cause of Placidia and, after four months, Rome
saw the usurpation of the throne by Ioannes, who was supported by the new
rising star among the officers, Flavius Aëtius.[46] As a result, civil war erupted and
imperial forces battled each other in Africa, Dalmatia, and Italy. Bonifatius prevailed on his home ground against an Italic field army, while the eastern Roman
armies forced their way to Ravenna and crushed the rebellion there. One contemporary chronicler noted that: "Ioannes' defenses were weaker because he
was trying through force of arms to take Africa, which was held by Bonifatius."
(Prosp. *s.a.* 424). That Bonifatius was able to resist the usurper's army is a
very noteworthy fact. Previous rebellions in Africa featuring the regional field

---

44   Oost (1968) 169–171. O'Flynn (1983) styled the over-mighty *magistri militum* who domi-
        nated the western Roman court during the late fourth and fifth century anachronistically
        but aptly as '*Generalissimos*'.

45   For Hydatius' chronicle, I follow the edition of Burgess (1993).

46   *PLRE* 2 'Fl. Aetius 7' 21–29; O'Flynn (1983) 74–103; Zecchini (1983); Stickler (2002).

'WARLORDISM' AND DISTINTEGRATION OF THE WESTERN ROMAN ARMY 199

army, including those by Firmus (372–375), Gildo (398) and Heraclian (413), had always ended in failure when confronted by field armies from the continent.[47] Bonifatius' *buccellarii* probably played a decisive role in his defence.

Valentinian was crowned emperor in 425 but, being only six years old, he could not yet exercise his authority properly. Bonifatius, who had played a pivotal role in the restoration of the Theodosian dynasty in the West, was merely given the position of *comes domesticorum* (commander of the imperial household troops). The hitherto unknown Felix was given the senior generalship of the western field army, the position that Bonifatius had craved.[48] Lastly, Castinus was exiled while Aëtius struck a deal with the new regime through which he was given a command in Gaul in exchange for redirecting several thousands of Hunnic auxiliaries, originally recruited to support the usurper Ioannes (Prosp. *s.a.* 425). The precarious balance between the three commanders did not last long. Already in 427, Felix made the first move by declaring Bonifatius an enemy of the state, after the latter had refused to appear at the court to account for a trumped-up charge.[49] An army under three generals was sent that managed to besiege the *comes Africae*. During the siege mutual discord broke out and one of the generals murdered his colleagues only to be killed in his turn by Bonifatius (Prosp. *s.a.* 427). A new force was sent under the command of the Goth Sigisvult.[50] It captured Carthage, while Bonifatius retreated to the interior of Numidia. However, no fighting occurred during this campaign in 428. It has been argued, therefore, that Sigisvult's mission was a reconciliation mission between Placidia and her dissident general. However, the imperial envoy Darius confessed in a letter to Augustine that there was still potential for armed conflict—indicating that this was a truce established at arms length.[51]

Despite having to face three imperial armies in less than five years, Bonifatius still remained unbeaten. At the same time, the last civil war demonstrated the limits to his private power. His fall-out with Ravenna meant that Bonifatius could only rely on his Gothic retinue. In return, they exercised pres-

---

47   On Firmus and Gildo, see Blackhurst (2004). On Heraclian, see Oost (1966) 236–242.

48   *PRLE* 2 'Felix (14)' 461–462.

49   Procop. *Vand.* 3.3.14–21 tells a fantastic story how Aëtius was responsible for a nefarious plot against Bonifatius which ultimately led to his conflict with the western court. However, the former spent most of 425–430 fighting various campaigns in Gaul against Visigoths and Franks. It is safer to accept the testimony of Prosp. *s.a.* 427 that Felix engineered the war against Bonifatius.

50   *PLRE* 2 'Fl. Sigusvultus', 1010.

51   On the peaceful mission, see Mathisen (1999) 176–183. On the potential for armed conflict, see Aug. *ep.* 230.3.

sure on him for rewards for their service. The only way he could satisfy their demands was to let them provide for themselves by looting the local population (Aug. *Ep.* 220.6). At the same time, Bonifatius' daughter was baptized by an Arian priest. In these circumstances, Bonifatius could only acquiesce, but it would cost him the friendship of Augustine. Negotiations followed and Bonifatius was reconciled with Ravenna in 429 and re-established in his office, just in time to deal with the Vandal invasion. For fourteen months they would besiege him in the town of Hippo Regius that was defended by his Gothic warriors (Poss. *V.Aug.* 28). Augustine would not see the end of the siege, passing away in the summer of 430. A relief army was sent from the combined forces of East and West under the command of Aspar, who had also commanded the eastern army against Iohannes. After a second defeat, Placidia recalled her former ally to deal with Aëtius, while Aspar remained in Africa to keep the Vandals in check.

During the Vandal war, sinister power games had continued in Italy. Aëtius had Felix and his wife murdered during a mutiny at Ravenna, allegedly to prevent a plot against him (Prosp. *s.a.* 430; Hyd. 84 (94); Marcell.com. *s.a.* 430.2; Joh. Ant. *fr.* 201). When Bonifatius landed in Italy, he was invested with the title of *magister militum* at Rome (Prosp. *s.a.* 432). The two respective warlords came to blows near Rimini in 432 (Addit. Prosp. Haun. *s.a.* 432; Hyd. 89 (99); Marcell.com *s.a.* 432; Joh. Ant. *fr.* 201). When the battle was over, Bonifatius had prevailed, but he would not enjoy his victory long. A couple of months later he died from a battle wound. His son-in-law Sebastian briefly took over his position and attempted to liquidate Aëtius, who had retreated to his fortified estates. This failed and Aëtius fled to the Huns, from whom he acquired a new army before returning to Italy (Prosp. *s.a.* 432; Chron. Gall. 452, 112). He subsequently forced Sebastian into exile and took over his position, forcing the latter into a decade-long Odyssey all over the Mediterranean.[52] Pelagia became Aëtius' new wife and, for the next twenty years until his death, he would be the virtual ruler of the imperial west.

Conclusions

Between 435 and 460, the supreme commanders Aëtius, Ricimer, and the (future) emperor Majorian were still able to score a series of victories over barbarian armies whilst retaining control over a core empire centered on Italy,

---

52   *PLRE* 2 'Sebastianus 3' 983–984.

southern Gaul, Dalmatia, and eastern Spain. But their armies and the nature of their command were significantly different from the centralized forces encountered in the *Notitia Dignitatum*. The career of Bonifatius, taking place less than a few years after the death of Constantius III, already shows the evolution taking place in the military upper classes. Both Stilicho and Constantius had wielded enormous power beyond their official rank of *magistri utriusque militiae*, exercising *de facto* control over the government of the imperial West. Yet both men had followed a traditional career path inside the officer's class and at no occasion did they break with the dynastic government.[53] Bonifatius achieved his position and power inside the western Roman army in substantially different ways. It was claimed in actions which, from a legitimist point of view, can only be identified as insubordinate and violently opposing the expressed wishes of central authority. Bonifatius was able to establish local dominance in Africa based on his private access to forces, which he could muster successfully on three different occasions against Castinus, Felix and Aëtius. The latter started his career serving a usurper, murdered his superior, violently resisted his deposition, and blackmailed the government into restoring his office at arm's length.

The fundamental importance of the careers of Bonifatius and Aëtius does not lie in them being "the last of the Romans" as Procopius famously but deceivingly called them (Procop. *Vand.* 3.3.14).[54] Instead, Bonifatius and Aëtius were the first ones who succeeded in to directly challenging and resisting central authority, without resorting to the traditional means of usurping the imperial office—a course still taken by other contenders in both men's early careers. That is not to say they were the first western Roman officers to do so: the reign of Honorius had seen several men maneuvering in similar fashion but ultimately failing.[55] It does not suffice to limit a study of such officials to

---

53 Stilicho did have the opportunity, during Olympius' coup in 408, to revolt against Honorius' government since he still carried the allegiance of his barbarian auxiliaries. Some of his officers urged him to do so, yet he declined and submitted himself to Honorius' guards at Ravenna and was duly executed (Zos. 5.33–34). Constantius' career shows nothing but a legitimist approach, eventually culminating in him being appointed co-*Augustus* of Honorius in 421.

54 Bonifatius and Aëtius only feature in Procopius' history as a digression on how Africa was lost to the Romans and conquered by the Vandals. The episode mainly highlights Aëtius' plotting against Bonifatius and the latter's desperate fight for survival.

55 The *comites Africae* Gildo and Heraclian clearly belong to this category. The Goth Sarus, one of Stilicho's officers, also showed himself quite adapt at operating as a free agent during Alaric's sieges of Rome (408–410). However, all three men ultimately failed in their autonomous endeavors and met a violent death. In the counter-factual scenario

individuals' policies or careers to comprehend the evolution of the western military aristocracy in the fifth century. Their military forces, whether it be *buccellarii* or the imperial army as whole, clearly possessed agency, which needs to be taken into account. The reciprocal nature of the *buccellarii* meant that Bonifatius was forced to allow them to pillage the provincial population during his battle to survive with imperial forces in the late 420s. After their patron's murder, Aëtius' *buccellarii* personally killed Valentinian III on the *campus Martius* and one contemporary could not help but note that Valentinian died "as his army stood around him" (Hyd. 154 (192)).

Bonifatius and Aëtius' martial abilities owed much to their extensive experience with the barbarian world.[56] Aëtius had served as a hostage in both Alaric's camp and among the Huns during his youth.[57] The Huns provided him with an external source of military power which he tapped into on various occassions between 424–439. Similarly, the Goths form a red line throughout Bonifatius' career: he launched his career by wounding Athaulf in 413, married a Gothic noblewoman, and commanded Gothic soldiers in Africa almost continuously between 417–432. Both men were noted for their prowess in single combat, and reputedly challenged each other during their showdown at the battle of Rimini (Marcell.com. *s.a.* 432). In the process they broke the western imperial government's monopoly on violence, paving the way for future warlord-commanders and the disintegration of the western army in the second half of the fifth century.

By virtue of his Gothic alliance, Bonifatius had become the first Roman in the western Empire to successfully challenge and resist state authority. When the central government tried to depose him, he took an autonomous stance in his province without declaring himself emperor, which had always been

---

that Honorius had yielded to Alaric's demands and appointed him *magister militum* in 409–410, the latter could have earned a place of pride in this new power-brokering paradigm. Even then, however, it would have been a significant issue for the western Roman government to deal with his dual position as king of his Gothic followers. Given Alaric's ultimate failure, these considerations cannot be drawn out further.

56   That is not say that both commanders' armed forces were exclusively barbarian. Bonifatius in Africa, and Aëtius in Gaul, will have commanded what was left of the remaining field armies. They are also attested with Roman officers amongst their personal staff (e.g. Aug. *ep.* 7* mentions the *domesticus* Florentinus and the *comes* Sebastian for Bonifatius. Sid. Apoll. *Carm.* 5 mentions Majorian, and the latter's father Domninus, for Aëtius). Barbarian retainers significantly enhanced his power, but Bonifatius was equally able to retain the allegiance of the regular African field army in his conflicts with Ioannes and Felix.

57   Gregory of Tours, *Hist.* 2.8 (citing the lost history of Renatus Profuturus Frigeridus).

the traditional method used by rebels to legitimize their positions. He carried the personal loyalty of his troops and was able to direct them as he saw fit during these struggles, in return for providing them with patronage—even at the expense of Roman provincials. All these elements made Bonifatius the first successful exponent of what can be described as 'Warlordism' in the western Roman army.

CHAPTER 11

The Significance of Insignificant Engagements: Irregular Warfare during the Punic Wars

Louis Rawlings

The conflicts between Carthage and Rome in the third century BC display considerable complexity. The broad regional and temporal span, the sheer number of campaigns, and the variety of engagements fought make this a rich area of study for military historians. While there were many major battles and sieges in both wars, there were also periods of relative stalemate and lethargy. Hannibal's last few years in Southern Italy, for instance, appear to stand in stark contrast to his audacious campaigns of 218–216.[1]

The surviving historical narratives regularly report periods of seeming inactivity in various theatres of operation, often despite considerable forces being present. Thus Livy (25.1.5) noted that in the summer of 213, with around 20,000 men, the consul Ti. Sempronius Gracchus "fought a number of minor engagements in Lucania, but none of them worth recording; he also stormed and captured a few unimportant towns." Whole campaigning seasons were described in terms such as "the Romans accomplished nothing of note in this year" (259, Polyb. 1.24.9) or the land forces of both sides "achieved nothing worthy of mention, spending time on minor operations of no significance" (257, Polyb. 1.25.6).[2]

---

1   Except where noted, all dates in this chapter are BC.

2   Other examples of little happening in a theatre: Livy 22.9 ("a few minor skirmishes" in Cisalpine Gaul, 217), Livy 23.48 (in Campania after Hannibal withdrew to Apulia, there were some minor skirmishes between Campanians and Fabius (23.46) after which "little happened", 215), Livy 24.49 (Spain, 213), Livy 25.32 ("little worth recording had been done during the past two years, being waged more by stratagems (*consilia*) than war", 214–212; on *consilium* see Wheeler (1988) 53–55), Livy 28.12 (two consular armies were allotted Hannibal, however as "he did not offer battle, the Romans were content to leave him alone as long as he remained inactive," 206; cf. App. *Hann.* 54a, Zon. 9.11), Livy 28.46 ("Nothing of importance occurred that year in Bruttium" despite the presence of the two armies, now under a consul and proconsul, 205, cf. Livy 29.6), Dio 17.70 ("In Italy nothing important was accomplished in the war against Hannibal; for though Publius Sempronius was defeated by Hannibal in a trivial battle, he later won a victory over him," 204, cf. Livy 29.36.4–12, with some narrative elaboration); cf. Rosenstein (2004) for study of Roman military activity in the period

THE SIGNIFICANCE OF INSIGNIFICANT ENGAGEMENTS   205

Our sources tend to display a studied indifference to what they regard as minor operations and report them, when they bother to include them at all, in partial and compressed ways. Polybius' reasons for not detailing what amounted to *five years* of warfare conducted by Hamilcar Barca against the Roman forces in Western Sicily, first at Heircte and then Eryx (247–242), were that:

> The causes or the modes of their daily ambuscades, counter-ambuscades, attempts and assaults were so numerous that no writer could properly describe them, while at the same time the narrative would be most tedious as well as unprofitable to the reader.
>
> POLYB. 1.57.1–3

The sheer volume of operations created stylistic concerns for the narrative, but Polybius' didactic purpose of relating only that which he thought might benefit the reader in moral or practical terms allowed him to severely compress this period.[3] Our sources often give the impression that such 'low-level' military activities were rarely worth reporting, but they also admit that they might be considerable. This is reflected in the frequent, albeit passing, mention of skirmishes, ambushes, raids, and harassments in the surviving historical accounts of the period. Indeed, these operations might be considered the 'dark matter' or the 'background noise' of campaign narratives. As Livy (24.3) indicated: "even the winter was not wholly without activity," in Apulia in 215/4, "the consul Gracchus was wintering at Luceria, Hannibal near Arpi, and skirmishes took place between them as the opportunity arose and as one or the other side spied an advantage." Despite the lack of detail, the references to such 'minor' actions suggest that, collectively, they may have had a substantial, if not necessarily positive, contribution toward the nature of campaigning and to the course of the Punic wars. This chapter will argue that the protagonists made extensive use of such minor actions in a variety of forms and, in order to do so, they employed considerable forces whose identity, motives, and military roles were often complex and mutable. Despite the many 'regular' engagements of armies and fleets in large-scale battles and sieges, and the undoubted importance of such actions as 'moments of decision', this chapter will argue that many

---

200–168, esp. 120 table 3, showing considerable numbers of years where no or only minor action occurred.

3   Other writers, judging by the surviving fragments of Diodorus (24.6, 9), appear to have related at least some of the individual operations, but our knowledge of them is very limited; Lazenby (1996) 148–149. On Polybius' moral didacticism see Eckstein (1995).

'irregular' and individually 'insignificant' actions, such as raids, skirmishes, ambuscades, and harassments, performed important military functions. Collectively these had tactically and strategically significant physical, psychological, and temporal effects.

Irregular Operators

The attitudes of our ancient sources to such operations appears to be, in part, bound up both with the types of activity pursued and the origins of the men undertaking them. They are subject to the rhetorical and ideological agendas of the authors (or their sources), and sometimes linked to ethical constructions of the standards of warfare. For instance, Livy characterized the warfare in southern Italy in the last few years of the Second Punic war as:

> The struggle in Bruttium had assumed the character of brigandage much more than that of conventional warfare (*latrociniis magis quam iusto bello in Bruttiis gerebantur res*). The Numidians had commenced the practice, and the Bruttians followed their example, not so much because of their alliance with the Carthaginians as because it was their traditional and natural method of carrying on war. At last even the Romans were infected by the passion for plunder and, as far as their generals allowed them, used to make predatory incursions on the enemy's fields.
>
> LIVY 29.6

The application of terms for banditry (Greek *leisteia* and Latin *latrocinia*), served the pejorative purposes of ancient writers, denying the protagonists political motivations by emphasizing their predatory aspirations.[4] The ethnic stereotypes of barbarians, such as Numidians and Bruttians, reverting in Southern Italy to their traditional approaches to warfare, allowed Livy to suggest the erosion of military conduct and the perceived standards of *bellum iustum*.[5] In this respect, Livy even described Hannibal's later campaigns as being waged through brigandage (Livy 28.12.9 *per latrocinia militiam exercere*). Such terminology might even be deployed to suggest a questionable martial ability. In

---

4 Grünewald (2004) 39–40.

5 Bruttian innate love of brigandage cf. Livy 28.12. *Bellum iustum* as conventional warfare see Grünewald (2004) 40. For the labelling of irregulars as bandits and criminals in other cultures see e.g. Beckett (2001) 49–50, 62–64, 101; Boot (2013) xxiii; Brice (forthcoming).

THE SIGNIFICANCE OF INSIGNIFICANT ENGAGEMENTS 207

his portrayal of the military situation in Spain in 206, Livy (28.32) gave the younger Scipio a speech in which he dismissed Spanish forces as "brigands and brigand-leaders" (*latrones latronumque duces*), good only for raiding neighbouring farms, but not for standing up to a regular army in pitched battle.[6]

Yet many of the predatory practices described by our sources in the Punic Wars clearly had important military dimensions. What Livy considered to have been banditry, we would understand as irregular warfare. Indeed, he used the terms *latrocinia* and *latrones* to describe raiding parties in the Hannibalic war in contexts that suggest guerrilla tactics. Gallic *latrones* harassed Hannibal's crossing of the Alps with hit and run attacks (*parva furta*) on the rear of his column (218, Livy 21.35), while Massinissa conducted "stealthy night raids and then open *latrocinia*" against Syphax, king of the Masaesyli, in Numidia (c. 205/204).[7] Indeed Massinissa is described as living as a *latro* (Livy 29.31–2, 33). Appian's description emphasizes the tactics of the Numidian chieftain:

> Syphax and the Carthaginians were much the more numerous. They marched with wagons and a great load of luggage and luxuries. On the other hand, Massinissa was an example in all doing and enduring and had only cavalry, no pack animals and no provisions. Thus he was able the more easily to retreat, to attack, and to take refuge in strongholds. Often, when surrounded, he divided his forces so that they might scatter as best they could, concealing himself with a handful until they should all come together again, by day or by night, at an appointed rendezvous. Once he was one of three who lay concealed in a cave around which his enemies were encamped. He never had any fixed camping place. His generalship consisted especially in concealing his position. Thus his enemies never could make a regular assault upon him, but were always warding off his attacks. His provisions were obtained each day from whatever place he came upon toward evening, whether village or city. He seized and carried off everything and divided the plunder with his men, for which reason many Numidians flocked to him, although he did not give regular pay, for the sake of the booty, which was better.
>
> APP. *Pun.* 12

---

6 cf. Livy 22.19, 28.22.

7 Livy 29.31, *inde nocturnis primo ac furtiuis incursionibus, deinde aperto latrocinio infesta omnia circa esse.* On *furta* see Wheeler (1988) 66–67; cf. Livy 24.29 (Hippocrates conducts stealthy raids on pro-Roman Sicily *furtiuis excursionibus uastare coepit*).

Appian's account presents us with a classic example of irregular 'guerrilla' tactics.[8] Massinissa employed harassment rather than frontal engagement, relied on mobility, concealment, and dispersal of forces to evade a more numerous, yet slower moving, enemy. His tactics illustrate the use of local populations to sustain his forces, and predation and rewards to enhance the loyalty and commitment of his forces. Whatever Massinissa's political motives might have been for conducting these raids, Appian emphasized the predatory motives of his supporters, which chime closely with Livy's representation of Numidian *latrocinia*, both in the campaign against Syphax and in Bruttium.

Brigands, Guerrillas, Warriors, and Soldiers

While one might understand why patriotic Roman writers might describe the Carthaginians' barbarian allies and neighbours in derogatory terms as bandits and predators, the picture is more complex. Livy admitted, not without some distaste, that in Bruttium the Romans had also adopted the pursuit of plunder (29.6, quoted above). Other Roman forces were also characterized as latrocinial: the Samnites, for instance, complained to Hannibal that they were subject to the banditry of Roman forces operating out of Nola (Livy 23.42). A broader examination reveals that the status of some of the Roman forces in the Punic wars was somewhat complex. Their operations indicate that some *latrocinia* went beyond purely predatory motives, being state-sanctioned irregular warfare (cf. Zon 8.16.8).[9]

Perhaps the clearest example of the Roman use of irregular forces occurred in 210. By then, operations in Sicily were over. The Carthaginian invasion of the island had failed. Punic allies, Syracuse and Agrigentum, had fallen. At Agathyrna, large numbers of men had taken to brigandage and posed a threat to the restoration of peace and the resumption of agriculture in that part of the island. This was a matter of strategic significance. The Romans desperately needed corn, since the war, still raging in Italy against Hannibal, had created shortages and threatened famine. The Roman consul in Sicily, M. Valerius

---

8   Modern definitions and examples of guerrilla tactics abound, e.g. Dupuy (1939); Kutger (1960) esp. 116; Janos (1963) 643; Asprey (1975); Lacqueur (1998); Beckett (2001); P. Rich and Duyvesteyn (2012) esp. 2–5; Boot (2013). It should be noted that some definitions link guerrilla warfare to the concepts of insurrection or insurgency, which are not relevant to the discussion at hand—cf. Brice (forthcoming). Beckett (2001) vii argues that revolutionary ideology has only taken centre stage in the practice and discourse of guerrilla warfare since the 1930s and 40s. It should be clear that the guerrilla operations of Massinissa express no ideology of popular uprising, but are a strategy of mobility and evasion, cf. Beckett (2001) 1–2.

9   Bleckmann (2011) 179.

THE SIGNIFICANCE OF INSIGNIFICANT ENGAGEMENTS   209

Laevinus, rounded up 4,000 of these men and sent them to Rhegium, on the mainland, the population of which, as Livy stated, "were in need of a band experienced in *latrocinia*" to help plunder Bruttium (Livy 26.40.18, cf. 27.12.4, Polyb. 9.27.11a, Plut. *Fab*.22).

The situation at Agathyrna suggests the dangers to the countryside and to the war effort of such armed and desperate free agents, but it also indicates a means by which such 'irregular' forces might be used by state actors in a major war. Their status mutates from desperadoes to privateers, serving Rhegium in return for rations and booty (Polyb. 9.27.10–11), and raiding pro-Hannibalic Bruttian communities under a Roman marque. *Latro*, in third and second century usage, could also sometimes mean pay-taker or mercenary, and may suggest a contemporary perspective on the status of these predatory privateers.[10]

We are informed of a number of other 'irregular' operators in southern Italy during the war. T. Pomponius Veientanus conducted a number of raids in Bruttium and Lucania with an irregular army (*tumultuarius exercitus*) until, as Livy hostilely remarked, Hanno defeated his "disorderly mob of peasants and slaves" (*inconditae turbae agrestium servorumque*, 212, Livy 25.1, 3). Their predatory activities, however, seem official since Pomponius was *praefectus sociorum* (Livy 25.1). It is also possible that some Lucanian cohorts in 214 were acting as a similar force of irregulars. They were sent by Gracchus under an unnamed *praefectus sociorum* to raid enemy territory (possibly locally in Lucania), where Hanno likewise intercepted and mauled them "while they were spread out in all directions" (Livy 24.20).[11]

The brigands of Agathyrna arrived in Rhegium and were augmented by Bruttian deserters from Hannibal's cause, who had been made "audacious and daring by necessity" (Livy 27.12.5). Their combined force numbered 8,000 according to Livy. They are referred to as a "garrison" (*praesidium* Livy 27.12.4), implying that they had security and defensive functions, including the interception or harassment of Bruttian raiding parties. They also appear to have been charged with ravaging enemy farms and crops and were even ordered to

---

10   *Latro*: Plaut. *Mil*. 4.1.3; *Poen*. 3.3.50; Enn. *ap. Non*. 134, 29 (*Ann*. v. 528 Vahl.); cf. Varr. *LL* 7.52; Festus p. 105. *Latrocinor*: Plaut. *Trin*. 2.4.198; *Mil*. 2.6.19 cf. Enn. *ap. Non*. 134, 28; Varro *LL* 7.52; Grünewald (2004) 5, 168 nn. 10–11 for further references. Fronda (2010) 273–274 regards the *latrones* of Agathyrna as mercenaries, Goldsworthy (2000) 235 as irregulars.

11   This episode may be a doublet of Pomponius' operation: Kahrstedt (1913) 255; Seibert (1993) 273 n. 22, but there is no compelling reason to think that such raids were not a feature of the warfare of this time, or that Hanno's counter-operations could not have intercepted several forays in Lucania.

assault Caulonia by Fabius Maximus (Livy 27.12.6, Plut. *Fab.* 22). On Hannibal's approach they withdrew to high ground, but could not escape and soon surrendered to him (27.15, 16). Whether they were killed, enslaved, or re-employed by Hannibal is not specified. The loss of so many men passes without further comment in Livy, though he had prepared the reader for their military insignificance and lack of formal status by remarking on their careless rapacity and calling them an irregular (or disorderly) crowd (*incondita multitudinis* 26.40.16). Other forces embarked on irregular operations might also be described in terms that emphasized their indiscipline, unruliness, or lack of formation. Livy regularly used terms such as *tumultuarius, incondita turba*, or *multitudinis* and perceived these as negative qualities.[12] Yet the men of Agathyrna had acted like a small army; clearly they were thought capable by a Roman general of threatening an entire town as part of a broader operation. They distracted Hannibal while Fabius seized Manduria and captured the key prize of Tarentum (27.12).[13] A significant parallel can be drawn with the army of centurion M. Centennius Paenula in 212, consisting of around 8,000 Romans and allies, plus volunteers collected on the way to Lucania. Livy (25.19) described it as "irregular and half-armed" (*tumultuarium ac semermem*). Paenula had apparently gained his command because of his close familiarity with the area and had promised to use "the stratagems (*artes*) invented by the enemy" against Hannibal himself.[14] Whatever his intentions may have been, his force was caught by Hannibal and

---

12 Cf. 21.57 *incondita turba* (a sally by townsfolk and refugees of Emporium), 22.45 *incondita turba* (a disorderly mob of water bearers from the Roman camp near Cannae), 24.29 *incondita multitudinis*, (mercenaries and deserters sent from Syracuse to raid in support of Leontini), 25.14 *inconditam inermemque aliam turbam* (unarmed crowd of Campanian wagoneers and rural "foreigners" in Hanno's camp, causing military discipline to breakdown), 25.15 *incondita turba* (Thurians routed in battle by Hannibal), 27.32 *inconditam inermemque multitudinem* (unarmed rustics and their cattle in a fortress), 29.1 *inconditam turbam tironum* (the Roman army in Spain after Scipio, acting as an untrained raw mob). Elsewhere Livy uses the term *tumultuarius* for a force of irregulars, 22.21 *tumultuarius manus*, 22.44 *tumultuaria auxilia*; for hastily raised levies in a pair of similes: 29.28 (where, in comparison to Scipio, Hasdrubal Gisco is like the irregular levies of his army) and 30.28 (in a speech designed to illustrate the magnitude of Scipio's challenge, Hannibal in Africa will not be like Hasdrubal Gisco or an irregular army drawn from a semi-armed mob of farmers *tumultuariis exercitibus ex agrestium semermi turba subito conlectis*), cf. 25.19. Similarly an irregular action is often described as a *tumultuaria*: 21.7, 8, 25.34, 27.42, 29.37.

13 Lazenby (1978) 175; Seibert (1993) 344, n. 14, 347.

14 On the meaning of *ars*: Wheeler (1988) 57 and n. 27.

THE SIGNIFICANCE OF INSIGNIFICANT ENGAGEMENTS 211

destroyed; yet it had drawn the Carthaginians away from supporting Capua, where the consular forces were concentrated. In both instances a case can been made that the irregular forces served to augment a broader Roman strategy.

In 212, Hannibal, at Herdonia, eliminated another force that had been extensively plundering in Apulia, but, unlike the brigands of Agathyrna and the *tumultuarii* of Pomponius Veientanus and Centennius Paenula, this force appears to have been a 'regular' two-legion army led by the praetor Cn. Fulvius Flaccus.[15] Livy suggested that it was their success in raiding that made the Romans unruly and led to the breakdown in discipline that caused their defeat at Herdonia.[16] By contrast, raiders from Nola in 214, although characterized by their Samnite victims as "operating not in maniples, but roaming more as bandits" (23.42 *iam ne manipulatim quidem, sed latronum modo percursabet*), were portrayed as being more disciplined. Drawn from the two urban legions assigned to garrison Nola, their predatory successes appear to have had no detrimental impact their role of safeguarding the city from Hannibal (Livy 23.31, 32, 44–46). Indeed, although the Samnites claimed that Hannibal's forces could deal with these *latrones* easily, Livy noted that close oversight by the praetor M. Claudius Marcellus resulted in effective preparation, organisation and caution in all their operations (Livy 23.41–3). Although perhaps contrived to enhance the reputation of Marcellus, the representation draws on the idea that well-led forces could prove formidable in irregular warfare.

That Roman armies might be capable of predations and irregular operations suggests regular forces had the potential to send out contingents to conduct such warfare. Consideration of the numbers, types, and capabilities of troops present in Roman and Carthaginian field armies underpins our understanding of the prevalence and impact of these operations.

Light Infantry and Cavalry
The field armies of both sides included substantial numbers of light infantry, who, besides their participation in battle (e.g. Polyb. 3.113.6, 73.7), appear to have undertaken a variety of other roles on campaign. They are recorded as being sent out in raiding parties, or to harass or skirmish with enemy forces, and were especially effective over rough ground, where they might be used to seize and hold positions or lay in ambush.[17] Hannibal's army employed Spanish,

---

15 Livy 25.21, probably raised in 214 see 24.11, 25.3; cf. 22.20 (detailing the raids of the Roman fleet on Onusa, New Carthage, Loguntica and Ebusus).

16 Livy 25.20, cf. 26.2, 28.24.

17 Raiding: Polyb. 3.69.5, Livy 21.52, 24.20. Terrain: Polyb. 3.43, Livy 21.31, 55, 22.18, 24. Ambush:

African and Gallic tribesmen in these capacities.[18] At Trebbia, Hannibal had about 8,000 light infantry in addition to his 20,000 line infantry (Polyb. 3.72.7–8), and possibly a greater number at Cannae, depending on how many of his recently recruited Gauls were armed in this way.[19] Unfortunately, it is often unclear how many light troops were mustered for other Carthaginian armies, though a third of the 20,000 Iberian infantry raised by Indibilis and Mandobius in 206 were light troops (Polyb. 11.33.6, cf. Livy 28.31, 33).[20] This is, broadly speaking, a ratio (33.3%) not too dissimilar to that of Hannibal's light infantry at Trebbia (28.7%).

The Romans raised light infantry from their own citizen body as an integral part of the creation of legions (Polyb. 6.21–22). There is some debate about changes to the nature and numbers of light infantry in the legions as the Punic wars progressed, with some scholars suggesting an increase and/or re-organisation of the *leves* in around 212.[21] Others, based on Polybius' figures (6.21.7–10), have settled on around 1200 light-armed in a full strength legion of 4,200 infantry. This number represents 28.5% of the infantry; a significant percentage of the legion's fighting capability and strikingly similar to the proportions in the Carthaginian and Spanish forces discussed above.[22] On such cal-

---

    Livy 21.55, 28.11, Polyb. 3.74, 104–105, 11.32.2; cf. Livy 28.3.2–6, App. *Hann.* 42. Surprise: Livy 21.57. Harassment: Polyb. 3.69, cf. Livy 21.51, Polyb. 3.102, Livy 22.3, 22.21, 41, Plut. *Marc.* 25.2, Livy 28.11. Pursuit: Polyb. 3.84.14.

18   Hannibal's light infantry was often collectively termed by Polybius simply and unethnically as *longchophoroi*, "spearmen"; Daly (2002) 108–111, often distinguished from Balearic slingers, but note Dominguez Monedero (2005). Italian recruits may have included light troops, especially in the latter stages of the war to replace losses, cf. Livy 23.42, 27.42, 28.12; Hoyos (2003) 128–129, 264–265 nn. 13–14; Rawlings (2011) 316.

19   Daly (2002) 31–32; Lazenby (1978) 81 suggests 11,400 light and 28,500 line infantry at Cannae.

20   cf. Livy 28.1–2 reports a force of 9,000 Celtiberians, consisting of 4,000 shielded line infantry, 200 cavalry, and the rest described as light armed. However, an overall ratio cannot be calculated since the Celtiberians were part of a larger Punic army (though camped separately), the numbers of which are unknown.

21   For change Gabba (1949) 177 ff., 181 ff.; (1976) 5–6; Toynbee (1965) 1.517; Brunt (1971) 402–404. Criticisms: Walbank (1957) 698; J. Rich (1983) esp. 294–295, 305–312; Lazenby (1996) 178. It is possible that Livy (26.4.4–7), in a confused passage, describes a local innovation to address a particular problem posed by Campanian cavalry during operations around Capua: Oakley (1998) 40; Daly (2002) 72–73; Walbank (1957) 701; Rawson (1971) 29; Rawlings (2007b) 56; Anders (2012) 65–68.

22   Keppie (1998) 34–35; Connolly (1998) 129; Walbank (1957) 703; Daly (2002) 56–57; Quesada Sanz (2006) 246.

THE SIGNIFICANCE OF INSIGNIFICANT ENGAGEMENTS   213

culations, Rome's six legions in 218 would have included a total of 7,200 citizen
infantry skirmishers. In 216, at Cannae alone, there may have been more than
10,000 light armed citizen soldiers, with a further 7–8,000 deployed in other
forces.[23] By 212/11 *velites* could have numbered 25–30,000 of the 90–100,000 citizens theoretically committed to twenty-five legions. Of course, paper strengths
probably rarely reflected reality, where attrition through combat and illness, as
well as desertion, would have made their mark. It is also unclear whether the
proportions of different legionary troop types were rigorously maintained during the realities of campaign. Scipio's part of the combined consular armies at
Trebbia, for instance, may have had little opportunity to make good the losses
to skirmishers from the mauling they had received at Ticinus (Polyb. 3.65.10,
72,2). Nevertheless, it is clear that the majority of Roman legions, like the forces
of their opponents, had a significant capacity and potential to use such troops
for light-armed tactics and irregular actions.

The capabilities of Roman irregulars were considerably enhanced by the
contributions of allies. We have already noted forces of Italian allies operating under *praefecti sociorum* to raid Bruttium and Lucania. That Livy described
some of these forces as *semiermis* "half-armed", if it is not merely a derogatory dismissal of their military value and instead reflects the nature of the
equipment or irregular status, hints at a preponderance of light-armed fighters
who might be adept at mobile hit and run activities.[24] In any case, the Italian
allies supplied some light-armed *socii* to the field armies of the Second Punic
War.[25] At Trebbia, as many as a third of the 6,000 light troops were allied.[26]
Rome also used non-Italian troops in certain campaigns. 500 Cretan archers
and 1,000 Syracusan light troops were sent to Rome by Hiero II in 217 (Livy
22.37, Polyb. 3.75.7). The Spanish and Numidian allies, present in their local theatres in the Second Punic War, probably included a proportion of light infantry.
A significant amount of Massinissa's 6,000 infantry, deployed on the left at
Zama, would have been skirmishers.[27] Of course, most, if not all of Massinissa's

---

23 Cannae: Lazenby (1978) 79–80; Daly (2002) 57, *contra* Brunt (1971) 402. Seven legions
    additional to the army of Cannae: Brunt (1971) 418–419.
24 Livy 25.19.14, 30.28.3, cf. Tac. *Ann.* 3.39.2, 43.3, 45.2; Konrad (1997) 41–42.
25 Lazenby (1978) 13; Anders (2012) 34.
26 Polyb. 3.72.2; Lazenby (1978) 79–80. The original, theoretical, maximum of 4,800 citizen
    leves (1200 × 4 legions), were probably reduced by the casualties sustained at Ticinus. A 2:1
    citizen: allied proportion applied to Cannae would give a skirmisher total c. 15,000. Daly
    (2002) 228 n. 9 suggests that if the ratio was normally nearer 1:1, then the total could have
    been 21,280.
27 Polyb. 15.5.12–13, 15.9.8, Livy 30.29; Ait Amara (2007) 115.

cavalry at Zama would have been light horse and the Romans also benefited from Numidian deserters in Italy, Sicily, and Spain during the conflict (Livy 23.46, 27.8).

The Carthaginians generally possessed a good supply of light cavalry, particularly from Numidia, whose talents for guerrilla warfare have already been noted.[28] The army that marched to relieve Agrigentum in 262/1 included Numidians, who were instrumental in causing the defeat of Roman cavalry in the skirmishes outside the city (Polyb. 1.19.2, 4, 9). Numidian forces were part of the Carthaginian armies of Spain from 237 and of the *Grande Armée* that invaded Italy in 218. The numbers that were raised suggest the importance of such troops. Naravas' 2,000 were regarded as particularly effective additions to Hamilcar Barca's forces during 'the Truceless War' (Polyb. 1.78.9, 11, 82.13), while perhaps around 4,000 of Hannibal's 10,000 cavalry at Cannae were Numidians.[29] There are frequent reports of Numidian detachments sent to raid, harass, and ambush the enemy during the Hannibalic war.[30] While the Romans generally lacked large contingents of light cavalry, their citizen and social horse were also capable of a variety of operations and were often accompanied by light infantry, which enhanced their operational flexibility and gave them a measure of resilience in the face of the numerical and technical advantages sometimes displayed by the enemy horse.[31] Carthaginian, Gallic, and Spanish cavalry were adept at irregular operations, particularly in skirmishing, raiding, and providing tactical cover for other forces.[32] It was the capability of cavalry to travel, strike, and retire quickly that enabled them to project power across open country. In raids, they were best suited to rounding up livestock and humans and in some acts of physical destruction, such as firing ripened crops.[33] In Italy, Hannibal's cavalry tended to have a considerable advantage, being both more

---

28   Ait Amara (2007) but note the supposed superiority of Spanish cavalry over Numidians in at least one skirmish in Spain, Livy 23.26.

29   Daly (2002) 92.

30   Polyb. 3.69.5, 117.7, 8.26.4–27.1, Livy 21.45, 52, 55, 22.13, 15, 23.1, 24.20, 25.9, 26.9, 28.11, App. *Hann.* 40, 50. Livy 27.20 (Massinissa given 3,000 of the "best cavalry" in the Punic army to rove through Hither Spain, helping allies and raiding enemy towns and farms).

31   Polyb. 3.65.3, Livy 21.26, 29 (against cavalry at Ticinus), Polyb. 3.69.5–8, Livy 21.52 (countering Punic raiders pre-Trebbia), Polyb. 3.102.3–4, Livy 22.24 (harassing foragers at Gerunium), Livy 26.4 (against cavalry at Capua), Polyb. 11.22.5–6 (harassing the camp of Hasdrubal); McCall (2002) 73–74; Anders (2012) 147–148, 155–156.

32   e.g. Polyb. 3.69.5, 9.7.5, Livy 21.47, 52, 61, 22.15; 23.25.

33   Erdkamp (1998) 219; e.g. Livy 24.20 (214, 4,000 remounts seized for Hannibal in Lucania and Apulia).

THE SIGNIFICANCE OF INSIGNIFICANT ENGAGEMENTS

numerous and experienced. Thus it was Hannibal's dominance in cavalry that was the cornerstone of his freedom to move, secure forage and terrorise the enemy.[34]

Heavy Infantry

Even line infantry were capable of some irregular operations. According to Polybius (3.104–105) Hannibal divided 500 cavalry and 5,000 *psiloi* and *pezoi* (clearly meaning light infantry and *other infantry*) into groups of two-three hundred, who lay in ambushes around a bare hill, to be used against Minucius' army near Gerunium. The Carthaginians in particular made a military virtue of the mutability of tribal troops, who seemed comfortable both in irregular warfare and as line infantry in more conventional engagements. The destruction of Postumius' army in the Litana forest in 216 suggests the capabilities of Gauls in difficult ground, while Spanish tribesmen, despite Scipio's unfavourable characterization of them as "bandits", could both fight pitched battles and be effective *guerrios* in Spain.[35] This was a talent transferred to field operations in Italy at Mount Callicula, where Livy stated:

> These men were more accustomed to the mountains and in better training for running amongst rocks and precipices, and being both more lightly made and more lightly armed they could easily by their method of fighting baffle an enemy used to fighting on a plain, heavily armed and more static.
>
> LIVY 22.18.3

Livy's characterization of the Romans at Callicula as preferring the plains belies their use on difficult terrain in many operations.[36] The decision of Fabius to

---

34   Erdkamp (1998) 127–128.

35   Livy 28.32, cf. Diod. 5.34.6 (Spanish bandit-guerrilla lifestyle and upbringing); Grünewald (2004) 38. Strabo 3.4.18 calls Celtiberians "peltasts", indicating a dual role. Livy's uses the term *caetrati* (21.21.12; 23.26.11) for Spanish infantry, while in Greece he employs it to mean peltasts (31.36; Anders (2012) 38). Quesada Sanz, (2011a) 227–228 emphasises the peltast-like flexibility of Iberian infantry, see also (2011b).

36   cf. Livy 28.2, the Romans moved swiftly through rough country and were not discomfited by it in a battle against a Celtiberian camp. Roman success at Baecula involved ascending a strong hill, held by Carthaginian and Iberian infantry (Livy 27.18). The success of Regulus at Adys turned on attacking the Carthaginians on steep terrain, where their elephants and cavalry were ineffective (Polyb. 1.30.9–11). Note also the Roman use of difficult ground at Pydna to break up the Macedonian phalanx Plut. *Aem.* 20, Livy 44.41.

shadow and harass Hannibal, but remain in terrain where Hannibal's cavalry would be less effective, is a famous, though not the only, example (Livy 22.12, cf. Polyb. 1.39.11–12). Indeed Polybius emphasized the flexibility of Roman legionaries, noting that,

> Every Roman soldier, once he is armed and ready, is able to meet an attack from any quarter, at any time or place. He is equally prepared and in condition to fight as part of the whole army, or a section of it, or in maniples, or singly.
>
> POLYB. 18.32.10–12

It should not be forgotten that most legionaries had, in their earliest campaigns, probably spent some time as *leves*, before progressing to the *hastati* (Polyb. 6.21.7–9), and would not have simply forgotten the skills they had acquired. It was these skills that enabled them to contend with Hamilcar's troops for five years around Heircte and Eryx (247–242). The qualities of Hamilcar's veterans were demonstrated later at Utica, during their mutiny, where despite being forced from their camp they regrouped in rough ground and retook it from the Carthaginian forces (Polyb. 1.74.4–12). After their defeat at the battle of Bagradas (c. 240), they resorted to 'Fabian' tactics, keeping to terrain unfavourable to the superior Carthaginian cavalry and elephants (Polyb. 1.77.1–2, 4).

The status of combatants in the Punic wars was not always clear-cut or, indeed, consistent. Irregulars might be non-state forces: warriors, bandits, and *ad hoc* levies, acting out of personal interest or for those of their warlords. Alternatively, they might be enlisted troops: mercenaries, allies or levied citizens expected to perform irregular roles as part of military operations commanded by state-appointed officers and furthering the broader strategy of the state. These combatants were often required to participate not only in irregular actions, but to take their place in regular battles and sieges. The large numbers and tactical flexibility of many troops gave both sides the means to conduct 'irregular' activity on a large and protracted scale.

The rest of the discussion will consider the tactical and strategic contribution and impact of these operators, and of the operations that they undertook during the Punic Wars. It will survey the physical and psychological aspects of irregular acts of force, and will also consider the temporal effects on the pace of campaigning and, indeed, the duration of the wars.

THE SIGNIFICANCE OF INSIGNIFICANT ENGAGEMENTS 217

An Arena of Stratagem

Irregular warfare allowed talented commanders the opportunity to manipulate the perceptions of their enemies through stratagem and trickery, to bedazzle, misdirect, and entangle them. Wheeler's analysis of stratagem vocabulary reveals that, despite the Punic reputation for deviousness and trickery, both sides embraced stratagem, deception, and *rusé de guerre*, and could be smart operators.[37] Cicero (*Off.* 1.108, Wheeler 1988, 68 trans.) stated that "we have heard that among the generals of the Carthaginians Hannibal, and from our own, Fabius Maximus, were *callidus* (cunning): ready to hide, to keep silent, to dissimulate, to ambush, and to perceive the enemies plans in advance ..." Indeed Polybius compared Hamilcar Barca and his Roman adversaries in the latter stages of the First Punic war to boxers:

> ... both distinguished for pluck and both in perfect training, [who] meet in the decisive contest for the prize, continually delivering blow for blow, neither the combatants themselves nor the spectators can note or anticipate every attack or every blow, but it is possible, from the general action of each, and the determination that each displays, to get a fair idea of their respective skill, strength, and courage, so it was with these two generals. The causes or the modes of their daily ambuscades, counter-ambuscades, attempts and assaults were so numerous that no writer could properly describe them ...[38]
>
> POLYB. 1.57.1–3

Polybius elsewhere portrayed Hamilcar as a skilful games-player, emphasizing how he out-manoeuvred his enemies in minor actions, cutting off large numbers and forcing their surrender, surprising them with ambushes even during major engagements, and unexpectedly appearing during day or night (1.84.7).[39] Irregular warfare gave ample scope for improvised plans, opportunities, and

---

37 Wheeler (1988) 56, 68.

38 Walbank (1957) 121; Traill (2001).

39 Walbank (1957) 147; Austin (1940) 260–263. Latin authors also extended the vocabulary of sport or gaming to irregular warfare, Wheeler (1988) 80–81. E.g. *ludificatio* (make sport of, tease, exasperate through trickery) described the strategies of Fabius in 217 (Livy 22.18.9, Val. Max. 7.3 ext. 8) and L. Porcius Licinius in 207 (Livy 27.46.6), but also of stratagems in battle: thus Metellus' use of *leves* at Panormus in 250 (Front. *Strat.* 2.5.4). Note the elite bias in Hamilcar's games-playing superiority over upstart rebel leaders at 1.84; Eckstein (1995) 176–177; cf. Hannibal against Paenula, Livy 25.39.

ad hoc solutions that would demonstrate the creativity, agility, and adaptability of the generals, officers, and the men involved (Polyb. 1.57.6, 58.3–4). When confronted with Fabius' tactics, Hannibal, "that sly boots, that wolf" (Lucilius *Satires* Frg. 952–953), responded by using his own artifices to try to undermine him, constantly shifting his camp, setting ambushes for Fabius' men, or provoking them through raiding and devastation (Livy 22.12).[40]

Like battles, however, irregular operations often required careful planning, particularly if co-ordination with other contingents or the main army was required. We have seen, for instance, that Marcellus' raids out of Nola were characterized as well-planned and executed. To be successful, such actions required subordinate officers to perform their tasks effectively, providing the appropriate leadership, command, and control, with a good dose of initiative thrown in. Carthaginian junior officers, such as Maharbal and Mago, and even Hannibal himself as a young man (under the command of his brother-in-law, Hasdrubal), were placed in command of detachments and minor operations, in which they acquired much experience.[41] Allied and subordinate officers, such as Massinissa and Muttines, were also given such roles. Indeed, it might be suggested that Hannibal's officers were so well trained in these types of operations that they could implement them regardless of scale—from handfuls of Numidians on a raid to thousands in a battle. The Carthaginian command structure included an "officer of supplies" (*ho epi ton leitourgion tetagmenos*, Polyb. 3.93.4) who is known to have organised and protected foraging parties. The versatility of such a position is indicated by a certain Hasdrubal, who not only performed this over-watch role at Gerunium (Polyb. 3.102.1, 5), but executed the stratagem of the oxen with flaming branches tied to their horns at Callicula (Polyb. 3.93.4), and, significantly, also commanded the Spanish and Gallic cavalry on the left at Cannae, which was instrumental in routing both Roman cavalry wings and falling on the rear of the infantry (3.114.7, 116.6–8, Livy 22.46, 48). Such command flexibility allowed Hannibal to transfer to the grand scale the "Punic" deceptions often practiced in insignificant engagements. Hannibal's ambushes and stratagems in pitched battles were sometimes on an audacious scale, as the battles of Trebbia, Trasimene, Cannae, and First Herdonia vividly demonstrate. Thus the influence of irregular operations and

---

40   Cf. Livy 27.2, 12 for Hannibal's evasions of Marcellus.

41   Livy 21.4; App. *Iber.* 6, *Hann.* 10; Nepos *Hann.* 3.1. The same could be said of younger members of the Roman elite who served as cavalry *decuriones* and military tribunes, McCall (2002) 79–81; e.g. Marcius who, under Scipio the Elder, had "developed all of the arts of war" Liv 25.37.

THE SIGNIFICANCE OF INSIGNIFICANT ENGAGEMENTS 219

tactics on the military thinking of Hannibal was, arguably, a factor in his greatest 'regular' victories on the field of battle. As we have seen, it is an attribute that Polybius indicated Hamilcar also had possessed; his games-playing generalship embraced both minor actions *and* ambushes in pitched battles (1.84.7).

'Grinding the Legions' and 'Kicking the Stomach'

Any use of force has both physical and psychological elements. An act of violence operates on the mind of the perpetrator and the victim, producing psychological effects (which in the context of irregular warfare will be discussed in a later section). The present discussion considers the physical ramifications of irregular warfare in terms of the casualties inflicted on soldiers and civilians, the damage caused to assets, and the logistical impacts on supply and movement.

A fragment of Naevius' epic poem, *Bellum Punicum*, exclaims: "arrogantly, scornfully, he grinds away the legions" (*superbiter contemtim conterit legiones*).[42] The statement has been linked to the years of bitter operations of Hamilcar at Heircte and Eryx, and aptly suggests the attrition inflicted on armies of the "daily ambuscades, counter-ambuscades, attempts and assaults," which Polybius (1.57.2) found too numerous to recount.[43] Such operations could, through the fatalities and injuries caused, reduce the numbers of the enemy, particularly of specialist troops, such as cavalry, who were often involved. Harassment, particularly of foragers or those on the march, might cut off small detachments or stragglers (Livy 21.35–36, 24.41, 25.34, 28.22), while skilful ambuscades might inflict bloody defeats on contingents, divisions, or whole armies.

The casualties caused to armies by irregular engagements are, however, often hard to quantify. Rarely are figures of losses in 'minor' episodes given and, where casualties are recorded, we are unsure how reliable such figures are.[44] According to Polybius, losses could be minimal if the troops could gain

---

42 Naevius frg. 38, Warmington (1936) 64.
43 Hoyos (2003) 238 n. 12.
44 E.g.: Livy 21.29 (at the Rhône, 200 out of 500 Numidian cavalry fell, while 160 of 300 Roman and Gallic cavalry were killed), 22.31 (c. 1,000 raiders lost, including quaestor Sempronius Blaesus, when intercepted by local African forces), 22.41 (an escalated minor engagement checked Punic raiding parties, 1,700 Carthaginians to 100 Roman and Gauls lost) 24.41 (Roman marching column attacked by cavalry, 2,000 stragglers lost), 25.18 (1,500 Romans raiders killed in Campania), App. *Pun.* 14 (over 1,000 picked Carthaginian cavalry and

safety quickly, without the bloodletting that normally accompanied a pursuit of the defeated (Polyb. 1.57.7–8). Larger armies could stand to lose more men before their fighting capacity was significantly degraded, but a severe reverse or reverses might force an army to remain inactive or even attempt to break away. Thus Hannibal appears to have been forced to keep his troops in camp after a reverse against Minucius, later withdrawing to a stronger position at Gerunium (Polyb. 3.101.8, 102.8–11, Livy 22.24). The attrition of the enemy through skirmishes and harassment probably contributed to a gradual decline in the size of armies, which would have required replacements to be recruited.

Despite, or because of, their limited capacity to eliminate the enemy, irregular operations were sometimes the only sensible and available course of action. It is difficult not to find compelling Polybius' assessment of the strategic imperatives that determined Fabius' strategy of delay:

> Fabius was not able to meet the enemy in a general battle, as it would evidently result in a reverse, but on due consideration he fell back on those means in which the Romans had the advantage, confined himself to these, and regulated his conduct of the war thereby. These advantages of the Romans lay in inexhaustible supplies of provisions and men. He, therefore, during the period which followed, continued to move parallel to the enemy, always occupying in advance the positions which his knowledge of the country told him were the most advantageous. Having always a plentiful store of provisions in his rear he never allowed his soldiers to forage or to straggle from the camp on any pretext, but keeping them continually massed together watched for such opportunities as time and place afforded. In this manner he continued to take or kill numbers of the enemy, who despising him had strayed far from their own camp in foraging. He acted so in order, on the one hand, to keep on reducing the strictly limited numbers of the enemy, and, on the other, with the view of gradually strengthening and restoring by partial successes the spirits of his own troops, broken as they were by the general reverses.
>
> POLYB. 3.89.9–90.4

Central to Fabius Maximus' strategic use of irregular warfare was an appreciation of the need to preserve the Roman field army from further defeats,

---

additional African forces killed or captured), cf. Liv 29.34 (3,000 lost at Salaeca, probably the same engagement), App. *Pun.* 15 (a failed Punic ambush on Scipio, 5,000 killed and 1,800 captured), *Pun.* 36 (African supply forces attacked by Thermus near Zama, over 4,000 killed or captured).

THE SIGNIFICANCE OF INSIGNIFICANT ENGAGEMENTS   221

while harassment of foragers was applied to rub away at Hannibal's more limited numbers and inhibit his ability to obtain supplies. Along with the tactics of 'scorched earth' (Livy 22.11), Fabius directed his action against the enemy sources of supply, "kicking the stomach," as later Roman military slang had it (Plut. *Luc.* 11.1).[45] The targeting of transports and foragers, and the interception of raiding parties, were designed to exert leverage on the enemy. Harassing forces and raiding parties might inhibit the free movement of troops and also civilians, restricting their access to the countryside; in some cases even preventing them from going beyond their own ramparts (Livy 21.55, 57). Confinement within or close to camp could be a particular problem for the grazing of horses and other livestock with the army (Polyb. 3.101.9–11). The harassment of foragers and water carriers brought general discomfort to an army (Polyb. 3.112, Livy 22.45). The effects can be seen in Spain in 211, where Hannibal's protégé, Massinissa,

> ... first sought to check Scipio's advance with a body of Numidian horse, and he kept up incessant attacks upon him day and night. He not only cut off all who had wandered too far from camp in search of wood and fodder, but he actually rode up to the camp and charged into the middle of the outposts and pickets, creating alarm and confusion everywhere. In the night he frequently upset the camp by making a sudden rush at the gates and the stockade; there was no place and no time at which the Romans were free from anxiety and fear, and they were compelled to keep within their lines, unable to obtain anything they wanted.
>
> LIVY 25.34

Prolonged harassment could bring "every kind of privation" on the enemy (Polyb. 1.58.4), potentially weakening them and sapping their energy, sometimes forcing them to give up their position. By targeting the sources and transportation of rebel supplies, Naravas' Numidians were able to force Mathos and Spendius to raise the siege of Carthage (Polyb. 1.82.13, 84.2, cf. 1.18.10). Disengaging in such cases might be very dangerous, since the harassment of moving forces slowed them, potentially allowing them to be caught by superior forces, as happened to the armies of Publius and Gnaeus Scipio in Spain and to the Carthaginians after Ilipa (Livy 25.35, 28.16). The other alternative was to fight. The logistical pressure on the Romans at the siege of Agrigentum in 262/1 was made acute by the constant harassment and skirmishing for almost seven

---

45   Goldsworthy (2000) 193.

months by the Carthaginian relief force, though for the most part "without any action more decisive than shooting at each other every day" (Polyb. 1.196). In fact, both sides became so hungry that they risked battle; the Carthaginians to save the city, the Romans to save themselves (Polyb. 1.19.6). Some forces could neither engage nor disengage. Having been out-manoeuvred by Hamilcar's games-playing, the rebels pinned at Prion ("more than 40,000" according to Polyb. 1.84.9), were so restricted that they became too weak to fight and offered to surrender.

Foraging, raiding and banditry are operations on the same continuum, in that they aim to obtain, through force, enemy assets for the operators' use, be it supply or enrichment (Livy 25.39).[46] Even when we are told by our sources that forces 'devastated' a region, it is implicit that there was plenty of looting occurring in the midst of the destruction. Thus the Roman raid on Consentia, which Livy (28.11) described as devastation (*depopulor*), brought back much plunder (*praeda*, cf. 24.20). Foraging, devastation and looting perpetrated by field armies generally appear to have been conducted by parties spreading out from the main force, and are regularly described in the sources.[47] Considerable numbers of men might be involved in these operations. Around Gerunium, in the late summer of 217, it is said that two thirds of Hannibal's army was engaged in foraging, while the other third acted as an over-watch, to ensure their protection.[48]

A collateral effect of foraging was often the destruction of assets that were surplus to the army's needs, but ravaging could be a strategy of deliberate destruction of crops, farms, agricultural buildings, and tools, or the killing of inhabitants and animals.[49] Though it was temporally constrained to the month or so up to harvest, when the crops were most flammable, burning could be effectively perpetrated by small and/or widely dispersed raiding parties, especially of cavalry.[50] Muttines, for instance, led Numidian cavalry on raids to burn the crops of Rome's Sicilian allies (Livy 26.21.15, cf. Front. Strat. 1.5, 16). Such operations allowed ancient writers the powerful image of ravaging as "fire and slaughter" (*caedibus incendisque*, Livy 22.3.6) or "fire and sword" (*ferro ignique*,

---

46   Erdkamp (1998) 123–124.

47   Polyb. 3.110.10, Livy 21.55, 57, 22.13, 15, 24.20, 28.11, etc.

48   However, on the approach of Minucius Hannibal deployed two thirds of the army to cover the rest as they continued to forage, Polyb. 3.101.4.

49   Erdkamp (1998) 123, 216 nn. 26–27.

50   Erdkamp (1998) 215–216 more positive on the impact than Harvey (1986) or Spence (1990) 101.

THE SIGNIFICANCE OF INSIGNIFICANT ENGAGEMENTS   223

Livy 23.41, 23.46).[51] Such activity denied the enemy the use of those resources. Consequently the raiding and ravaging of territories inhibited the operations of field armies by restricting forage or the capacity of local supporters to collect sufficient supplies for wintering. Thus the Roman devastation of Samnium and Campania in 215 and 214 made it difficult for Hannibal to winter there, requiring his army to withdraw south.[52] The repeated ravaging of Campania in the years to 211 meant that, even during summer, Hannibal, whose army tended to live off the land, was unable to spend much time supporting Capua. Indeed, his own presence intensified the pressure on his allies.[53] The irregular warfare in Bruttium in the latter stages of the war put increased restrictions on where Hannibal might winter and increased the burden on his supporting communities, which Livy regarded as leading some inhabitants to take up *latrocinia*, further exacerbating the problems of food supply.

The physical depredations of fields and farms thus inflicted collateral damage, creating poverty, food shortages, refugees, and even more brigandage.[54] Livy aptly summed up the impact of such warfare:

> The small holders had been carried off by the war, there was hardly any servile labour available, the cattle had been driven off as plunder, and the homesteads had been either stripped or burnt ... deputations from Placentia and Cremona came to complain of the invasion and wasting of their country by their neighbours, the Gauls. A large proportion of their settlers, they said, had disappeared, their cities were almost without inhabitants, and the countryside was a deserted wilderness.
>
> LIVY 28.11

The "sword" must not be overlooked. The targeting of local inhabitants is exemplified by Polybius' description of Hannibal's march into Picenum, where "just like at the capture of cities by assault, the order had been given to put

---

51   Erdkamp (1998) 216 n. 27.
52   Erdkamp (1998) 182–183.
53   Erdkamp (1998) 283–284.
54   Economic impact of war: e.g. Polyb. 1.59.1, 6–7; Livy 23.48; see Cornell (1996); Ñaco del Hoyo (2011). Refugees: e.g. Polyb. 1.31, Livy 25.1. Food supply: Erdkamp (1998) esp. 280–289. Livy 28.12 represents Bruttium as denuded of men to work the land, because many had been recruited into Hannibal's forces or resorted to brigandage. This created shortages and starvation in Bruttium and for Hannibal's army, Hoyos (2003) 128–129. Livy 23.41–42 represents Roman predations in 215/14 as creating significant hardships for the Samnites, cf. 27.4 (restoration of exiles and farms in Sicily after the war).

224   RAWLINGS

to the sword all adults who fell into their hands" (Polyb. 3.86.11),[55] while in Hannibal's advance on Rome (211), many refugees were killed or captured by the Numidian cavalry (Livy 26.9). Scholarship has noted the disproportionate violence inflicted on civilians in irregular warfare throughout human history.[56] Appian (*Pun.* 134) reported that Hannibal had destroyed 400 towns and killed 300,000 Italians in his invasion, but his figures may not be trustworthy and claim to relate only those killed in engagements. Assessing the demographic cost of the warfare on the peoples of Italy remains controversial, resting, in part, on the surviving Roman census figures, and is irrecoverable for Sicily, Spain, and other Carthaginian territories. Civilian losses in central Italy through irregular action probably does contribute to explaining the massive decline, by 140,000, of Roman adult male citizens recorded in the 209 census figures (compared to those of 234).[57] However, problems with safely conducting the census, combined with under-registration of refugees and of soldiers in service, are also likely to have been factors in the low figure reported.[58]

The Theatre of Violence

Because of the generally limited physical damage that irregular operations can inflict, it is the psychological aspect of such force that potentially has the greater impact.[59] It can have an immediate influence on the emotions and morale of combatants and victims, with longer term effects on the determination and vigour by which war is conducted, and, potentially, even upon the enemy will to continue.

As we have seen, Massinissa's operations against the Elder Scipio's forces in Spain caused physical inconveniences of supply and movement for the Romans, who appeared to have become more or less confined to camp. They also had an impact on the mental state of the soldiery, creating "alarm and confusion", and leading to "anxiety and fear" (Livy 25.34, cf. Front. Strat. 2.5.25, Polyaen. 6.38.6, Zon. 9.3). Evidently, the morale of troops could be affected by the success, or otherwise, of irregular engagements (e.g. Polyb. 3.110.5–7, 111.1). Livy (23.26), with ungenerous even-handedness, described the warfare

---

55   Rawlings (2007b) 24.

56   Boot (2013) 11–12, cf. Dwyer and Ryan (2012) xvii–xxiii.

57   Livy 27.36 (137,108); cf. Livy *Per.* 20 (270,713 in 234); Livy 29.37 (214,000 in 204).

58   cf. Erdkamp (1998) 289–291; (2011) 63–67; Toynbee (1965) 1.473.; Brunt (1971) 417–422; Rosenstein (2004).

59   Janos (1963) 645–646.

THE SIGNIFICANCE OF INSIGNIFICANT ENGAGEMENTS   225

between Hasdrubal Barca's forces and the Iberian Tartessi as: "several skirmishes [that] took place between the two sides, who were alternately frightening and fearing each other." Defeat in skirmishes tended to intimidate or make the troops and their commanders more cautious, while success often enhanced the morale of troops, making them more confident (Polyb. 3.102.11, Livy 25.18, 28.13).

It was an education to troops and officers to engage in skirmishes and harassment actions. The experience gained by troops improved their basic skills, cohesion, and ability to adapt and improvise, which all enhanced both their fighting capabilities and their morale. As part of the strategy of delay, Polybius claimed that Fabius sent his troop to engage enemy foragers "with the view of gradually strengthening and restoring by partial successes the spirits of his own troops, broken as they were by the general reverses" (Polyb. 3.89.4, cf. Livy 22.10.12). Anders has argued that such operations were a type of "stress inoculation", where, in the relatively low-risk environment of the skirmish, soldiers became less liable to be intimidated by combat.[60] This seems evident in the orders given to the pro-consuls at Gerunium, in the spring of 216, "to skirmish vigorously and unintermittently, so as to train the youths and give them confidence for a general battle" (Polyb. 3.106.4). At this time, new levies had been raised and sent to the force already blooded by repeated engagements with the Carthaginians there. Hannibal too had many relatively new Gallic recruits who benefited from the seasoning provided by the frequent and hard fought skirmishing at Gerunium. Prolonged contact allowed troops to become more familiar with the enemy. Thus, by Cannae, the Roman forces included the veterans of Fabian delaying operations and "these men had not only seen the arms, order, and numbers of the enemy, but had engaged in almost daily fights with them over the past two years" (Polyb. 3.109). Livy (24.3) claimed that Gracchus' men benefited from winter skirmishing with Hannibal's forces (215/14): "their efficiency improved; they became progressively more cautious and less liable to be caught off guard." Indeed, some troops became extremely experienced over time in the ebb and flow of irregular actions. Polybius (Polyb. 1.74.9) described the qualities of rebellious veterans as "men schooled in the daring tactics of [Hamilcar] Barca and accustomed from their fighting in Sicily to make repeated withdrawals and fresh attacks in one day." At Utica in 241, they appeared equally able to use the cover of overgrown high ground to evade pursuit and of level terrain to assault an enemy held camp. By contrast, the opposing Carthaginian forces were men more accustomed to defeating

---

60   Anders (2012) 236; Grossman and Christensen (2008) 34–36.

Numidians and Libyans, who usually fled for several days, and had assumed the veterans would act in the same fashion. Their lack of appropriate experience caused them swiftly to be surprised and defeated by the rebel counterattack (Polyb. 1.74.7, 10–12).

Skirmishes were a 'theatre of violence', allowing troops to demonstrate their *virtus* and acquire praise (*laus*) and glory (*gloria*) under the gaze of fellow soldiers in a context of competitive peer-rivalry.[61] The display of *virtus* and quest for *gloria* in the fluidity of skirmishing made some combats into personal expressions of prowess (Liv 23.47, 25.18). Appian (*Hann.* 37) stated that

> The space between the enclosing wall and Capua was about two stades, in which many enterprises and encounters took place each day and many single combats, as in a theatre surrounded by walls, for the bravest were continually challenging each other.

The predisposition of Romans and Campanians to such displays of prowess is well known, and the warrior ethos of Gauls and Iberians was equally well served by such engagements.[62] Polybius reported that some of the sorties involving Gauls (1.43.4) during the siege of Lilybaeum were so fierce and individualistic that they resembled *monomachiai* (1.42.13, 45.8–10, cf. Livy 28.2 Romans against Celtiberians). Marcellus, in his forties, had claimed the *spolia opima* for slaying the Gallic commander in single combat in 222 (Plut. *Marc.* 7–8, Polyb. 2.34–35, Propertius 4.10.39). It is interesting to note that his death in 208 came not in battle, but in a skirmish.[63] The story of the death of Flaminius during the ambush of Trasimene, apparently at the hands of Gauls who recognised him from his earlier campaigns against them, carries the resonance of personal display and of grudge settling (Livy 22.6, Sil. Ital. 5.132, Polyb. 3.83.6).

The award of decorations and prizes for valour endorsed the culture of aggression in skirmishes. Hannibal offered rewards to the bravest Numidians for their attempt to lure out the Romans at the Trebbia (Polyb. 3.71), while Scipio rewarded his bravest cavalrymen after their successful ambush of Hanno's force at Salaeca (Livy 29.34–35). Polybius emphasized that, for the Romans:

---

61   McCall (2002) 83–86; Lendon (2005) 187.

62   Oakley (1985); McCall (2002) 85; Rawlings (1996).

63   cf. Livy 25.16 (Gracchus also killed in a skirmish with ambushers), 24.41 (Publius Scipio with a light-armed party also surrounded while scouting, but Gnaeus was able to bring relief. However, in App. *Iber.* 16.1 Publius perishes during scouting—probably a conflation with later events).

THE SIGNIFICANCE OF INSIGNIFICANT ENGAGEMENTS 227

This [award of decoration] does *not* take place in the event of their having wounded or stripped any of the enemy in a set engagement or the storming of a town; but in skirmishes or other occasions of that sort, in which, without there being any positive necessity for them to expose themselves singly to danger, they have done so voluntarily and deliberately.

POLYB. 6.39

The thirst for distinction was probably the reason why a few young *velites* covered their helmets with pieces of wolf or other animal skin, in order to be identified by their officers when they fought bravely in skirmishes and other operations (Polyb. 6.22.3).[64]

Officers too gained recognition and reputation in their conduct of irregular operations. According to Polybius, during the winter of 217/16, at Gerunium,

numerous skirmishes and minor engagements took place, in which the Roman commanders [proconsuls M. Atilius Regulus and Cn. Servilius Geminus] gained a good reputation, their conduct of the operations being generally thought to have been brave and skilful.

POLYB. 3.106.11

The successes of the Liby-Phoenician Muttines and his Numidians in raiding across Sicily from the base at Agrigentum, earned him personal fame and popularity.[65] Peer-rivalry for honor in warfare emerges in this case, for the success of Muttines ignited the envy of Hanno, the commander in the region. His jealous removal of Muttines from the cavalry command was a slight on the man's reputation and led the Numidians to mutiny and to the betrayal of Agrigentum by Muttines (Livy 26.40). Subsequently the Romans honored Muttines sufficiently (Livy 27.4–5) and he served them with distinction thereafter (Livy 38.41).

Yet, success in skirmishing and raiding, when linked to a culture of aggression and praise for displays of *virtus*, could breed overconfidence. Sempronius Longus' success against raiders prior to Trebbia stoked his pride and passion and convinced him that luck would be with him in a major engagement (Livy 21.53). According to Livy (22.41), when in the early summer of 216, Hannibal's raiders suffered 1,700 casualties in a defeat near Gerunium; he regarded them as "a goad to the rashness" of the newly arrived consul, Varro. Sources tended

---

64 Lendon (2005) 187.
65 Livy 25.40, 26.40; cf. Livy 21.4, App. *Iber.* 6, Nepos *Hann.* 3.1 (young Hannibal); App. *Hann.* 10 (Maharbal).

to emphasize how irregular operations had an adverse effect on discipline and had the moral repercussion of unleashing human greed.[66] Hit and run operations in particular were often *too easy*; raids targeted non-combatants and their property, and were rarely contested by the enemy, who struggled to assemble and respond before the raiders had escaped. Consequently the expectations of raiders could become distorted by the one-sidedness of these operations, leaving them vulnerable to self-delusion, the breakdown of discipline or manipulation by skilful enemy commanders.[67] The legionaries at First Herdonia were so emboldened by their extensive pillaging as to demand their commander let them engage Hannibal's army (Livy 25.21, cf. 22.31), but Hannibal had prepared an ambush, which he sprang upon them during the battle.

Irregular operations often exhibit performative aspects for psychological purposes. In antiquity, this 'theatre of violence' had an immediate audience, directly witnessed by the troops and commanders involved in skirmishes or looking on from the camps, contributing to peer-group competition for *virtus*, enflaming passions and intimidating the enemy.[68] Raids too had their audiences, Livy (22.13) claimed that Hannibal devastated around Beneventum in order to provoke to battle an enemy incensed at the insults and suffering inflicted on its allies. In 256, according to Polybius (1.30), the Punic generals in Africa, "decided on marching to the assistance of the country and no longer looking on while it was plundered with immunity" by the forces of Regulus. Similarly, Hannibal's devastation of Etruria drew Flaminius to seek decisive battle, despite his officers' recommendation that employing light troops and cavalry would contain the pillaging (Livy 22.3). Both decisions led to major defeats, at Adys (256) and Trasimene (217). Perceptively, Polybius claimed that Hannibal turned his operation in Campania in 217 into a performance:

> The Carthaginians, then, by quartering themselves in this plain made of it a kind of theatre, in which they were sure to create a deep impression on all by their unexpected appearance, giving a spectacular exhibition of

---

66 Diod. 24.9.9 (Vodostar disobeyed Hamilcar's orders not to plunder at Eryx, Diodorus emphasises the indiscipline of the looting infantry, but the good order of the 200 cavalry who retrieved the situation), Livy 21.48 (Numidians plunder the Roman camp instead of maintaining pursuit), Livy 28.24 (seditious legionaries plundered the peaceful Spanish countryside, 206), cf. 29.6.

67 Liv 23.27 Spanish warriors' success in raids and skirmishes make them overconfident and *incompositus* (disorderly).

68 Davidson (1991) 12.

THE SIGNIFICANCE OF INSIGNIFICANT ENGAGEMENTS 229

the timidity of their enemy and themselves demonstrating indisputably
that they were in command of the country.

POLYB. 3.91.10

Hannibal sent out raiding parties to harry the countryside, particularly the *ager Falernus*. Livy emphasized the visual element of these operations: of an exasperated Roman army witnessing the columns of smoke caused by the depredation (Livy 22.14, cf. Polyb. 3.82.3). Both Polybius and Livy emphasized how Hannibal's raids were a psychological demonstration of the enemy's inability to protect its civilians, allies, wealth and farms.[69]

The broader impact of the acts resonated beyond the combatants, through report and retelling, to affect the mood of civilians and home authorities. Hannibal hoped that the news of the death of Marcellus in ambush would strike "great terror in the enemy" (Livy 27.28). Intimidation and psychological damage were clear motives in raiding and devastation, directly affecting the morale of the victims.[70] According to Polybius (1.31.2–5), after Adys (256), the combined pressures of Roman devastation and Numidian pillaging (which apparently inflicted even more damage than the Romans), brought terror-stricken refugees into Carthage, forcing the Carthaginians to make peace overtures to Regulus. In the devastation of the *ager Falernus* in 217, "The work of destruction extended to the Baths of Sinuessa; the Numidians inflicted enormous losses, but the *panic and terror* which they created spread even further ... everything was wrapped in the flames of war" (Livy 22.13). It was often the fear of further attack that drove farmers to seek refuge in defended settlements, impacting upon the harvesting or sowing and inducing hardship and even starvation in affected communities (e.g. Livy 28.11, Polyb. 9.11a).

A subtler, though perhaps unintended manifestation of the stress of such warfare, was a fear for the moral damage inflicted on the character and belief of the general populace. By 213, after years of continuous warfare, there were concerns about the rise of superstitions and foreign rites in Rome (Livy 25.1.6–9), the numbers of prophets and priestlings had increased because of the influx of needy peasants sheltering in city, whose farms were too dangerous to live in and had become uncultivated due to nature of the war. Indeed, the Senate censured the aediles for failing to control professional priests and conmen, and the *praetor urbanus* had a *senatus consultum* read out requiring the surrender

---

69   Davidson (1991) 16–17.
70   Bragg (2010) 52, 61–62; on the relationship between this strategy and terrorism as a tactic
      in the ancient world see Brice (forthcoming).

of private books of prophecy and foreign rites. Livy reported numerous other religious responses to the stresses of the war made by the Romans. For example, in 217 the Romans performed a number of extraordinary religious offerings, dedications and rites. They vowed a *ver sacrum*, an offering of the firstborn animals and children five years hence, "if the war went well and the Republic remained in the same condition" as before hostilities commenced (Livy 22.9). A period of public prayer was decreed and "not only the population of the City but the people from the country districts, whose private interests were being affected by the public distress, went in procession with their wives and children" (Livy 22.10). These rites, coming in the aftermath both of the disaster of Trasimene and the depredations of Hannibal through Etruria, Umbria, and Picenum, indicate the fears and psychological stresses being felt by the urban and rural populations at the time.

On the other hand, Bragg has observed that, during both the First and Second Punic Wars, Roman coastal raids on Africa bolstered home morale.[71] This useful observation could be applied also to Hamilcar's raiding of the Italian coast during the 240s, which would have had a beneficial impact on morale in Carthage and on pro-Punic elements in Sicily.[72] As the naval raids on Africa played well to a Roman audience, allowing them to feel that they were striking back at the enemy, so the pressure on Rhegium of Bruttian raiding persuaded Laevinus to send the *latrones* from Agathyrna to protect, but also to bolster, his allies by providing them the means to strike back (Livy 26.40, Polyb. 9.27.10–11).

Clearly irregular operations also attracted the gaze of the allies and neutrals. The transformation of Campania into a theatre of violence by Hannibal showed the inhabitants of Italy the impotence of the Romans. The strategy of using irregular warfare as coercive diplomacy to influence hearts and minds of communities, is indicated by Maharbal's raiding of identifiably pro-Roman Gallic farms, while sparing others, in order to encourage the tribes to join Hannibal (Livy 21.45). Refraining from devastation was the other side of the coin. Hannibal ravaged neighbouring territories, but spared the Tarentines in order to encourage defection (Livy 24.13, 17, 20). It could be seen as an inverse terror tactic with an implicit message for the beneficiaries, since on a number of occasions ravaging took place only after a community had rejected diplomatic overtures (e.g. Livy 23.14.5–7, cf. 23.44).[73] On the other hand, Marcellus

---

71 Bragg (2010) 57–64.

72 Cf. Polyb. 1.46.13, 47.3 (the daring blockade running of Hamilcar the Rhodian kept up the morale of the besieged at Lilybaeum, emboldening others to pursue the same activity); Rawlings (2010) 278–279.

73 Fronda (2010) 248.

THE SIGNIFICANCE OF INSIGNIFICANT ENGAGEMENTS 231

sent out counter-raiding parties seemingly to deter defection (Livy 23.44.3–8, cf. 24.33), while the defeat of the irregular force under T. Pomponius Veientanus by Hanno caused some wavering communities in Bruttium to remain with Hannibal (Livy 25.1). Indeed, the capture in 210 of several thousand Bruttian exiles and deserters, who had joined the privateers of Agathyrna (27.15, 16), strengthened Hannibal's control of Bruttium; much of it remained loyal for the rest of the war.

The Long Game

Despite its capacity to spread fear and misery, irregular warfare rarely breaks the morale of the enemy by itself. During the First Punic War, the years of protracted campaigning in western Sicily brought neither side nearer to victory until the Romans cut off supplies to Hamilcar by sea after the naval victory of the Aegates. In the second war, Hannibal's initial depredations did not persuade the Italian allies to change sides. It was only when his theatrics in the *ager Campanus*, which suggested Roman weakness, was compounded by the success at Cannae that it convinced some (though not all) Campanians to join him. Yet, despite reports of fear and panic in the city of Rome itself (Livy 22.54–56, cf. 22.7, 26.7,9), the senate resolved to take military and religious measures to diminish Roman anxiety and to continue the war.

Nevertheless, the combination of irregular and regular warfare could be a powerful tool. Against purely irregular forces, it gives a decided advantage to the hybrid operator who can play the enemy at their own game while benefiting from the stamina and potency of a regular field army.[74] It is worth noting that, despite Massinissa's qualities as a guerrilla leader in 205, it was Syphax and his Carthaginian allies who had gained the upper hand in Numidia. But for the arrival of Scipio in Africa, it is possible that Massinissa may have ended his days as a *latro*. On the other hand, the application of irregular and regular operations against a power without counter-guerrilla capacity allows the psychological impact of the former to compound the menace of the latter. Perhaps the closest to a complete breakdown in morale caused by such a combination of regular and irregular forces came in the aftermath of the battle of Adys in 256 when, having crushed the Carthaginian army in the field, the simultaneous ravaging

---

74 Huber (2002) 2 employed the term compound warfare to analyse campaigns in which irregular and regular force is used against an enemy; contributors to Murray and Mansoor (2012) tended to prefer the term hybrid warfare.

of Regulus and the Numidians created the conditions for a Carthaginian capitulation, which was only wasted by Regulus' mishandling of the armistice.[75] A blend of irregular and regular forces also almost ended the war in Spain in 211. Massinissa's harassment of the armies of the Scipios prevented them from retreating to safer territory, and each in turn was delayed by the Numidians and smashed by several Carthaginian and Spanish field armies (Livy 25.34–36). Interestingly, however, Roman fortunes appear to have been partly restored by an irregular operation mounted by the Roman survivors. A talented junior officer, Marcius, "adopting Punic tactics" through a combination of ambush and surprise, launched an unexpected attack upon one of the Carthaginian camps (Livy 25.39, with variants). The success disinclined the Carthaginians to press their still considerable advantage, allowing time for Roman reinforcements to arrive in Spain.

Marcius' success shows that irregular warfare can be used to retrieve defeat. Fabius' strategy of delay was clearly intended to create 'breathing space' to allow the rebuilding of Roman forces after the defeat of Trasimene. The temporal impact of irregular warfare can be considerable, but it was not necessarily always a direct strategy. Irregular operations tended to be manifestations of the inertia that armies in proximity experienced. Battles were such high risk enterprises that armies spent days, weeks, even months, tentatively skirmishing, attempting to obtain positional or psychological advantage, to protect or threaten sources of supply, and to provoke or dissuade a general engagement.[76] The longest periods of such proximate skirmishing that we encounter during the Punic Wars came when both sides had dug in. This is exemplified by the warfare waged by Hamilcar and his 'boxing' rivals at Heircte and Eryx between 247 and 242, which gave the impression that little decisive or important happened, while paradoxically, compelling our source, Polybius, to admit that countless operations were undertaken (Polyb. 1.57.1–3). Elsewhere, protracted instances of skirmishes and the harassment of foragers occurred during sieges (e.g. Polyb. 1.19, Livy 29.6, App. *Hann.* 37) or where both sides were wintering in close proximity (e.g. Livy 21.57, 24.3). Breaking away in such cases could represent strategic setbacks, intrinsically dangerous in execution, and yielding the

---

75   Polyb. 1.31. Whether Regulus co-ordinated his actions with the Numidians is unknown, but Huber (2002, 1–2) notes that it is not necessary for the protagonists consciously to understand that they are waging compound war, for its effects to come into play. The Carthaginians nevertheless felt the compound impact of Romans legions and Numidian predators.

76   e.g. Liv 22.12, 23.26, 23.29.1, 23.40, 27.41, 28.14; cf. Polyb. 11.21.7; App. *Iber.* 25, 65, *Hann.* 16; Goldsworthy (2000) 56, 371 n. 36.

THE SIGNIFICANCE OF INSIGNIFICANT ENGAGEMENTS 233

positional and psychological advantage to the enemy. Where pitched battles could not force the quick ending of the conflicts, months and years were given over to actions that prepared for, exploited or attempted to retrieve the results of battles. This had a debilitating effect on the protagonists. Indeed, Polybius noted,

> We may compare the spirit displayed by both states to that of game cocks engaged in a death-struggle. For we often see that when these birds have lost the use of their wings from exhaustion, their courage remains as high as ever and they continue to strike blow upon blow, until closing involuntarily they get a deadly hold of each other, and as soon as this happens one or the other of the two will soon fall dead. So the Romans and Carthaginians, worn out by their exertions owing to the continual fighting, at length began to be despairing, their strength paralyzed and their resources exhausted by protracted taxation and expense.
>
> POLYB. 1.57.7–9

Conclusions

There are several ways to account for the pervasiveness of irregular operations in the Punic wars. For the most part, generals derived some practical tactical and strategic value from them. As well as acting as an arena of stratagem, where commanders might demonstrate their cleverness in manipulating the enemy and earn a positive reputation for it among their peers, most irregular operations had more prosaic functions: providing opportunities for the destruction of life and property, enrichment through plunder, and acquiring or contesting essential supplies. They were relatively low risk compared to pitched battle, defeat in which could cost thousands of lives, and might even lose a theatre or the war. As such they also provided opportunities for enhancing the training and experience for the troops, instilling confidence, applying terror, and undermining the enemy's will to fight. There appears to have been an understanding, therefore, at least among the more able commanders, of the potential benefits such operations brought in terms of training, logistics, morale, and intimidation. There was an appreciation that these operations demonstrated "skill, strength and courage" (Polyb. 1.57.2) as part of a broader process of warfare. The recourse to irregular operations was also facilitated by the makeup of the armies themselves. Both Romans and Carthaginians had considerable capacity for prosecuting such warfare, having large numbers of troops capable of both regular and irregular operations, albeit in different degrees and capacities, depending on the types and qualities of troops that they had available. Some of these troops may have been ideologically predisposed to irregular operations, reflecting their native pursuits, but most warriors and soldiers appear able and willing to undertake these pursuits when the opportunities arose.

It was rarely the case that an individual irregular action or engagement had a decisive impact; as such it was natural that our sources regarded them as having had minor or insignificant impacts. Nevertheless the accumulation of actions could affect the course of a campaign or war. Irregular warfare was a long game in which both sides struggled to land a knockout blow, despite the occasional major victory on the battlefield. The First and Second Punic wars, at 23 and 16 years respectively, were protracted conflicts, in which neither side could find an entirely adequate answer to the approaches of the other. Irregular operations may not have been able to decide these wars, but they did much to prolong them.

Fortifications and Sieges

∴

CHAPTER 12

'Siege Warfare' in Ancient Egypt, as Derived from Select Royal and Private Battle Scenes

Brett H. Heagren

Fortifications have two essential characteristics. First, they should present an obstacle to potential attackers (physically and/or psychologically), and second, they should offer some form of protection for their defenders.[1] With respect to the first point, the Egyptians employed fortresses in order to protect not only their traditional borders,[2] but also newly acquired foreign territory.[3] Furthermore, as their empire expanded, fortresses were also required to secure long, and in some cases quite exposed, lines of communications.[4] Indeed, in their defence against internal and foreign aggression, the Egyptians employed what could be best described as a 'multi-layered network' which possessed both offensive and defensive capabilities. The lowest level of this network was seen in its most tangible form with the tactical battle camp, which provided both protection and security for the army at the furthest reaches of the empire. Ultimately, this camp was also a form of protection for Egypt itself, for if the camp and army were ever lost, the repercussions could reverberate up to the higher levels with the final result being a major strategic defeat. The next layer of protection was provided by operational bases located at key geographical and strategic points throughout occupied foreign territory, as well as within the borders of Egypt. These bases were essential both to guard against foreign

---

1 Jones (1987) 10.

2 For the major border fortress of Tjaru, see: Abd el-Maksoud (1998a) 61–65, and Abd el-Maksoud (1998b) 9–18. For Tjaru and other New Kingdom border fortresses, see: Morris (2005), Thomas (2000), and Kitchen (1998).

3 For the Nubian fortresses, publications are extensive but see (for the Middle Kingdom): Emery (1965) 64–81; Trigger (1982); S. Smith (1991); Williams (1999); and Vogel (2004). For the Libyan fortresses, see: Snape (1998); Snape (1997); and Snape (2003). For Egyptian fortresses established in Asia and in Nubia (New Kingdom), see Morris (2005).

4 For the strategically important 'Ways of Horus', see: Gardiner and Peet (1920); Oren (1987); Oren and Shershevsky (1989); Cavillier (2001a); and Oren (2006). An important strategic corridor also existed in Nubia with the Nile and this too needed protection; for this, see: Kemp (1986) 133–134, and S. Smith (1995).

© KONINKLIJKE BRILL NV, LEIDEN, 2016 | DOI: 10.1163/9789004284852_013

incursions and to maintain the integrity of the empire, especially when the main army was to be absent from a particular strategic theatre for an extended period of time.[5] The final layer of defence was with Egypt itself (the prime strategic base for all military activity). In addition to being protected by the above-mentioned devices, the Nile Valley possessed natural defences in terms of favourable geography which served to isolate it somewhat from external threats. Egypt's key border entry areas were also protected by major (strategic) military installations, which served to effectively 'seal' the country off from hostile intrusions.[6]

These three layers of protection were, however, only part of the overall defensive system. The Egyptians also relied on *offensive* action against enemy fortresses and cities in order to ensure the security of their empire (in other words, they employed a form of active defence). This meant that the ability to conduct successful 'siege' warfare was of paramount importance, not only in the initial construction of their Asiatic empire, but more importantly in maintaining it. By using such warfare, the Egyptians were able to capture key strategic locations, which was a necessary precondition, for logistic reasons, for further military expansion. Capturing and controlling such locations also denied their use to any potential enemies and as an extension (either through direct or indirect control) enabled the creation of a viable defensive zone made up of subservient buffer states.[7] This was an important consideration as enemy cities and fortresses had the potential to be used as jumping off points by foreign intruders, and pre-emptive strikes to capture such locations were a more desirable option than awaiting a possible attack, especially if it were to come at an inopportune time.[8]

---

5 Furthermore, these bases (whether they be cities or fortresses) would have served as jumping off points for military campaigns, see, for example, Kemp's (1986) 133–136 analysis of the layout of the Second Cataract Forts, and especially fig. 6.

6 On the function of Egypt's *ḥtm* bases, see Morris (2005).

7 The creation of buffer zones during the New Kingdom was not limited to the Asiatic theatre but was also undertaken in Nubia; see Morket (2001).

8 An attack or insurrection occurring while the army was heavily engaged elsewhere was a situation to be avoided. Yet, the Egyptians faced such a scenario when a rebellion took place in Nubia soon after the Year 5 Libyan invasion during the reign of Merenptah. It has been argued that the close timing of these two events was more than just a coincidence, that is, it is probable the Nubians were prompted into action by the Libyans, see Kitchen (1977) 222. A similar incident was narrowly avoided when the Hyksos king Apophis unsuccessfully attempted to incite an uprising in Nubia, which would have drawn Egyptian forces south and away from the northern frontier; see Habachi (1972) 54–55.

'SIEGE WARFARE' IN ANCIENT EGYPT

The defensive network of the Egyptians was not unique to them, but was also a feature of the other Asian superpowers.[9] That is, against the major superpowers like Mitanni and Hatti, the Egyptians had to contend with their respective networks. Thus, by examining the siegecraft capabilities of the Egyptian military we are not only able to ascertain the extent of their ability to conquer fortified locations but also, on the flip side, gain an insight into how they may have perceived such an attack against their own fortified bases. This chapter will look specifically at the visual representations of Egyptian siegecraft and city assault techniques while avoiding any detailed discussion on the physical nature of the targets under attack. This is because, as recent archaeological evidence has indicated, certain cities (witnessed in battle representations and textual accounts) may not have possessed some of the features generally associated with fortresses or fortress-cities.[10] This fact alone poses a dilemma in our attempt to reconstruct Egyptian siege tactics and undermines, somewhat, the historical accuracy of the representations—especially those which are known for certain to be fictional.[11] Yet, as will be argued, even unwalled cities could still be considered formidable military targets. In addition, the absence of city walls and other fortification devices may in fact explain why we no longer find representations (at least from the New Kingdom period onwards) of certain pieces of sophisticated siege equipment found in earlier periods. Regardless of these obvious difficulties, it will be argued that the battle reliefs do indeed provide us with a valuable insight into Egyptian siege warfare tactics.

According to Yigael Yadin, there are five main ways to successfully assault a fortified location.[12] These methods were not always mutually exclusive and

---

9   Neither Kerma nor the Libyans had such an equivalent system. The former, while a powerful political entity lacked any form of 'strategic depth' in terms of geography and as a result found itself within reach of the Dynasty XVIII Egyptian armies. The Libyan theatre on the other hand possessed considerable 'strategic depth' yet lacked the high degree of urbanism and organisation found within the territories of the powerful Asiatic Empires.

10   To what degree certain Late Bronze Age cities were fortified (if at all) is a question of particular importance; see for example: Hasel (2006) 96 and 114 n. 5; Kempinski (1992); and Gonen (1984). In addition, one must also consider that the locations for a number of these fortified cities (as depicted for instance in the Ramesside reliefs) is still the subject of debate; see Na'aman (2006). Furthermore, difficulties also arise when later generations made use of earlier fortifications (MB II fortifications proved to be considerably durable); see Gonen, (1984) 62, Baumgarten (1992) 145. For the presence or absence of other fortification devices, see especially Oredsson (2000).

11   There is also the added difficulty that these images, in particular the earlier ones, were not intended to serve as historical documents, see especially: Shaw (1996) 241–244.

12   Yadin (1963) 16–18.

in many cases a combination of techniques were employed in order to achieve the desired result. The first three methods involve overcoming the walls of the city or fortress. This was accomplished by going over the top with the aid of siege towers and ladders (scaling), by breaching the gate or penetrating the walls themselves (using battering poles, rams or even hand weaponry), or by penetrating the walls from below (sapping). Sapping was quite difficult and time consuming to carryout, and was not always practical. However, unlike the first two methods, it did offer the attackers protection from enemy fire in that soldiers worked from the relative safety of a mine. Alternatively, if the attacker was reluctant to risk an active, and potentially costly, assault, then a fourth option was available for them—the siege. This was, however, a lengthy process and the attackers (now the passive force) would have had to ensure that they had enough resources to carry it through. Not only did they have to maintain the blockade of the city, but they also had to be prepared to fight off relief forces or any attack originating from the city itself. Finally, the fastest and least expensive way (in terms of resources and casualties) of capturing a city was the fifth method, trickery, but this was probably the most difficult to accomplish.[13] As well as the above mentioned methods, we may add two more. The first involves intimidation, that is, threats of wanton death and destruction if resistance is offered but countered with the promise of favourable conditions for a quick surrender, preferably before the first shot is fired. The second, on the other hand, is far more circuitous but it too could be considerably effective. Instead of attacking a fortress or city directly, other more vulnerable locations could be attacked thus forcing the enemy to abandon their strongholds in order to deal with this new threat.[14]

---

13   During Dynasty XVIII, General Djehuty under Thutmose III, supposedly smuggled 200 Egyptian soldiers hidden in baskets (which were carried by other disguised soldiers) into the city of Joppa; see Peet (1925). While this may be a fictional tale, it still highlights the fact that the Egyptians understood that it was theoretically possible to capture a city by such means. Other notable historical examples include the capture of the supposedly impregnable Krak des Chevaliers in 1271 by the Sultan Baybars, and the fall of British Singapore in 1942 to Japanese forces.

14   Avoiding the main concentration of enemy strength and attacking less defended but vital areas was an 'operational' concept that the Egyptians were familiar with: see in particular Sety I's operation against the ruler of Hamath, where the Egyptians attacked three separate targets simultaneously while avoiding the bulk of the enemy force (which was laying siege to another city); see KRI. I, 11:11–12:14. Thutmose III in his Year 22/23 campaign performed a similar action by taking the Aruna pass (the most difficult route) to reach Megiddo; see URK. IV 649:13–652:11. Whether or not the Egyptians had a basic understanding of what

'SIEGE WARFARE' IN ANCIENT EGYPT

Without a doubt, we can say the Egyptians knew how to successfully assault a fortified location, and we can even say with some confidence that they also knew how to lay down an effective siege—although evidence for the latter is more limited.[15] Among the earliest known representations of city assaults are two scenes found in the tombs of Khaemhesy at Saqqara and Inty at Deshasheh, both of which are dated to Dynasty VI.[16] From the end of the First Intermediate Period to the early part of Dynasty XII we find more pictorial evidence of city assaults namely from private tombs found at Beni Hasan.[17] Tombs 14, 15, and 17 are dated towards the end of Dynasty XI and the beginning of Dynasty XII whereas tomb number 2 is dated to the reign of Sesostris I. As well as the Beni Hasan scenes, fragments found at the Deir el Bahri temple of Mentuhotep II appear to show Egyptian soldiers assaulting a city.[18] Yet it is from the tomb of Inyotef (386) at Thebes that we find one of our best examples of Egyptian assault warfare, again dated to Dynasty XI.[19]

Unfortunately, it is not until late in Dynasty XVII and early Dynasty XVIII that we once again find detailed references to city assaults. The struggle to liberate northern Egypt from Hyksos rule involved a number of assaults against Hyksos controlled cities culminating with the capture of their capital at Avaris.[20] The Hyksos were subsequently driven out of Egypt by Ahmose, and their last stronghold was taken after a siege of three years.[21] Although city assaults certainly occurred during the first half of Dynasty XVIII especially as the Egyptians extended their control into Asia, we only find the briefest mention of the tactics that were used, none of which, incidentally, appear in pictorial form. Most

---

is now termed 'Operational Art' is another matter entirely and cannot be discussed here. For a comprehensive study on this concept, however, see Vego (2000).

15   Schulman (1982) 179. Schulman rightfully makes the distinction between a siege "surrounding of a fortified place by an army attempting to take it by a continued blockade and attack" and an assault. See also Schulman (1964).

16   Schulman (1982) 165; W. Smith (1965) figs. 14 and 15. For a recent analysis of early siege scenes, see Schulz (2002), and also Vogel (2004) 41–44.

17   Newberry (1893–1894) I for tomb number 2 (pls. XIV and XVI) and number 14 (pl. XLVII); and Newberry (1894) II for number 15 (pl. V) and number 17 (pl. XV). The scenes from tomb number 14 are for the most part lost.

18   W. Smith (1965) fig. 185; Vogel (2004) 54 and fig. 9. In addition, see Naville (1907) pls. XIV ('D', in particular) and XV.

19   Arnold and Settgast (1965) fig. 2; and Vogel (2004) 50–54.

20   See the account of Ahmose son of Ibana: *URK.* IV 1:16–4:13.

21   *URK.* IV 4:14–15:2. Hans Goedicke however argues that the siege of Sharuhen was not the investment of just one city in one specific location but possibly a series of sieges carried out within a specific *district*, see Goedicke (2000).

of the written evidence comes from the campaigns of Thutmose III, and it is also from the records of this king that we find only our second reference to an Egyptian siege.[22] Our first New Kingdom pictorial evidence for a city assault comes from the reign of Tutankhamun. From a number of fragmentary talatat, Raymond Johnson has reconstructed two battle scenes, one of which appears to depict an Egyptian attack on a walled Asiatic city.[23]

The situation improves dramatically for Dynasty XIX, and from the reigns of Sety I, Ramesses II, and Merenptah we find numerous reliefs depicting Asiatic cities under attack.[24] We also see for the first time the aftermath of a successful assault: the abandoned fortress with breached doors surrounded, perhaps, by devastated fields.[25] Only in a few of these scenes, however, do we actually see a full assault in progress.[26] The written records for this period are less informative as to the techniques used and, in addition, we find very little reference to proper siege warfare from both the texts and the reliefs. For Dynasty XX, most of our information comes from the Medinet Habu temple of Ramesses III.[27] Recorded here are a couple of scenes which are among the best preserved examples of Egyptian assault tactics against fortified locations. The historical validity of these scenes, however, has been the subject of much debate.[28] Taken as a whole, the pictorial representations provide a wealth of information regarding the tactics likely used by the Egyptians in order to attack a fortified position (whether it be a fortresses or a fortified city). Indeed, from the pictorial representations alone we find that the Egyptians were seemingly capable of employing a number of, apparently quite effective, assault weapons in order to overcome the defences of their target.

Egyptian Tactics for Assaulting Fortified Locations

Once the enemy forces (if any) defending the city had been defeated in open battle,[29] the Egyptian soldiers could then begin their assault on the city itself.

---

22 *URK.* IV 660:4–662:6.

23 Johnson (1992).

24 Heinz (2001) 121–122, for a discussion of the various 'city types'.

25 See especially Wresz. II, pl. 65.

26 That is, an assault where Egyptian soldiers, armed with siege weaponry, actively participate.

27 *Medinet Habu* II, pls 87–90 and 94–95.

28 See Morris (2005) 698–707 for a useful summary of the arguments.

29 See below for a more detailed discussion of this visual element of the battle images.

'SIEGE WARFARE' IN ANCIENT EGYPT 243

Although the cities, as seen in the representations, are not accurate depictions and there exists the added difficulty as to just what fortification devices they did actually possess, these images highlight the important defensive features which—traditionally or stereotypically—had to be overcome: outer and sometimes inner enclosure walls, moats, towers, and so forth. The range of assault weapons and techniques used are listed as follows:

Assault Ladder

The ladder appears among the earliest assault tools used by the Egyptians. In the scene from the tomb of Inty, Egyptian soldiers are depicted attacking a fortified location, which is shown in plan view. The fortress, which is oval shaped and appears to possess rounded salients,[30] has been identified as possibly being located in southern Palestine.[31] The ladder has been placed against the 'side' of the fortress and it is being held in position by a single Egyptian soldier.[32] Our next occurrence of an assault ladder is found among the small number of Dynasty XI fragments uncovered at the temple of Mentuhotep at Deir el-Bahri, which depict an assault on a city. While the fragments show very little, we do see part of an assault ladder being climbed by at least two Egyptian soldiers (one of which is armed with a socketed battleaxe). Next to them, enemy soldiers tumble towards the ground.[33] Little can be ascertained concerning the fortress (shown in elevation) in this scene, but the wall appears to have been supported by a buttress. Schulman, in his analysis of the fragments, believed the assault that was pictured here was not too dissimilar to the Beni Hasan scenes (see below), namely, a fortified city (in this case Asiatic) being surrounded and attacked on all sides by Egyptian soldiers supported by Nubian archers.[34]

---

30   As proposed by Lawrence (1965) 71. On the other hand they could simply represent buttresses. For a detailed discussion of the visual elements of this scene, refer to Schulz (2002) 29–34, and Shaw (1996) 256–257.

31   W. Smith (1965) 149; Shaw (1996) 243, 256. This would tend to support the assumption that the Egyptians were indeed militarily active in this region. The interior of this fortress is depicted as divided into five registers in which we see the inhabitants reacting to the attack in a display of various emotions, see Gaballa (1976) 31 and Schulz (2002) 30–31.

32   Schulz (2002) 31.

33   Reconstructed by W. Smith (1965) fig. 185, with analysis by Schulman (1982) 170–176. Schulman identified the defenders as Asiatic because one of them sports a short beard. See also Schulz (2002) 35.

34   Schulman (1982) 172–176: The attackers probably occupied two registers on each side of the city. Schulman argues further that this scene (as well as the Inyotef scene) was supposed to represent the fall of the city of Herakleopolis.

The basic ladder proved to be a favoured siege tool featured in reliefs, although it is not until Dynasty XVIII, among the various talatat dated to the reign of Tutankhamun, that it reappears.[35] By the Ramesside period, with the sudden explosion of pictorial representations, we find it again depicted and this time in substantial numbers. In a scene from the Ramesseum, which depicts Ramesses II's assault on the city of Dapur, Egyptian soldiers use a ladder in their attack against the city.[36] Also present in this battle are four of the king's sons, who actively take part in the assault, each protected by his own personal mantelet.[37] During Merenptah's one and only Asiatic field campaign, he attacked the city of Ashkelon, and in the scene depicting this battle, the Egyptians have positioned two ladders on opposite sides of the city.[38] An Egyptian soldier is scaling one of the ladders armed with a dagger, and he is offered some protection by the shield strapped to his back, perhaps secured in such a way that if he were to duck, the shield would protrude far enough forward to protect his head. This is a departure from earlier times, when the assault troops were shown scaling ladders often without shields. In the scene depicting Ramesses III in his (fictional) assault against the city of Tunip two ladders are used, each being scaled by two soldiers with shields on their backs.[39] Some Egyptian soldiers are shown having already reached the top of the outer wall and are depicted engaged in hand-to-hand fighting with the defenders.

Wheeled Ladder

Another version of the simple ladder, this time equipped with wheels, was also depicted as being used in city assaults.[40] In the scene in question, from the tomb of Khaemhesy, three soldiers scale the ladder while two others near the top use their axes against the walls of the fort or town.[41] The way that the Egyptian soldiers scale the ladder appears to be a bit haphazard, although perhaps this is because the wheeled ladder may have been some kind of forerunner to

---

35   Johnson (1992) 158, fig. 12.

36   Wresz. II, pls. 107–109.

37   It is believed the mantelets originally contained battering ram crews but these have been subsequently obscured by the addition of the princes; Schulman (1964) 17.

38   Wresz. II, pls. 58a.

39   *Medinet Habu* II, pls. 88–89.

40   Senk (1957); Schulz (2002) 27.

41   W. Smith (1965), fig. 15. Although the use of such weapons against fortifications may appear suspect, Schulman (1964) 14, on a visit to Egyptian Nubia in 1962, commented on the effectiveness of such 'primitive' weapons when employed against mudbrick structures.

'SIEGE WARFARE' IN ANCIENT EGYPT 245

the siege tower. In any case, we do not see this particular assault weapon used again.[42] Smith identified this town as Asiatic, and in his description he believes that the inhabitants and their animals may be entering into underground shelters in order to escape the Egyptian attack.[43]

Personal Weapons

The frequently depicted, straightforward attack upon an enemy fortification using personal hand weaponry alone was undoubtedly the least effective, in real terms, of the siege techniques discussed here. One of the earliest representations of this occurs during Dynasty VI (from the tomb of Khaemhesy),[44] but we also see soldiers as late as Dynasty XIX employing this technique. In scenes from Ramesses II's temples at Beit el Wali and Amara,[45] we see soldiers armed with axes assaulting the walls of Asiatic cities. The Amara soldier is fortunate enough, however, to work under the protection of a mantelet. In the same scene, another soldier appears to be armed with a long spear, although it is difficult to make out if this is an actual siege weapon. In the scene depicting the assault on the city of Ashkelon, a lone soldier is seen carrying his axe at the base of the walls and it is possible he is moving into position in order to attempt to breach one of the gates of the city. He has also strapped his shield to his back which provides him with some cover. In the Tunip assault scene of Ramesses III, three Egyptian soldiers attempt to breach the main gate with their axes. With their shields tied to their backs, they are able to wield their weapons with both hands. It appears that the axe was indeed the

---

42   Of all the assault scenes that we know of, this particular example is in many ways the most problematic. This is the only time that we see a ladder equipped with wheels used in an assault, and it is difficult to believe that such an unwieldy device could have been employed in a hostile situation. One would also expect that even a ladder with wheels would have been depicted resting *against* the wall and not standing parallel with it.

43   W. Smith (1965) 149. The interior of this 'fortress' is divided into four registers—the upper two depict the inhabitants engaged in some sort of frenzied activity, in the third register we find the farmer directing his cattle and sheep through an opening, while in the lowest register, and possibly inside the shelter proper, we find women and children. For a detailed description of this scene and its composition, see Schulz (2002) 25–28. Schulz (2002) 26–27, also argues for an Asiatic identification for the inhabitants.

44   In addition to the soldiers attacking the side of the fortress with hand axes, we also see an Egyptian above doing likewise but armed with a mattock: Schulman (1964) 14. Adjacent to this soldier are two additional figures (apparently leaning on staves) who have been identified as officers: Gaballa (1976) 31.

45   Wresz, II, pl. 163; Spencer (1997) pl. 34 a–b.

246   HEAGREN

personal weapon of choice when attacking a fortress—and we will discuss below whether such a weapon could have been truly effective if used in this way.

Siege Tower

From the tomb of General Inyotef, Egyptian and Nubian soldiers are shown assaulting an Asiatic city with the aid of a moveable siege tower.[46] The tower is presented as being open-sided in which the soldiers have to climb up inside its interior using both hands to reach the fighting platform at the top. There also appears to be some sort of protruding bridge which may have enabled the soldiers to cross from the tower and over to the enemy walls. The fortress under assault has been identified as Asiatic, as the defenders are depicted with beards and shoulder length hair. In addition, their style of dress serves to set them apart from their Egyptian and Nubian attackers.[47] This scene is notable in that it is the only time we see an Egyptian siege tower depicted.

'Battering Ram'

The city assault(s) depicted in the Beni Hasan tombs (excluding tomb number 14) are notable for the fact that the Egyptian attackers make use of a type of 'battering ram.'[48] In the scenes from the tomb of the nomarch Khety (number 17) and of Baket III (number 15), the ram is manned by three Egyptians who are protected by a mantelet. These engineers do not appear to possess any personal weapons. In the scene from the tomb of Amenemhet (number 2, dated to the reign of Sesostris I), the ram with mantelet is also present, but this time it is manned by only two Egyptians. As we have seen above, mantelets could also be employed, without the battering ram element, as additional protection for Egyptian engineers. Bruce Williams convincingly argued, however, that a ram used in such a way would have had minimal impact on the walls of

---

46   Schulman (1982) 168 and n. 23; Vogel (2004) 52; and Schulz (2002) 36–40. In addition, Schulz covers various arguments for alternative localities for the battle including Nubia, Schulz (2002) 40.

47   Gaballa (1976) 38–39; Schulman (1982) 168–170. Schulman (1982) 182–183, has suggested that this was not an Asiatic town being attacked, but rather a distorted representation of the capture of Herakleopolis. This, he argues further, tends to better fit the narrative that he has reconstructed based on the battle-scenes from this tomb. This argument, however, has not received much support, see Shaw (1996) 247–248. See also Schulz (2002) 40, who argues against the possibility that this scene depicts a possible Asiatic campaign.

48   The long presumably wooden shaft of this weapon appears to have been capped with another unidentified material Williams (1999) 440.

'SIEGE WARFARE' IN ANCIENT EGYPT   247

the fortress.[49] Instead of a weapon for breaching walls, this device may have been used to probe for imperfections along the wall and then to pick out hand and foot-holds.[50] This could have been accomplished relatively quickly and at multiple points along the walls of the fortress. Once completed, the attacking infantry would begin their assault. In order to guard against such an assault, countermeasures included stone lined ditches (two or more metres deep) with low walls lining the inner parapet in order to prevent easy picking at the curtain.[51]

All the Beni Hasan scenes are believed to represent the one same event and have generally been discussed as one unit,[52] but it would be worthwhile to quickly cover some points of interest from each scene. As well as the above-mentioned differences, the fortress from tomb number 15 possesses one door and nine visible defenders (see below). The fortress also features buttresses in addition to turrets and a crenellated parapet. These same features are also found in the fortress from tomb number 2 (although in this case only six defenders are visible). Rather than buttresses, however, Williams saw this fortress as resting on a splayed base of brick or mud plaster.[53] The fortress from tomb number 17 appears to be of a different sort. Although it possesses buttresses, turrets and a parapet with crenellations, the entire structure, this time with two doors, is clearly depicted as resting upon a sloping earthenwork platform. The fortress and its defenders in all three extant scenes have been identified as Egyptian and, as Schulman has argued, what we have represented here may have been the final assault against the city of Herakleopolis, although this is not definite.[54] Our only other known occurrence of the use of the 'battering

---

49   Williams (1999) 440.

50   Williams (1999) 440–442. Williams points out that the walls of these mud brick fortresses were not solid constructions. While the mass would have been brick, they also contained additional material as well as open space (poles and beams for strength, mats to control moisture, vents and so forth). These imperfections would have been hidden from view by layers of mud plaster but an attacking force would eventually have been able to pick out these weak spots. Indeed, given the nature of these walls, the attackers could potentially have used fire to weaken the walls although this is not depicted in any of the extant siege scenes, Williams (1999) 442.

51   Williams (1999) 442, Williams adds that such defences were not insurmountable and would have only slowed down a determined attacker. But the fact that most of the fortresses in Nubia employed such defences indicates their effectiveness.

52   See in particular Gaballa (1976) 39–40, and for a fuller discussion Schulman (1982) 176–178.

53   Williams (1999) 440.

54   Schulman (1962) 182–183. If this battle took place between Egyptians and not a foreign enemy, this may explain the absence of civilians Shaw (1996) 257.

ram' is possibly found among the Deir el-Bahri fragments. Schulman, from his analysis of fragment 'A', believed that the butt-end of a battering ram (similar to the ones pictured in the Beni Hasan scenes) can just be made out.[55]

'Battering Poles'

Battering poles apparently were the predecessor to (or at least the more portable version of) the battering ram. In the Inty scene, three soldiers (or two soldiers and an officer) armed with these poles are attempting to break open the gate of the city.[56] From the Khaemhesy scene, two soldiers are again depicted armed with long poles while their colleagues use the mobile ladder to assault the city. Their actions, however, have been the subject of differing interpretations. The reaction of the two male figures in the lowest register *inside* the shelter is particularly interesting. They may possibly be listening to the two Egyptians, which, if so, may indicate that the latter are engaged in sapping or at the very least are using their battering poles to breach the walls. Yet while what we have here is very similar to the activity taking place in the Inty assault scene, it is more likely that our two Egyptians are instead attempting to secure the mobile ladder in order to stop any backward or forward movement.[57] Also of interest, next to the top of the ladder, we find a soldier who is possibly transporting additional weapons to the battlefield.[58]

Supporting Fire and the Use of Auxiliaries

As the assault of the city or fortress takes place, covering fire for the engineers and attacking soldiers is often provided by Egyptian or Nubian archers, and in some cases light infantry armed with sling bolts. Indeed, the use of Nubian archers was a common trope. The scene from the tomb of Inty likely depicted archers in the topmost register (now partially damaged) as we see a number of

---

55 For fragment 'A' see Naville (1907) pl. XV; Schulman (1982) 173; this interpretation is however open to debate.

56 As described by Schulman (1964) 14. He has also suggested that the soldiers in this particular scene may be engaged in sapping. See also Gaballa (1976) 31 who follows this view. This interpretation is supported somewhat by the reaction of one the inhabitants— he appears as if he is listening intently to the sound of digging while at the same time hushing (or beckoning) those behind him. Shaw (1996) 256, and Schulz (2002) 31, however, believe that they are merely attempting to breach the walls, which would be a more valid interpretation.

57 As convincingly argued by Senk (1957) 210, and Schulz (2002) 27.

58 The weapons may be either spears or arrows. A similarly equipped soldier is also seen in one of the Beni Hasan scenes (tomb number 14), but see n. 61 below.

'SIEGE WARFARE' IN ANCIENT EGYPT 249

the fortresses' defenders falling to the ground having received multiple arrow hits. Later, from the tomb of Inyotef, Nubian archers are also depicted among the assaulting troops.[59] Schulman, in his analysis of the Deir el-Bahri fragments, determined that Nubian archers likely participated in the assault of the city,[60] while the archers in the Beni Hasan scenes are also Nubian, and from tomb number 17 we can see how they prepared themselves for battle. First, they would lay their arrows on the ground and string (or check) their bow. This was done at a safe distance from the fighting. Next, they would move up closer, carrying their arrows in their free hand, and once in range they would thrust their arrows into the ground. Once they had set themselves up in this fashion, with arrows in front of them and in easy reach, they would then fire their complement. Support troops in the meantime would bring up fresh supplies.[61] The bowmen depicted in tomb number 2, rather than thrusting their arrows into the ground, have instead set their arrows up in a more organised way. In most of the Beni Hasan scenes (and indeed in other scenes as well), the bowmen shoot their arrows standing, but from tomb number 15, one of the Nubian archers, in very close proximity to the enemy fortress, is shown kneeling while firing.

From Sety I's Kadesh campaign, enemy soldiers occupying the defences of Kadesh are shown being attacked by arrows (yet no Egyptian archers are depicted).[62] In yet another assault against an Asiatic city, this time from the reign of Ramesses II, the defenders are again attacked by arrows. One of the defenders has been hit twice and is toppling over the wall.[63] A similar fate has befallen another unfortunate Asiatic from a different scene who is starting to topple over the wall having been hit by an arrow, presumably fired by the king.[64] In another attack against an unnamed fortress, the defenders outside the city, along with the garrison inside, appear to have been overwhelmed by a heavy arrow bombardment.[65]

---

59   Schulz (2002) 36.
60   The archers are providing covering fire for the advancing Egyptian troops Schulman (1982) 175. The presence of archers at this battle is further supported by fragment 'D' in which we see falling enemy soldiers that have succumbed to multiple arrow wounds; see Naville (1907) pl. XIV.
61   Hayes (1953) believed these soldiers were carrying javelins (within a case) and actually used the term 'ammunition parties' to describe them.
62   *Epigraphic Survey*, 1986: pl. 23.
63   Wresz, II, pl. 25b.
64   Wresz, II, pl. 183.
65   For this scene from Karnak see: S. Heinz, *Die Feldzugsdarstellungen*, 269.

The city of Dapur also appears to have suffered a heavy arrow bombardment when it was assaulted. The city's standard was hit four times,[66] and in another scene of this same assault,[67] a number of the defenders, who are depicted falling to the ground below, are also struck and killed by arrows. We also see the four arrows that have pierced the enemy standard, and the positions of the arrows suggest that our vantage point has now changed. The same city assault is now depicted from the opposite side. Judging from the two scenes, it is possible the attack commenced with an assault against one side of the city in the presence of the king.[68] The purpose of this assault may have been to act as a diversion, as it appears that no serious attempt was made to breach or scale the wall at this stage.[69] Rather, the goal was merely to force the enemy defenders to concentrate in this part of the city in order to fight off the attackers. It is notable that the troops used for this dangerous task included foreign auxiliaries as well as regular Egyptian troops.[70] While this diversionary attack was taking place, the real assault would have begun on the other side.[71] The king mounted a chariot, allowing him to quickly reach the other side of the city in order to be present for this part of the battle. The Egyptians, by attacking at the rear(?) of the city, caught a number of enemy soldiers and civilians out in the open. Some of the latter escape by being hoisted up over the wall by their colleagues while others are less fortunate and are massacred along with the soldiers. With the enemy defenders clearly distracted (they are, for the most part, firing not at the Egyptians but at the auxiliaries who are attacking on the other side), the real assault, undertaken primarily by Egyptian soldiers, would have been able to take place.[72]

The combination of arrow bombardment and use of auxiliaries is also seen in the Ramesses III war scenes. In the assault against a city in Amurru, the king is depicted firing arrows at the defenders while at the base of the city,

---

66   Wresz, II, pl. 78.

67   Wresz. II, pl. 107–109.

68   Wresz, II, pl. 78.

69   For example, there is no visible siege equipment.

70   The soldiers have been identified as Sherden warriors: G. Gaballa, *Narrative*, 111.

71   Wresz. II, pl. 107–109.

72   See for example Yadin (1963) 23–24. Yadin states that the main weakness of perimeter fortifications was the magnitude of their circumference (from around 700 metres for an average size city to several kilometers for a large city). Thus, the assaulting force had the initiative here as they could launch diversionary attacks anywhere along the perimeter forcing the defending garrison to spread its forces. Once this was achieved, the attackers could then strike with their main force against a vulnerable point.

'SIEGE WARFARE' IN ANCIENT EGYPT

and in a quite vulnerable position, we see Sherden auxiliaries armed with their characteristic daggers and round shields.[73] Two of the defenders are shown as having already succumbed to the King's arrows and are toppling to the ground. In addition, the standard of the city has been pierced by two Egyptian arrows. Sherden warriors are also used in the assault against the city of Tunip, but it is the Egyptian regular troops who, having crossed the moat, are leading the assault. In this particular example, their covering fire is being provided by another batch of foreign troops. The cities in the three remaining scenes at Medinet Habu are all also subjected to heavy arrow bombardment.[74]

Fall of the City

In the majority of the battle scenes where a city is under actual attack, the instant the artist has often chosen to represent occurs just prior to the penetration of the outer enclosure. But in a couple of scenes we actually see Egyptian soldiers who have successfully penetrated the city's first wall. This is seen, for example, in the assault of Tunip, where the outer defences had been scaled and Egyptian troops, armed with sickle-shaped swords and shields (one soldier still has his strapped to his back), are engaged in hand-to-hand combat with the defenders. In another example (again from Ramesses III), the city's outer enclosure wall has been breached via one of its gates and Egyptian soldiers, armed with staves (but without shields), are in the process of slaughtering the defenceless inhabitants.[75]

The image of a conquered, and presumably abandoned, city is frequently found in Ramesside representations. For example, the city of Mutir is depicted under fierce attack in a scene from the Luxor temple,[76] but in the Karnak temple we see the same city now empty and with its doors breached.[77] Other fallen

---

73 *Medinet Habu* II, pls. 94–95; They appear to be 'urged on' by the three stick wielding princes behind them. This scene is quite similar to the Dapur scene of Ramesses II. Could these particular soldiers be utilized as 'cannon fodder' under certain circumstances?

74 *Medinet Habu* II, pls. 87 and 90. As further testament to the power of this weapon, archers (in particular Nubian archers) are frequently mentioned in the Amarna Letters, specifically in requests for military aid, see for example: *EA* 53; *EA* 70–71; *EA* 73; *EA* 90–91; *EA* 93–94; *EA* 102–103; *EA* 108; *EA* 111; *EA* 114; *EA* 117; *EA* 121; *EA* 123; *EA* 127; *EA* 129.

75 *Medinet Habu* II, pl. 87.

76 Wresz, II, pl. 71.

77 Wresz, II, pl. 55.

cites depicted in such fashion include Akko,[78] Krmjn,[79] [ ]an,[80] and [ ]t.[81] Possibly some of these were indeed abandoned following their capture and there is a wonderful example of such a dilapidated Asiatic city, with its smashed doors, broken windows and crumbling walls.[82] A less dramatic scene, but still worthy of note, depicts a similar city with a number of birds emerging from its ruins as if they were startled.[83] Other cities appear to have escaped such a fate and are shown in fairly good condition,[84] but it is from the Ramesseum that we find the most curious collection of conquered cities compiled in such a way to resemble a bizarre trophy collection.[85] The complete destruction of enemy cities is also referred to both directly and indirectly in certain textual accounts. King Kamose mentions destroying the cities of his enemies and burning down their dwellings,[86] while in the Gebel Barkal Stela we find a mention of Thutmose III destroying by fire and making into mounds the various cities and tribes belonging to Mitanni.[87]

As well as devastating the city itself, the Egyptians also describe engaging in a bit of logistic and economic destruction in order to ensure that the city remained uninhabitable.[88] In real terms, this would have no doubt involved the killing or seizure of livestock. From the tomb of Khaemhesy, we see cattle been driven into a shelter for safety. Similar images are found in Sety I's assault against the city of Kadesh, and also with Ramesses II's attack on Mutir. In another scene, an enemy bowman flees for his life while at the same time attempting to save the livestock.[89] In addition to this, the Egyptians are depicted destroying the natural resources of a city, as is seen most explicitly

---

78   Wresz, II, pl. 55a.

79   Heinz (2001) 271; Kitchen (1964) 56–61 and Scene 'E'. This city is also shown under attack in a scene from Karnak: Wresz, II, pl. 54a.

80   Wresz, II, pl. 54a.

81   Kitchen (1964) 59–62; Heinz (2001) 271.

82   Wresz, II, pl. 65.

83   Kitchen (1964) 56.

84   For example, *Jkt*: Wresz. II, pl. 56; *J[n]d* and 'Dibon' in Kitchen (1964) 51f. The identification of *Tbn* as Dibon has since been questioned, see especially Na'aman (2006) 65. This is an important consideration—as excavations conducted at the site of Dibon have not uncovered evidence for Late Bronze city walls.

85   Wresz. II, pls. 90–91.

86   See the following translations: Habachi (1972) 38; Redford (1997) 14.

87   *URK*. IV 1231.

88   The level of destruction inflicted may have been considerable Shaw (1996) 253. For the 'Destruction of Life Support Systems' see especially Hasel (2006).

89   Wresz. II, pl. 183.

'SIEGE WARFARE' IN ANCIENT EGYPT

in Ramesses III's assault of Tunip. While the main attack is taking place, a detachment of troops is in the process of cutting down trees nearby.[90] The textual accounts also provide us with numerous instances of the employment of such a strategy. From the Dynasty VI 'Autobiography of Weni', mention is made of sacking strongholds and the cutting down of figs and vines.[91] Indeed, the written texts are far more informative with respect to logistic and human destruction, and it is worthwhile to look at some other examples. From the Wadi Halfa Stela of Sesostris I we find evidence of the destruction of crops by the Egyptians who threw barley belonging to the enemy into the river,[92] while from the Semna Stela of Sesostris III mention is made of corn being pulled out from the ground and then burnt, and cattle being slaughtered by the wells.[93] An alternative fate for the cattle was for it to be plundered, as is evidenced in the 'Instructions of Merikare'[94] and also in the booty list of Thutmose III following the capture of Megiddo.[95] In addition, the Annals of Thutmose III contain a couple of excellent examples of the repercussions for resisting the might of Egypt. Following the destruction of the city of Ardat, the king ensured that the harvest was destroyed and all the fruit trees were cut down.[96] Later, the city of Tunip fared no better, having its trees cut down and corn pulled out,[97] and the same fate befell another city mentioned in the Gebel Barkal Stela: corn being pulled out, trees cut down, all the resources destroyed, and the entire land being turned into a place where "no trees will ever exist".[98]

In addition to the destruction that occurred outside of the city, one expects that there must also have been the usual instances of plunder (whether condoned or not), and indeed it does appear that the Egyptian soldiers were permitted to partake in the 'spoils of war' at least to some extent. Kamose, after a

---

90 Schulman (1964) 18. On the other hand, they may be chopping down the trees in order to make additional siege weaponry.

91 *URK.* I 103:11–14.

92 From an inscription of General Mentuhotep dated to Year 18 of Sesostris, see H. Smith (1976) 40 f.

93 Lichtheim (1975) 119.

94 Lichtheim (1975) 104.

95 The large amount of livestock captured is listed in separate categories (for example, cows: 1,929; goats: 2,000; and sheep 20,500): *URK* IV 664:12–14. Cattle not taken back to Egypt would have undoubtedly been consumed by the troops. The meat from one decent sized ox would have been sufficient to provide one day's meat ration for 1,000 soldiers assuming a portion of 160 grams per man.

96 *URK.* IV 687:5–7.

97 *URK.* IV 729:15–730:1.

98 *URK.* IV 1231.

254 HEAGREN

successful military operation against a Hyksos-controlled town, related how he allowed for the sharing out of captured booty to his troops, and no doubt this was done in an attempt to maintain order and discipline.[99] Occasionally, the urge to plunder did overwhelm the discipline of the troops and the one notable case of this occurring followed the Battle of Megiddo. The Egyptian soldiers had successfully defeated the opposing army in open battle but became distracted with plundering the enemy camp and thus failed to follow up their victory with an immediate assault on the city.[100]

Finally, what became of the inhabitants of the city itself? For the answer to this question there is ample evidence both from the battle scenes and the written accounts. It is clear that a percentage of the inhabitants ended up as prisoners and were transported back to various locations within Egypt. Others would have been killed (either in battle or executed after), and if the population of the city was not in its entirety transported away, the remaining inhabitants (now the 'loyal subjects' of the Egyptian king) would have remained to carry on with their lives. Returning to the battle images, the carrying off of captured city folk as spoils of war is seen in some of our earliest scenes. In the bottom register of the Inty scene we see a line of bound captives consisting of men, women, and children. One of their guards, armed with an axe, also carries a young child over his right shoulder. Asiatic prisoners (for the most part women) are also seen in the Deir el-Bahri fragments being led away from their doomed city.[101] From Dynasty XVIII, images from the Memphite tomb of Horemhab depict lines of prisoners, male and female, being carried off into captivity,[102] and from Dynasty XIX, in a damaged scene from Ramesses II's Amara temple, we see a procession of prisoners being led away from a conquered Asiatic city.[103] As one would expect, if it was believed that the city was about to fall, some of the inhabitants or the defenders may have tried to flee, and in the battle scenes we find examples of this. From the Deir el-Bahri fragments, we see Asiatics fleeing their besieged city, while from the battle reliefs depicting Sety's attack on Yenoam there is a wonderful image of Asiatic soldiers (who are supposed to be protecting that city) hiding among nearby trees.[104] In Ramesses II's assault on

---

99   Redford (1997) 14.

100   *URK.* IV 658:8–10.

101   Schulman (1982) 175. For these fragments, see also n. 18 above.

102   Martin (1989) pl. 105. But out of interest, no Hittites are among the captives, see Darnell (1991).

103   Spencer (1997) pl. 36 c–d.

104   For a recent study of such images and their comparison with similar occurrences in Asian art, see. Gilroy (2002).

Mutir we can just make out the head of an Asiatic soldier who is also hiding in trees close to the city. Another memorable image is the hapless individual who fled the city of Satuna,[105] and although he escaped the armies of Ramesses II, he ran afoul of a wild bear. In certain cases however, it proved impossible to flee a city as was the case with Megiddo. Thutmose III had it surrounded with a wall, constructed from local materials, making it impossible for anyone to escape.[106] This mirrors the textual accounts when describing the fate of the inhabitants. Those who were lucky enough to survive the Egyptian assault often found themselves as prisoners of war, as seen for example in the texts of 'Ahmose son of Ibana' and 'Ahmose son of Penakhbet'. Such prisoners either ended up as servants to the soldiers who captured them or were handed over to that soldier's superiors for other duties. Along with the devastation wrought by Thutmose III against the towns of Mitanni, as described in the Gebel Barkal Stela, mention is made of the fact that the inhabitants (the surviving ones at least) were carried away.[107] Other prisoners were less fortunate and were killed either during the assault or afterwards as punishment for their crimes—but, in depictions at least, such harsh punishment was generally reserved for the chiefs or leaders and not for the rank and file.[108] As with the images, it is difficult to ascertain clearly whether the Egyptians engaged in the wholesale slaughter and mutilation of inhabitants of certain cities.[109] That they did do so against rebellious elements in Nubia is likely and evidence for this type of practice dates back to at least Dynasty XII,[110] but in Asia this does not appear to have been the case. Certain cities were likely devastated severely, and the logistic and economic destruction is evident, but their populations probably had more value to the Egyptians alive rather than dead.

---

105 Wresz. II, pls. 66–67.

106 *URK.* IV 660:14–661:1.

107 *URK.* IV 1231.

108 See for example, Der Manuelian (1987) 50.

109 For example, as with the Romans at New Carthage under the command of Scipio or the Thracians at Mycalessus during the Second Peloponnesian War, see: Polyb. X. 15 and Thuc. VII. 29 respectively.

110 Indeed, their punishment could be quite extreme, see for example the Amada Stela of Merenptah which records the extreme punishment meted out to Nubian rebels, see *KRI.* IV 35:5–36:5.

The Defence

A city that was faced with attack basically had two options available. First, they could attempt to repel the enemy force before it reached the city. Second, their forces could withdraw behind their defences and hope to fight off any assault or, at the very least, survive any logistical destruction that would result from not meeting their opponent in open battle. From the Egyptian battle representations and textual accounts, it appears that a number of threatened cities did attempt the first option.

Open Battle

From the tomb of Inty we see the Egyptian forces engaged in heavy hand to hand fighting with enemy troops from the fortress. The Egyptian soldiers, armed with axes, are shown defeating the enemy ranks which appear to have been hard hit by the Egyptian or Nubian bowmen, as a number of them appear to be suffering from multiple arrow wounds.[111] Combat is also depicted outside the fortress/shelter depicted in the tomb of Khaemhesy as we see two individuals engaged in an unarmed struggle. From the Deir el-Bahri fragments it also appears as if the Egyptians may have engaged in some type of hand-to-hand combat before advancing on the city itself. The scenes from the four tombs at Beni Hasan also depict open combat outside the city. In the scene from tomb number 15, the attacking Egyptian soldiers are armed with spears and are in the process of stabbing their opponents, and from tomb number 17, violent hand-to-hand combat is taking place with a pile of corpses depicted right in the middle of the battle.[112] The surviving fragments from tomb number 14 are of interest as we see some Asiatic auxiliaries who are armed with bows and slings.[113] A pile of corpses is also depicted and is being inspected by a Nubian bowman. Finally, from tomb number 2 we find two registers of fighting on the south side of the east wall. Schulman believed that what was represented here should be considered separate from the city assault scene on the north side,[114] but it is more likely that the two events were connected. The attacking Egyptian and Nubian troops advancing from the left in the upper register are a mixture of archers and infantry. The latter are armed with eye-axes, spears, throw-sticks, and shields.

---

111 One individual, an enemy bowman, has been hit by five arrows but only finally succumbs after being struck by an Egyptian wielding a battle axe.

112 Schulman (1982) 177.

113 The one exception is armed with an Asiatic style 'eye-axe' in addition to his bow, see Schulman (1982) 178.

114 Schulman (1982) 178.

'SIEGE WARFARE' IN ANCIENT EGYPT

Their enemy (consisting of a mixture of Egyptian and Nubian troops) are likewise armed, although they appear to lack archers. Instead, one individual to the extreme right appears to be armed with a club. The lower register depicts two somewhat confused melees and is notable for the inclusion of Asiatic troops.[115] Schulman is uncertain as to with whom these soldiers are allied, but one would expect that the Herakleopolitans (if they were indeed on the defensive) would not have had any major qualms about enlisting the support of Asiatic mercenaries.

The inscriptions of Ahmose son of Ibana and Thutmose III provide us with two detailed accounts of combat taking place before the commencement of the attack on the city. The soldier Ahmose recounts the numerous battles that were fought around the city of Avaris before it was captured.[116] Later, Thutmose III had to defeat an enemy coalition outside Megiddo.[117] In this particular case, although the Egyptians defeated the defending force, the army, due to a lapse of discipline, failed to follow up this victory with an immediate assault on the city.[118] What followed was a siege that lasted seven months. From Dynasty XIX, a number of the city assault scenes depict the Egyptians engaged in open battle with some type of defending army. Sety, prior to attacking Yenoam, defeated a force of Canaanites equipped with chariots,[119] and during his attack on Kadesh he overcame enemy defenders, again equipped with chariots plus a healthy number of bowmen.[120] Ramesses II also fought a number of such battles with one of the best depicted examples actually belonging to small number of documented instances where the Egyptians failed to achieve their strategic objective (in this case the capture of a specific city). In his campaign to capture Kadesh, Ramesses, due to an error in judgement, ended up facing a defending force that almost overwhelmed his entire army.[121] The Egyptians managed to extract themselves out of the situation reasonably well but they nonetheless failed to capture the city. Other more successful battles, such as those which are depicted as taking place near Mutir and Satuna, are depicted in great detail. The attack on Dapur is also of interest as Egyptian troops have decimated the

---

115 Schulman (1982) 178.

116 *URK*. IV, 3:2–4:13.

117 *URK*. IV, 657:2–660:1.

118 This appears to be the *modus operandi* of the Egyptians when assaulting fortified locations: defeat the defending force in open battle and follow up with a quick assault thus (hopefully) eliminating the need for a drawn out siege.

119 *Epigraphic Survey*, 1986: pls. 9–14.

120 *Epigraphic Survey*, 1986: pls. 22–16.

121 Wresz, II, pls. 82–89 (Luxor); 169–178 (Abu Simbel); 92–101 (Ramesseum).

ranks of the enemy and appear to have caught some civilians out in the open, while others are being pulled to safety by their colleagues inside the city.[122]

Defending the Perimeter

If the defending force lost the open battle (or did not offer one in the first place) then their next hope was to rely on the fortifications of their home base. In preparing themselves for the impending attack, the inhabitants of the fortress or fortress city had few available options. While their fortified location would offer them some protection (moats, bastions, outer and inner defensive enclosures) and also slow the attackers down, it was the weaponry they possessed and how well they used it that would ultimately decide their fate (along with an adequate supply of stockpiled food and water). The defenders were forced to rely on successfully using one or more of the following three weapons systems: archers, spearmen, or stone throwers in order to repel the assault. From the pictorial representations, we see the defenders armed in such fashion. In some cases only one weapon system is employed, in other cases all three.[123]

The use of archers in the defence of a fortress is seen in the scene from the tomb of Inyotef, and in the same image we also see individuals armed with small hand held stones.[124] Both classes of troops are intermingled with each other, and indeed a combined defence force of archers and stone throwers appears to have been fairly common during this early period. In one of the Beni Hasan scenes (number 17), the defenders are likewise armed (four archers and six stone throwers). Two of the stone throwers are also equipped with shields. In another scene (tomb number 2), four stone throwers are present (one of which holds a shield) along with two archers, while from tomb number 15 we can make out a mixed force of nine soldiers consisting of archers and stone throwers.[125] Unlike in the attacking infantry, we never see slingers taking part in the defence. This was most probably due to the fact that they may have required considerable space on the wall in order to effectively deploy their weapon.[126]

---

122 Wresz, II, pl. 109.

123 In rare instances, we also find defenders armed with daggers (see below), but their use would have been limited unless the assault degenerated into street-to-street fighting.

124 Such projectiles, while crude, could be effective. Of the sixty or so soldiers uncovered at Deir el-Bahri, Winlock (1945) 15, noted that fourteen had received wounds consistent with being hit by stones thrown from a height. Many of these blows, while not immediately fatal, would have been disabling.

125 Williams (1999) 440. It is likely that the stones are to be thrown overhand.

126 Williams (1999) 446 n. 38.

'SIEGE WARFARE' IN ANCIENT EGYPT

Typical defensive practices meant that soldiers would have been in very close proximity ideally with one or two stationed for every metre along the curtain wall (if, that is, numbers permitted).[127]

In Sety I's assault against the city of Kadesh, the defenders are depicted armed primarily with bow and arrows, whereas defenders using spears appear for the first time in pictorial form during the reign of Ramesses II. In his attack against one unidentified Asiatic city, the king faces defenders armed with spears,[128] and in his assault against the city of Jj the defenders are again likewise armed. A lone spearman can just be made out in another city assault scene,[129] and the same can also be seen in a scene from Luxor.[130] The garrison of Dapur, which seems to be putting up quite a spirited defence against the Egyptian attack, also consists of spearmen in addition to archers and stone throwers. The latter are throwing rocks both small and large at the attackers, and indeed one of the stone throwers (stone held with both hands above his head) is shown facing us directly and not in profile. The garrison of Kadesh (during Ramesses II's Year 5 campaign) is depicted armed with a mixture of weapons albeit not all in the same scene. From Abu Simbel, the defenders are armed primarily with spears,[131] yet from Luxor the garrison appears to consist of a mixture of bowmen, dagger, and shield bearing troops.[132] In our two Rames-seum scenes, only one of the garrison troops in each appears to be armed (with a bow).[133] The rest of the troops, in both cases, appear to be content to watch the battle unfold.[134] Finally, in the scene of Ramesses III's (fictional) assault of Amurru, the defenders are armed to the teeth with spears. It is also from this king's reign that we have a rare representation of Egyptian troops defending their fortress.[135] In the scene in question, Egyptian archers shoot at Libyans who are in full retreat following the failure of their second invasion attempt. As for other representations, the inhabitants of the beleaguered city do not appear

---

127   Williams (1999) 446 n. 38.

128   Wresz, II, pl. 183.

129   Heinz (2001) 269.

130   Heinz (2001) 271.

131   Wresz, *Atlas* II, pls. 169–170.

132   As well as at least one spearman: Wresz, II, pl. 84.

133   Wresz, II, pls. 96–106.

134   In the Ramesseum (R 1) scene, a detachment of dagger wielding Hittites is, however, stationed just outside the city, Wresz, II, pl. 96a. A similar detachment (but armed with daggers, spears, shields and at least one bow) is also depicted in the Luxor scene, Wresz, II, pl. 84.

135   *Medinet Habu* II, pls. 69–70.

to possess any weaponry at all. Some individuals may have shields but for the most part they are no longer relying on conventional weapons to save them but have turned to a more spiritual defence.[136]

Conclusions

While it is important to remain mindful of the limitations of the artistic evidence, in particular the veracity of the events and activities depicted, this should not serve to distract us from the wealth of information provided by these battle scenes or the Egyptian conception of siege warfare and their capabilities in this sphere. As a result, although recent archaeological excavations have shown, for instance, that Megiddo may not have possessed city walls at the time of Thutmose III's famous attack—and this may have been the case with a number of other Asiatic cities as well—this should not cause us to discount the artistic depictions entirely.[137] Obviously this sort of disjunction between the archaeological evidence and the Egyptian account of the 'siege(s)' is problematic, although not entirely inexplicable. The absence of such walls would, of course, nicely explain the fact that certain pieces of Egyptian siege equipment fail to reappear during the New Kingdom. But more importantly, the depiction of various siege practices and equipment seem to have represented an accepted part of the Egyptian artistic vernacular for military scenes and likely an expected and recognized part of warfare during the period. Additionally, it must be stressed that even though a particular city may not have possessed defensive walls, it may still have represented a formidable target.[138] Any built-up city with a garrison would have posed a substantial obstacle to an invading army—especially considering they were often sited to dominate key geographical areas—and a defending force's ability to construct *ad hoc* defences should not be ignored. Even when an invading army had penetrated a city, street-to-street combat would likely have been just as deadly for the ancients as it remains today for a modern military force. In this respect, what siege equipment we do see represented (mainly ladders) could have been used to gain access to roof-tops and so forth. Personal weapons in the meantime, while

---

136   For a detailed discussion of the ritual performed by the inhabitants of numerous besieged cities, see Spalinger (1977–1978) 47–60.

137   Gonen (1987) 97; Kempinski (1992) 137.

138   For example, the walls of the outer layer of buildings on the perimeter of the tell would have provided some form of protection, see Baumgarten (1992) 145 n. 15.

'SIEGE WARFARE' IN ANCIENT EGYPT   261

probably useless against the main doors of a major fortress, would have been effective in gaining access to individual buildings.

All in all, reluctance on the part of the attacking troops to enter into a hostile city (one which contained a considerable body of enemy troops) is understandable and may explain why Megiddo was placed under a siege which was to last seven months despite its probable lack of city walls.[139] The fact still remains that these cities had to be captured and only by doing so would the Egyptians have been able to achieve tactical, operational, and ultimately strategic success. The acquisition or loss of an important centre or vital fortress would have had serious repercussions at the higher levels of war. For Egypt, the ability to conquer such targets played a significant part in determining the nature of their imperial expansion and subsequently impacted on their ability to defend their newly acquired Asiatic possessions. Indeed, it was crucial that these cities be either militarily conquered or won over through more diplomatic means in order to maintain the operational flexibility of the army and more importantly to ensure the safety of the Nile Valley itself.

---

139   Goedicke (1984) 126, however, argues for a significantly reduced siege of only one month and seven days. Nevertheless, the view that the Egyptians did not possess the necessary abilities in order to conduct siege warfare (see for example Gonen [1987] 97) does not really stand up to the overwhelming evidence to the contrary. On this point, see the comments of David O'Connor (2006) 31, who is critical of the idea that the New Kingdom Egyptian military was deficient in such areas. For a comprehensive study on the New Kingdom army in Asia, see Cavillier (2001b).

CHAPTER 13

Tissaphernes and the Achaemenid Defense of Western Anatolia, 412–395 BC[*]

John W.I. Lee

The Achaemenid Persian Empire had many frontiers to defend, from the Indus valley in the east to the steppes of Chorasmia in central Asia, from western Anatolia at the shores of the Aegean Sea to the Nile Delta and the Arabian Desert in the south. The empire's frontier in western Anatolia was neither the longest nor the most crucial for imperial success, but was arguably among the most difficult to defend.[1]

From the time that Cyrus the Great conquered western Anatolia around 545 BC, Persian officials had to handle a diverse and fractious native population that included local dynasties in Caria and elsewhere, Greek cities in the regions of Ionia and Aeolis, and an assortment of mountain tribes, as well as the remains of the once-powerful Lydian kingdom.[2] In contrast to the Empire's frontiers in Egypt, central Asia, and the Indus valley, on the western frontier land and sea intertwined. The sea offered easy travel for raiders and invaders, whether from the large islands of Samos, Chios, and Lesbos just offshore, or from mainland Greece across the Aegean. A fragmented Achaemenid administration, with competing major satrapal centers at Sardis in Lydia and Daskyleion in Hellespontine Phrygia, sometimes hindered imperial responses.

The uprising of Pactyes around 545 and the Ionian Revolt of 499–494 revealed the dangers posed by restive native populations, underscored the strategic importance of Sardis, and highlighted the difficulty of guarding a lengthy

---

[*] Over the course of four topographical study trips in 2008–2011 I was able to examine firsthand most of the places and routes discussed in this chapter. I am especially grateful to Cengiz İçten and Christine Thomas for their invaluable assistance during these trips. Many thanks also to Elspeth Dusinberre for sharing a chapter of Dusinberre (2013) before publication, to Frank Frost for commenting on an early draft, to Yingjie Hu for his expert cartography, and to the Press' anonymous reader. Grants from the UC Santa Barbara Academic Senate supported my research travel.

[1] Ruzicka (2012) makes a strong case for the strategic importance of Egypt.

[2] All dates are BC unless stated otherwise. Ancient geographical names are transliterated as they appear in the Barrington Atlas.

© KONINKLIJKE BRILL NV, LEIDEN, 2016 | DOI: 10.1163/9789004284852_014

TISSAPHERNES AND THE ACHAEMENID DEFENSE OF WESTERN ANATOLIA   263

coastline against aggressive, mobile, and seaborne opponents.[3] The magnitude of the challenge only increased after 480, when the Persians had to defend against the expanding Athenian Empire, which used its naval strength to great effect.[4] Starting around 450, though, a generation of relative calm settled over the coast of western Anatolia.[5] The end of open hostilities between Athens and Persia seems to have resulted in the dismantling of Ionia's coastal fortifications. The coastal cities remained unfortified into the early 420s (Thuc. 3.33), while the Persians based their defenses inland at Sardis and Daskyleion.[6] After the start of the Peloponnesian War, the Achaemenids attempted to capitalize on Athenian distraction, for example by making a grab for Colophon around 430 (Thuc. 3.34). After that, the revolt of Pissuthnes, which lasted from around 420 to 413, seems to have occupied Persian attention. While Athens did support Pissuthnes' illegitimate son Amorges against the Persians, it was only in 412, as the Empire returned to active hostilities with Athens, that the question of western Anatolia's frontier defenses again came to the fore.

This chapter takes a closer look at the Achaemenid defenses of western Anatolia during the years 412–395, with an emphasis on the regions of Lydia, Ionia, Aeolis, and Caria, and on the Persian commander Tissaphernes.[7] The military affairs of this pivotal era have often been examined from the Greek perspective, but less often from the Persian point of view.[8] As Achaemenid scholars have increasingly succeeded in revealing what once seemed an elusive imperial presence in western Anatolia, they have sometimes emphasized the overall long-term effectiveness of the empire's defenses.[9] While this broad picture of imperial security has been very important in counteracting stereotypes of Persian weakness or decadence, there is also much to be gained from employing a short-term perspective that helps reveal some of the complicated realities and difficult choices involved in guarding the western frontier.

---

3 Pactyes: Hdt. 1.154–160. Ionian Revolt: Hdt. 5.35–6.42.

4 On Persian defenses during 480–450, see Briant (2002) 497, 554–559; Raaflaub (2009) 103.

5 On the "Cold War" of these years, see Eddy (1973).

6 Amit (1975): 39–40; Hornblower (1996) 414–415; Raaflaub (2009) 110. See Hansen and Nielsen (2004) 1373–1374 for a list of fortified and unfortified sites in the 5th–4th centuries.

7 My analysis mostly excludes Pharnabazus and the defense of Hellespontine Phrygia; I hope to take up this topic elsewhere.

8 See for example Cartledge (1987); Kagan (1987); Krentz (1989a); and Lazenby (2004). For the Persian perspective see Briant (2002) 591–645; Westlake (1981).

9 For recent research on Achaemenid western Anatolia see Roosevelt (2009); Miller (2011); Khatchadourian (2012); Dusinberre (2013). Tuplin (1987a) is an essential starting point for the study of Achaemenid garrisons.

264   LEE

After assessing the situation around 412, when Tissaphernes first took power,
we investigate his successful response to the Athenian expedition of 409, then
discuss the changes wrought by his successor Cyrus the Younger between 407–
401. From there we trace how the civil war between Cyrus and his brother
Artaxerxes II affected the western frontier defenses, and end by examining
Tissaphernes' second period of command during 400–395 in the face of Spartan invasion. In many cases, the absence or ambiguity of evidence means we
can only offer possibilities or suggestions rather than definite conclusions.
Nonetheless, we can at least assess the conditions that Achaemenid commanders faced, the resources they could deploy, and the ways their defenses may
have developed over time.

Tissaphernes Takes Command, 412–411

Our story begins with the appointment of Tissaphernes as satrap at Sardis
and "general along the coast." Exactly when Tissaphernes took charge is uncertain, but he was probably relatively new to the position in 412. The new Persian commander had some significant resources. His capital at Sardis, with its
triple-walled citadel and fortified lower town, provided a centrally located base.
Sardis was within easy reach of the coast, but also far enough inland to maintain secure overland communications with the rest of the empire. Secondary
centers at Tralles and Magnesia-on-Maeander controlled the fertile Maeander
valley and the main routes north to the Cayster valley. Beyond these major
centers stretched an array of forts and outposts interspersed with rural communities of Iranian colonists and other military settlers.[10] These sites were probably densest on the ground around Sardis and progressively thinner toward the
Ionian coast, although coastal Aeolis possessed some important strongpoints
such as Larisa and Kyllene. To the south, rugged Caria had strong fortifications both inland and along the coast.[11] From his territories, Tissaphernes could
call up a wide range of troops, amongst them skilled Iranian horsemen and
other soldier-settlers who held land grants in return for military service, as

---

10   Sardis: Dusinberre (2003). Forts and outposts: Gezgin (2001); Roosevelt (2009) 117–121, 197–
       198; and Dusinberre (2013) 99–103. I have not yet been able to obtain Doğer and Gezgin
       (1998) or Doğer (2001). For an example of a minor fortified site, see Beden (2005) on
       Oroanna, inland from Teos. The *pyrgoi* ("towers") of Teos may have been incorporated
       into this network; on these *pyrgoi* see Hunt (1947); and Morris and Papadopoulos (2005)
       207.
11   Larisa and Kyllene: Hansen and Nielsen (2004) 1043–1046. Caria: Dusinberre (2013) 111–112.

TISSAPHERNES AND THE ACHAEMENID DEFENSE OF WESTERN ANATOLIA   265

well as professional Greek and non-Greek mercenaries.[12] To link these sites and troops together, Tissaphernes likely possessed a communications system organized along the lines found elsewhere in the empire. This system would have employed mounted messengers as well as possibly chains of fire beacons or other signal systems in some areas.[13]

Tissaphernes could also count on varying degrees of support from Greek cities along the coast, many of which had recently revolted from Athens. Ephesus and Miletus were the most firmly in the Achaemenid camp.[14] Ephesus, home to the famous sanctuary of Artemis, had joined with the Persians perhaps as early as 414; Persian forces were increasingly common sights in the city (Plut. *Lys.* 3.2–3). Built around fortified seaside heights, Ephesus was not only a major port but also controlled the lower Cayster River valley and the easiest land route south to the Maeander River valley, making it a linchpin for the defense of the Ionian coast.[15] Miletus, strategically located on a defensible peninsula with good harbors and fertile hinterland, dominated the Gulf of Latmos and its coasts.[16] The Milesians had subsidiary settlements in the lower Maeander valley and could field a strong hoplite force.[17]

Despite these resources, Tissaphernes faced several challenges. Some of his troops had battle experience against Pissuthnes, but they were still green compared to battle-hardened Athenians who had survived two decades of war. Tissaphernes and his officers were accustomed to suppressing land-based internal revolts, not defending against seaborne attack. Many of his men were soldier-settlers, so Tissaphernes had to be careful about ordering mass call-ups that would take them away from farming and repeated false alarms that

---

12   Tuplin (1987a); Dandamaev and Lukonin (1989) 229–232; Briant (2002) 599; Miller (2011) 334–337; and Dusinberre (2013) 85–93. Given his position, Tissaphernes probably had operational command of all forces in his realm, including royal troops. Without some authority over these troops, even if only for local defense, Tissaphernes' task would have been much more difficult. Or did Tissaphernes have to get the King's permission to use royal troops on specific missions? See Briant (2002) 66–67, and Waters (2010) 821.

13   Imperial communications: Lewis (1977) 56–57; Briant (2002) 364–378; and Briant (2012). On fire signals see Russell (1999), 145–149, and Colburn (2013).

14   Revolts of Ephesus and Miletus: Hornblower (2008) 799–800, 805.

15   Topography of Classical Ephesus: Hansen and Nielsen (2004) 1070–1073; Kerschner et al. (2008). Strategic position: Buckler (2003) 45.

16   Miletus' strategic location is difficult to appreciate today because silt from the Maeander River has filled in most of the Latmian Gulf, leaving the city landlocked; on the ancient coastline see Brückner et al. (2006).

17   On Milesian territory and settlements see Greaves (2002) 1–7, and Hansen and Nielsen (2004) 1082.

would make them progressively more lax to respond. And even in a genuine emergency, troops from dispersed rural settlements took time to assemble and dispatch. As for the Greeks, his Spartan allies would help with troops and ships, but he could not rely wholeheartedly on them, especially with the rival satrap Pharnabazus also seeking their attention. And, while the treaty between Sparta and Persia had formally returned the coast to the King, Tissaphernes had difficulty exerting effective authority over the Greek cities.[18] Having cast off the Athenian yoke, most coastal Greeks had little inclination to obey Persian orders. Moreover, the revolt of Pissuthnes had been disruptive and Pissuthnes' son Amorges remained at large, with a stronghold at Iasos in Caria.

The most significant challenge Tissaphernes faced was the geography of western Anatolia, which was unfavorable for a land-based defender facing a seaborne attacker. Tissaphernes had more than four hundred kilometers of coastline to watch. Much of this littoral was segmented into small plains, such as at Pygela and Teos, ringed by low coastal ranges that hindered movement to and from the interior. Further inland, rugged mountains separated the east-west valleys of the Hermus, Cayster, and Maeander rivers, making it difficult to transfer troops between north and south.[19] Marching along the valleys was easier, although the distances remained a challenge: from Sardis west to the sea it was more than eighty kilometers. And Tissaphernes had not only to guard urban centers, but also to protect the rural estates that provided essential revenue. Meanwhile, the Athenians could sail quickly and easily to any part of the coast they chose or, if needed, take refuge on nearby islands, especially Samos. The Athenians also held some walled strongholds on the mainland, amongst them Notion near Ephesus, and Pteleon and Sidousa on the peninsula of Erythrai (Thuc. 3.34, 8.24.2).

Under these circumstances, Tissaphernes could not successfully defend everywhere. He therefore avoided committing resources to guarding vulnerable places such as Teos. Though it was prosperous and occupied a defensible headland with good harbors, Teos was difficult for Persian forces to reach from the interior but easy for the Athenians to reach by sea. Furthermore, the Teian population was of particularly ambivalent loyalties. After dispatching his hyparch Stages to help pull down the walls of Teos, and then making a personal inspection tour, Tissaphernes completed the demolition work and left Teos an

---

18   On the terms of the treaty see Thuc. 8.5.4–5, 8.18 with Meiggs (1972) 445, and Hornblower (2008) 764–771. On Thucydides' portrayal of Tissaphernes, see Hyland (2007), and Munson (2012).

19   Buckler (2003) 42 considers that the geography favored the Persians, without accounting for the difficulty of transferring troops north-south between river valleys.

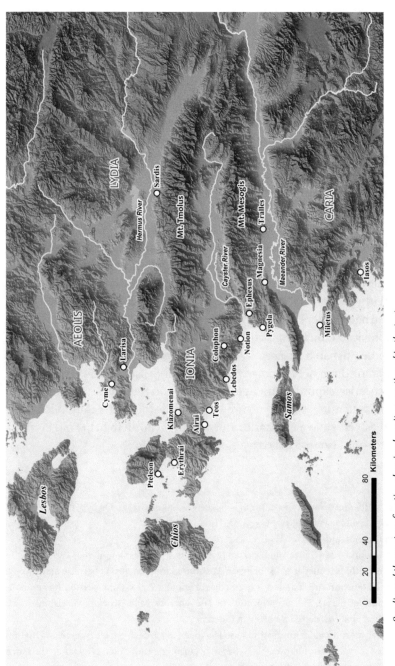

Sardis and the western frontier, showing key sites mentioned in the text
MAP BY YINGJIE HU, UC SANTA BARBARA DEPARTMENT OF GEOGRAPHY, USING DATA FROM THE ANCIENT WORLD MAPPING CENTER (HTTP://AWMC.UNC.EDU/WORDPRESS/)

open city (Thuc. 8.16, 8.20).[20] He also seems to have left alone nearby Lebedos and Airai, which were too small and isolated to be of much consequence.[21] Elsewhere, Tissaphernes apparently tried to solidify the power of anti-Athenian, and hopefully pro-Persian, factions without actually dispatching troops. For example, he sent another hyparch, Tamos, to Klazomenai to join an unsuccessful Spartan attempt to compel the pro-Athenian faction there to relocate to Daphnous, which lay inland in Klazomenian territory.[22]

Meanwhile, Tissaphernes attempted to consolidate control over more defensible regions. He may have set up additional outposts in the countryside, and he certainly placed garrisons in key coastal cities including Miletus, Knidos, and Antandros (Thuc. 8.35, 8.84, 8.108–109).[23] Miletus was an obvious choice given its strategic location. Knidos had good harbors and commanded the south Aegean Sea lanes.[24] Antandros was a shipbuilding center, with ready access to quality timber from Mount Ida.[25] In addition to protecting against Athenian attack, Tissaphernes may have hoped that these garrisons would forestall internal revolts. The presence of imperial troops, though, only antagonized the locals. Within a year, all three cities had ejected their garrisons. Antandros, which had been under Persian control for much of the fifth century, might have been more accommodating had it not been for ill treatment at the hands of Tissaphernes' hyparch Arsaces. Tissaphernes had more luck at Iasos, the last stronghold of the rebel Amorges. With Spartan assistance the Persians took Iasos by surprise, capturing Amorges himself (Thuc. 8.28–29). Tissaphernes fortified Iasos and established a garrison there, while the Spartans enrolled some of Amorges' erstwhile mercenaries and sent them north to help hold Erythrai. Even so, by 405 Iasos too may have ejected its garrison.[26]

---

20 For a different view on the status of Teos see Westlake (1979) 12. On hyparchs, see Tuplin (1987b) 120–121; Petit (1990) 152–154; and Briant (2002) 748.

21 On Lebedos and Airai see Hansen and Nielsen (2004).

22 On Daphnous see Hansen and Nielsen (2004) 1077, and Hornblower (2008) 840.

23 Greek authors tended to think of urban garrisons as most important because in the Greek city-state world the standard way to control another city was to garrison its acropolis (Tuplin 1987: 234). The exact location of Tissaphernes' Milesian *phrourion* is unknown, though it may perhaps have been on Kaletepe.

24 The location of Knidos during this period has been much debated: see Hansen and Nielsen (2004) 1123–1125; Hornblower (2008) 847–851. Ongoing excavations at Burgaz on the Datça peninsula may help clarify the question.

25 On Antandros see Xen. *Hell.* 1.1.26, 1.3.17, 2.1.10; Hanson and Nielsen (2004) 1004; Krentz (1989b) 101; and Tuplin (1987a) 210.

26 Diod. Sic. 13.104.7 says the Spartan Lysander sacked Iasos in 405 BC; if Diodorus is correct,

TISSAPHERNES AND THE ACHAEMENID DEFENSE OF WESTERN ANATOLIA   269

Tissaphernes narrowly passed his first major battlefield test in 411 when the Athenians tried to retake Miletus. The satrap himself joined the defense, leading cavalry and mercenaries. The veteran Athenians were too much for Tissaphernes' inexperienced troops, but the Milesians routed Athens' overconfident allies, and the Athenian commander withdrew on receiving news of approaching Spartan reinforcements (Thuc. 8.17, 8.25–27). Athens would never again control Miletus. After 411, as fighting shifted to the Hellespont, Tissaphernes had time for further defensive preparations. He probably built or restored walls at Ephesus, Miletus, and elsewhere, either by providing imperial funds and workers or encouraging local efforts.[27] Oddly, however, Tissaphernes did not fortify Magnesia-on-Maeander, which controlled the southern approaches to Ephesus from the Maeander. He may have deemed it unnecessary to fortify Magnesia while he already held Ephesus and Miletus.

The Defense against Thrasyllus, 409

The next test came in 409, when an Athenian expedition under Thrasyllus sailed for Ionia. The Athenians struck first at the town of Pygela, which lies on the coast southwest of Ephesus on the outskirts of modern Kuşadası. On an early summer day, the Pygelans spotted Thrasyllus' triremes approaching from the direction of Samos. The Athenian ships carried armed sailors along with peltasts, a thousand hoplites, and about a hundred cavalry (Xen. *Hell.* 1.2.1–3).[28] Disembarking his men on the coastal plain of Pygela, Thrasyllus set them to ravaging the countryside and ordered an assault on the city wall.

Thrasyllus had not chosen his target at random. Pygela had a protected harbor and overlooked the coastal road from Ephesus to the Panionion sanctuary on the Mycale peninsula. It also had access to several routes leading inland to the Maeander valley. The town offered views west back to the Athenian base at Samos as well as north across the gulf to Athenian-held Notion, but could not be directly observed from Ephesus. Reinforcements coming by land would

---

the city must have ejected Tissaphernes' garrison and allied with Athens. Xen. *Hell.* 2.1.15 describes Lysander campaigning further south in Caria, where he sacked Kedreai on the Gulf of Ceramus.

27   Miletus: Gorman (2001) 241. Ephesus: McKechnie and Kern (1998) 117, and Scherrer (2001) 61.

28   On Thrasyllus' campaign see McCoy (1977) 281–282; Kagan (1987) 270–273; and Lazenby (2004) 208–209. For convenience, I follow the low chronology for Xenophon's *Hellenika* (Thomas 2009).

have to wind their way through the hills ringing the coastal plain around Pygela, while the Athenians could easily withdraw by ship to Samos if necessary.[29] Taking Pygela would give the Athenians a useful springboard for future operations on the mainland.

Thrasyllus probably hoped to catch Pygela by surprise, as Tissaphernes had done to Iasos a few years before, but the townspeople seems to have been on their toes. The town walls were in good repair, and the Pygelans managed to shut their gates and repel the Athenian assault.[30] They also must have had the means to summon help, because after some time hoplite reinforcements arrived. Ephesus, just fifteen kilometers to the northeast, would at first glance seem the logical source of aid, yet according to Xenophon (Xen. *Hell.* 1.2.1–2) the troops came from Miletus, nearly forty kilometers away as the crow flies. Thanks to the silting of the Maeander valley, today it is possible to drive much of that distance straight across what was once the Latmian Gulf. In 409, however, the Gulf extended much farther inland.[31] To reach Pygela from Miletus, troops would have had to travel in an arc around the head of the Gulf, perhaps as much as eighty kilometers or a march of two to three days. Possibly Xenophon had incomplete information and the reinforcements came not literally from Miletus itself but from Milesian settlements or outposts in the lower Maeander valley, less than twenty-five kilometers away, or on the Mycale peninsula. Hoplites in battle order could have covered that distance in under a day.[32]

Wherever in Milesian territory they came from, why did Miletus rather than Ephesus send these reinforcements? The answer may be that Thrasyllus had hit a weak spot in Tissaphernes' defenses, where local politics trumped military practicalities. Pygela was geographically closer to Ephesus, but it had developed stronger political ties to Miletus, probably to counterbalance territorial threats from Ephesus and Samos. Indeed, inscriptions suggest that during the Classical period the Pygelans and Milesians had limited mutual citizenship, extended

---

29   Topography of Pygela and coastal plain: Keil (1908) 137–138; Strabo 14.1.20; and Livy 37.11. Access to the site these days is through the gate of the Havana Disco, unfortunately closed at the time of my 2008 visit.

30   Mellink (1976) 280. Excavations at Pygela in the 1970s cleared some of the city wall, though a brief report mentions only fourth century sherds in association with the wall.

31   For a reconstruction of the ancient coastline see Brückner et al. (2006).

32   Another possibility is that the Milesians ferried troops across the Latmian Gulf to the south side of the Mycale peninsula—a risky proposition with Athenian warships in the area—then marched from there to Pygela.

TISSAPHERNES AND THE ACHAEMENID DEFENSE OF WESTERN ANATOLIA   271

during the late fourth century into a formal *isopoliteia*.[33] In such circumstances, the Pygelans would be more likely to summon help from Miletus even if it would take longer to arrive. For his part, if Thrasyllus knew about these ties he may have hoped that by attacking Pygela he might lure the Milesians or even Tissaphernes out to fight on terrain of his own choosing.

In any case, the arriving Milesians fell upon some Athenians who had dispersed in search of loot and supplies, but then Thrasyllus' hoplites and peltasts counterattacked. Xenophon says the Athenians wiped out most of the Milesians, taking nearly two hundred shields (Xen. *Hell.* 1.2.3). Given that in 411 they had fielded just eight hundred hoplites to defend their home city against Athenian attack (Thuc. 8.25), the Milesians apparently sent a large proportion of their available force to Pygela. It is unlikely, though, that the Milesians sent every hoplite they had. They must have recognized that if they marched in full strength to reinforce Pygela, the Athenians could potentially re-embark, sail around Mycale, and fall on Miletus before the Milesian troops could get back to defend it.

Significantly, Persian reinforcements, which Xenophon surely would have mentioned had they made an appearance, did not show up at Pygela. Such forces could well have been available at nearby Magnesia-on-Maeander, a major Persian base which Tissaphernes himself sometimes used as a headquarters.[34] Several possibilities are worth considering. If the local Persian commander at Magnesia had the authority to send troops as he wished, they could have reached Pygela within a day, so the absence of imperial troops may mean the local commander had to consult Tissaphernes first. At top speed, a relay of mounted messengers could have required twenty-four hours or more for the roughly 300 kilometer round trip to and from Sardis.[35] By the time the orders came, the Athenians might already have departed. Then too, Thucydides reports that Tissaphernes was angry with the Milesians for ejecting his garrison in 412/411 and had sent an emissary to express his displeasure (Thuc. 8.85.2). If Miletus was unwilling to be integrated into his defensive network by accepting a garrison, Tissaphernes may have decided in advance to let the Milesians take care of their own. Even if he had not made such a decision and even if the Athenians stayed at Pygela long enough for Persian troops to get there, Tissaphernes no less than the Milesians knew the Athenians had the advantage of seaborne

---

33   Hansen and Nielsen (2004) 1094. Samos' dependent Anaia lay just south of Pygela, and by the early Hellenistic period Pygela belonged to Ephesus.

34   On Magnesia see Thuc. 1.138, 8.50.3; Strabo 14.1.39; and Hansen and Nielsen (2004) 1081–1082.

35   On the speed of the Persian fast messenger service see Minetti (2003).

272   LEE

mobility. If he rushed to Pygela before being sure of Thrasyllus' true objective, the Athenians could easily slip away and hit another vulnerable target. As long as local defenses held, Tissaphernes could conserve his own troops for later.

Thwarted at Pygela, the Athenians the next day sailed north across the gulf to Notion, the port of Colophon. Notion had briefly been under Achaemenid control in the early years of the Peloponnesian War, but since 427 a mixture of Athenian colonists and pro-Athenian Colophonians had held it. Notion offered a gently shelving beach for triremes, ample water, and a fortified acropolis, but unlike Pygela it was in direct view of Ephesus. Notion was difficult to approach by land along the coast from Ephesus, but the Hales river valley provided a natural route inland. About fifteen kilometers north up the Hales valley lay Colophon, still in the hands of a pro-Persian faction.[36]

From Notion, Thrasyllus led his men inland to Colophon, which came over without a fight.[37] Colophon was strategically vital, for it controlled access between the Ionian coast and the fertile inland plain south of modern Izmir. Yet, the apparent ease with which the Athenians took Colophon suggests there were no imperial troops in the immediate area. Given the history of civil strife between Colophonian pro- and anti-Persian factions, Tissaphernes may have—mistakenly, as it turned out—believed he could rely on the local defenders here as he had at Pygela.

After taking Colophon the Athenians launched a raid against the estates and villages of the inland plain—Xenophon calls it Lydia—perhaps striking north towards modern Cumaovası or east towards modern Torbalı.[38] It was harvest time in this fertile agricultural area and the Athenians took a rich haul of prisoners and booty. Xenophon says nothing about any local defenses, but it is noteworthy that Thrasyllus began the raid at night, suggesting that he was concerned about advancing into the open plain of Lydia where the Persians could best use their superior cavalry.

Thrasyllus' worries came true: the next day, Tissaphernes' hyparch Stages caught the Athenians in disorder while they were scattered in search of plun-

---

36   Notion and Colophon: Xen. *Hell.* 1.2.3–6; Thuc. 3.34; Hansen and Nielsen (2004) 1077–1080; and Sokolicek (2009) 109–110. The walls visible today at Notion are Hellenistic, but the site must have been fortified in 427 BC for a *diateichisma* (Thuc. 3.34.2) to have been useful, and it is unlikely the Athenian settlers demolished any fortifications.

37   On Colophon see Bruns-Özgan et al. (2011), who tentatively dates the visible city walls there to the fourth or early third century.

38   On the plains north and east of Colophon see Bean (1966) 151, and Aslaksen (2007) 34–35. Aslaksen investigated several mounds and worked stone blocks in this area, but was unable to provide secure indications of date.

TISSAPHERNES AND THE ACHAEMENID DEFENSE OF WESTERN ANATOLIA   273

der.[39] Stages had made a similarly sudden appearance at nearby Teos a few years before (Thuc. 8.16.3), perhaps an indication that Tissaphernes had placed him in charge of this portion of the coast.[40] Stages may have been on a roving cavalry patrol, touring his territory in emulation of the itinerant Achaemenid royal court. Or he could have been responding to reports of Thrasyllus' raid or to the sight of burning villages.[41] Whatever brought him to the scene, Stages and his Persians beat the Athenians in a cavalry skirmish, prompting Thrasyllus to retreat to Notion.

Having probed the Persian defenses at Pygela and Colophon, Thrasyllus at last decided to assault Ephesus.[42] Probably around this time Tissaphernes himself arrived at Ephesus, in order to better monitor events. While Thrasyllus inexplicably delayed more than two weeks at Notion, Tissaphernes got wind of the Athenian plan. How this intelligence reached him is unknown, but the satrap found it reliable enough that he decided to concentrate his forces at Ephesus. As he mustered his army, Tissaphernes reportedly sent horsemen "telling everyone to bring help to Artemis at Ephesus" (Xen. *Hell.* 1.2.7). This phrase could mean Tissaphernes appealed to the general population or that he summoned all available troops regardless of status or type.[43] The latter may have included contingents from nearby Magnesia-on-Maeander and as far away as the upper Cayster valley some seventy kilometers distant.[44] Whatever the precise nature of his call-up, Tissaphernes' action suggests that appealing to local religious feeling was a more effective way to foster allegiance than installing garrisons.[45]

Tissaphernes did not undertake a mass call-up lightly, especially since it was harvest time and men were needed in the fields. With the Athenian goal now clear, though, the time for local defense was over. The reinforcements made the

---

39   Xenophon does not give precise time indications, but does mention a camp (*Hell.* 1.2.6), suggesting the raid continued beyond a single night.

40   On Stages' position see Hornblower (2008) 798.

41   For cavalry patrols compare Xen. *Hell.* 3.4.13–14; Xen. *An.* 4.4.4; and Tuplin (1987a) 197. Itinerant court: Briant (2002) 187–189. Sub-court emulation: Miller (2011) 336. Mounted patrols would also help Persian commanders display their authority and military power, as Dusinberre (2013) 93–94 points out.

42   Battle of Ephesus: Xen. *Hell.* 1.2.7–10; Diod. Sic. 13.64.1; and *Hell. Oxy.* Cairo Fr. 1–3; Krentz (1995) 112–114; McKechnie and Kern (1998) 116–121.

43   Tuplin (1987a) 196–197.

44   Upper Cayster (Kilbian plain): McKechnie and Kern (1998) 119, and Roosevelt (2009) 43–44. Magnesians: Cuniberti (2009). A small number of Spartans may also have been present: McKechnie and Kern (1998) 117.

45   On Tissaphernes and Artemis, see Hornblower (2008) 1052–1053.

274   LEE

difference, and the Athenian assault failed. By correctly judging the situation, and waiting for the opportune moment to summon his forces, Tissaphernes had beaten Thrasyllus. Yet the satrap did not have long to enjoy this victory, for soon he would lose power to the teenaged prince Cyrus.

Cyrus the Younger, 407–401

Cyrus the Younger's arrival in spring 407 led to new developments in the defense of western Anatolia. Like Tissaphernes before him, Cyrus was appointed *karanos* or regional overlord, but Cyrus got along better with the locals than Tissaphernes did. Indeed, of the Greek coastal cities only Miletus preferred Tissaphernes, although the satrap did manage to maintain his support in Caria. Notably, Cyrus succeeded in installing city garrisons where Tissaphernes had failed, perhaps by making greater use of mainland Greek mercenaries. He was able to recruit these troops in large numbers thanks to his connections in mainland Greece. Such mercenaries may have been Greek enough not to not seem like foreign occupiers to the Greek coastal cities, but still outsiders enough that Cyrus did not have to worry about their loyalties.[46]

The city garrisons were just part of Cyrus' military buildup, which despite his loss of some of his territories only intensified after his brother Artaxerxes II came to the throne in 405/4. By now Athens had surrendered and troops were not needed to guard the coasts, but Cyrus found ways to camouflage his preparations against Artaxerxes. He deployed his troops to punish criminals and brigands, so that people could travel without fear in his territory, and launched punitive expeditions against semi-autonomous mountain peoples such as the Mysians and Pisidians (Xen. *An.* 1.9.13–14, 2.5.13, 3.2.23). Tissaphernes, meanwhile, built up his defenses in Caria, where he had acquired substantial estates.[47]

When Cyrus set out to usurp the throne from his brother in early 401, he stripped his domain of some twelve thousand mercenaries, leaving behind only the bare minimum needed to hold the acropolis of each garrisoned city (Xen. *An.* 1.2.1). He also collected local Greek and Anatolian troops, perhaps some twenty thousand altogether. Many of these local troops may have returned back west following Cyrus' death at Cunaxa in late 401 BC, along with a few

---

46   *Karanos*: Xen. *An.* 1.9.7, *Hell.* 1.4.1–4. Support for Tissaphernes: Xen. *An.* 1.1.6, 1.9.9; Xen. *Hell.* 3.1.3.
47   Ruzicka (1985) 208.

mercenary deserters. More than ten thousand of Cyrus' mercenaries, however, never returned to Achaemenid service, instead embarking on the arduous retreat across Anatolia famously chronicled in Xenophon's *Anabasis*. The five thousand or so Cyreans who survived this trek ended up back west by early 399, only to take up arms there for the Spartans rather than the Persians. The loss of this great number of troops would have a lasting effect on Achaemenid defensive capabilities in western Anatolia.

Defending against the Spartans, 400–395

Following Cyrus' death, Artaxerxes returned Tissaphernes to command along the coast (Xen. *Hell.* 3.1.3).[48] Tissaphernes now needed to rebuild the defenses Cyrus had denuded. True, he had made progress in fortifying Caria during the years Cyrus was in power, and Carian troops had become a distinctive element of his forces.[49] But even with sufficient funds it would take time to recruit more mercenaries and without Cyrus' overseas connections Tissaphernes never managed to muster anywhere near as many of them as Cyrus had. More insidiously, despite formal shows of reconciliation the aftermath of the War of the Brothers found former partisans of Cyrus living side-by-side in western Anatolia with those who had stayed loyal to Artaxerxes. Tissaphernes' shortage of troops only made the Greek cities of Ionia more willing to refuse imperial control. As the Ionians asked for and received Spartan help, Tissaphernes now had to defend against Spartan invasions of a frontier that was still coping with the legacy of civil war.

By 400, Tissaphernes had given up control of Ephesus to the Spartans and failed to retake the important port of Cyme in Aeolis (Diod. Sic. 14.35.7).[50] Worse, a Spartan force under Thibron landed at Ephesus and then campaigned into the Maeander valley, getting as far east as Tralleis. Thibron succeeded in capturing Magnesia-on-Maeander, which still had no walls. He relocated the city to a more defensible location on Mount Thorax, overlooking the direct route between Ephesus and the Maeander valley (Diod. 14.36.2–3, Strabo. 14.1.39).[51] Possession of Magnesia enabled the Spartans to shift troops unhin-

---

48 On this period see Westlake (1986); Briant (2002) 634–645; and Buckler (2003) 39–70.
49 These Carian troops were recognizable by their white shields: Xen. *Hell.* 3.2.15.
50 On the defenses of Cyme see Frederiksen (2011) 156–158.
51 Bean (1966) 207–208 identifies Gümüş Dağı as the site of Thibron's relocated city, but compare Bingöl (2007). Thibron later strengthened this position by taking the (still un-

dered between the Cayster and Maeander valleys. On the other hand, without Magnesia it was much more difficult for the Persians to observe Spartan movements out of Ephesus. Tissaphernes must have understood the strategic importance of Magnesia, yet only after Thibron had relocated the settlement and pillaged the plain did Tissaphernes finally appear with a cavalry force. This belated response suggests his local defenses were not as strong or swift as they had been a decade before against the Athenians.

There are other indications of weakness in Tissaphernes' defenses. Early in 399, while Thibron was still around Ephesus, the roughly five thousand surviving Cyrean mercenaries reached Pergamon in Mysia. There they met the dynast Hellas, matriarch of an Eretrian family whose ancestor had sided with Darius I during the Persian invasion of Greece almost a century before. Hellas suggested the Cyreans attack her neighbor, the Persian Asidates, tempting them with visions of rich booty at his estate.[52] Hoping to make up for their disappointing winter in Thrace, Xenophon and nine hundred men took up the invitation. They set off at night—as had Thrasyllus on his Lydian raid a decade before—but found Asidates' tower too strong to capture. Through the night Asidates and his men called for aid by shouting and lighting fires. As day came, more than a thousand cavalry, peltasts, and hoplites, including royal mercenaries and military colonists, arrived from around the neighborhood to help Asidates. Soon the Cyreans were under heavy pressure.

This episode has been taken as evidence that the Persians in the west possessed an efficient local defense network, populated with military colonists and garrisons, bristling with small forts, and capable of summoning substantial reinforcements in under a day's time, possibly through a formal signaling system.[53] Yet a closer look reveals otherwise, for not everyone in the neighborhood came to succor Asidates. Indeed, despite his mother's protests, Hellas' own son Gongylus actually led troops to assist Xenophon's raiders. So too did Procles, a descendant of the exiled Spartan king Demaratus, whose family had received land from Xerxes. Only with Procles' and Gongylus' assistance were the Cyreans able to escape.

This picture of neighbors inciting raids on neighbors and then coming to blows with each other more resembles low-level internecine war than efficient

---

located) fortress of Ionda (or Isinda) along with Mount Solmissos south of Ephesus (Diod. 14.99). Hansen and Nielsen (2004) 1076.

52 Attack on Asidates: Xen. *An.* 7.8.8–24. On the sites Xenophon mentions see Hansen and Nielsen (2004) 1041–1050.

53 Briant (2002) 376, 643; Kuhrt (2007) 824–825; and Dusinberre (2013) 91, but note the reservations of Tuplin (1987a) 176, 213–214, 221–222, and Sekunda (1985) 11.

TISSAPHERNES AND THE ACHAEMENID DEFENSE OF WESTERN ANATOLIA   277

local defense. Procles had taken up arms for Cyrus in 401 (Xen. *An.* 2.1.3), and though he seems to have formally reconciled with Artaxerxes, his former loyalties apparently re-emerged at this moment, as he came to help the Cyrean mercenaries against Persian royal troops.[54] Hellas is even more intriguing. She urged the Cyreans to attack her neighbor, even providing guides, yet tried to restrain her son from bringing aid when the Cyreans got in trouble. Her behavior has the look of someone trying to hurt a rival while keeping her own involvement hidden. Viewed in its full context, the Asidates episode suggests not a smoothly functioning, unified defensive network, but rather an imperial borderlands that two years after the death of Cyrus remained deeply divided between those who had favored Cyrus and those who remained faithful to the King.[55] Indeed, when Thibron arrived, Procles, Gorgion, and Gongylus all came over to the Spartan cause, at least as long as Thibron was in the area (Xen. *Hell.* 3.1.6).

To the Asidates episode we can add another fascinating bit of evidence. Polyaenus recounts the activities of a certain Alexander, who commanded the strongholds of Aeolis with a garrison force of both Greeks and "barbarians" (Polyaenus, *Strat.* 6.10). This passage has been used to suggest the presence of a "dense network of small forts" in Aeolis during the early fourth century, a conclusion that may gain support from recent archaeological survey work identifying a network of towers in the region.[56] Again, though, a closer look is revealing. According to Polyaenus, Alexander lured spectators to a public concert by four famous performers.[57] He then surrounded the audience with his troops and extorted ransoms for their release, before handing over his strongholds to Thibron and departing. Just as with the Asidates episode, viewed in full this tale can hardly be taken as unqualified evidence for solid imperial defenses in Aeolis. Rather, Alexander's actions suggest weakness and disarray. Indeed, Thibron successfully captured a number of poorly defended sites in the area by assault, another indication of disorganization. It was fortunate for Tissaphernes that the garrison of Larisa, inland from Cyme, mounted an energetic defense, leading the Spartan authorities to order Thibron out of Aeolis (Xen. *Hell.* 3.1.7).

---

54   Briant (2002) 631.

55   Westlake (1986) 411 calls it a "medley of loyalties."

56   Briant (2002) 643; compare Tuplin (1987a) 213. Survey: Gezgin (2001).

57   The four performers are all elsewhere attested as active in the time of Thibron: Callippides (Plut. *Alc.* 32.2, Xen. *Symp.* 3.11), Nicostratus (*OCD*³ 1044), Philoxenus of Cythera (*Marm. Par.* 69), and Thersander (Xen. *Hell.* 4.8.18–19, in textually challenged company with no less than Thibron himself).

278   LEE

Thibron's successor Dercylidas at first campaigned against Pharnabazus, seizing Achaemenid strongpoints in Aeolis and the Troad (Xen. *Hell.* 3.2.11). Internecine strife in the Troad, where a jealous son-in-law had just assassinated Pharnabazus' hyparch Mania, played a major role in Dercylidas' success (Xen. *Hell.* 3.1.14–16). Tissaphernes meanwhile was content to let Dercylidas discomfit Pharnabazus. In 397, though, the Spartan government ordered Dercylidas to attack Tissaphernes' lands in Caria. Crossing into the Maeander valley, Dercylidas found himself confronting the combined forces of Tissaphernes and Pharnabazus, who together had been shoring up the defenses of Caria. Dercylidas only managed to bluff his way out because Tissaphernes and Pharnabazus could not agree whether to risk battle (Xen. *Hell.* 3.2.14–20).[58]

After Dercylidas came Agesilaus, who spent two ultimately fruitless years marching here and there against both Tissaphernes and Pharnabazus. The details of these campaigns have been extensively discussed, but several aspects of Tissaphernes' response to Agesilaus deserve additional attention for the light they shed on Persian defensive capabilities during this period.[59]

To begin with, recall that during the defense against the Athenians in 409, Tissaphernes had received an accurate report of Thrasyllus' plan to attack Ephesus, enabling him to concentrate his forces in the right place at the right time. In 396, things were different. Agesilaus gave out that he planned to march from Ephesus for Caria, even sending phony orders to cities en route to provide markets for his army (Xen. *Hell.* 3.4.11–15). Had Tissaphernes controlled Magnesia-on-Maeander, he could easily have deployed scouts or outposts from there to get early confirmation of the actual direction of Agesilaus' march. Without holding Magnesia, though, it was much more difficult to observe which way the Spartans went when they left Ephesus. Tissaphernes ended up buying the deception and concentrated his forces in the Maeander valley, but Agesilaus instead marched swiftly northward, taking several cities by surprise. Meanwhile, Tissaphernes could not easily make up for his mistake because Spartan-held Magnesia blocked the direct route north to Ephesus.[60] Tissa-

---

58   Westlake (1986) 422. In narrating Dercylidas' march, Xenophon at one point (*Hell.* 3.2.14–15) mentions Greek scouts climbing up rural towers (in the Maeander valley?) to look for enemy forces. If these towers were not solely agricultural, on which see Morris and Papadopoulos (2005), they may again suggest an under-manned defense network, since there seems to have been no one around to hold them against Dercylidas' scouts.

59   Agesilaus' campaigns: Cartledge (1987) 208–218; Briant (2002) 637–445; and Buckler (2003) 58–69.

60   Westlake (1981) 266 suggests Tissaphernes should have threatened Ephesus, but does

phernes does not seem to have pursued the Spartans and not until Agesilaus reached Daskyleion far to the north did Pharnabazus' Persian troops offer resistance.

In 395, Agesilaus tried a variant of the same stratagem, announcing that he was going to attack "the best part of the country by the shortest route possible" (Xen. *Hell.* 3.4.20–22).[61] Again he succeeded in misdirecting Tissaphernes, who deployed his forces in the Maeander valley while Agesilaus marched northeast for Sardis. What exactly transpired after that is notoriously difficult to untangle, but whether one prefers the account of Xenophon or of the Oxyrhynchus historian, Spartan control of Magnesia would have prevented Tissaphernes' troops in the Maeander valley from easily pursuing Agesilaus once they discovered his true march route.[62] Moreover, it is striking that Agesilaus was able to traverse the Cayster plain and cross over Mount Tmolus unimpeded, whether by the Karabel pass, by a route in the vicinity of modern Gölcük, or by some other path. Fortified passes were common elsewhere in the empire, yet there is no indication the routes over Tmolus were similarly defended, even though the King had sent Tissaphernes reinforcements.[63] Nor does Tissaphernes seem to have had a functioning network of lookouts on Mount Tmolus.[64] Tissaphernes may have been careless, or perhaps he did not have the manpower to maintain such posts, because he could no longer draw on the large force of Greek mercenaries that Cyrus had employed.

---

not account for the importance of Magnesia. In 391 the Spartans still held Magnesia (Leukophrys), nearby Achilleion and Priene (Xen. *Hell.* 4.8.17), and other strongholds south of Ephesus (Diod. Sic. 14.99).

61　On the campaign of 395 see Gray (1979); Wylie (1992); McKechnie and Kern (1998) 140–148.
62　To pursue Agesilaus, the next available option for Tissaphernes' troops would have been to cross the low but rugged range of Mount Mesogis (Aydın Dağları) farther inland from Magnesia. Several narrow modern roads wind north over Mount Mesogis from the vicinity of Tralles (modern Aydın) to the Cayster valley, but the ancient routes here have not yet been identified.
63　Fortified passes: Briant (2002) 374–376.
64　Lookouts: Kel Dağ for example, on which see Roosevelt (2009), commands a view of most of the Cayster valley, yet Tissaphernes does not seem to have received early warning of Agesilaus' marches in either 396 or 395. One possibility for 395 is that during the first day of his march from Ephesus, Agesilaus hugged the north edge of Mount Mesogis (following the modern road to Tire), so that his advance would be screened by the low hills flanking the left of this route.

Tissaphernes and the Western Frontier

Agesilaus soon departed western Anatolia for Greece, but defeat at Sardis nonetheless meant the end for Tissaphernes: Artaxerxes had him arrested and executed (Diod. 14.80.6–8, Xen. *Hell.* 3.4.24–26). Despite this inglorious end, Tissaphernes' activities over the course of two decades reveal much about the challenges that the Persian Empire faced in defending western Anatolia.

In his early years, as exemplified by his defense against Thrasyllus, Tissaphernes showed a keen grasp of the strategic realities on the western frontier. Tissaphernes understood that he could not be strong everywhere. He abandoned vulnerable areas, garrisoned important cities when he could, and relied on local forces wherever possible—successfully at Pygela, not so at Colophon. As the Milesians showed at Pygela and Stages showed again near Colophon, local defenders could also take advantage of the lax discipline and insufficient logistics of Greek armies by attacking dispersed plunderers. Only after he had clear evidence of the attacker's main goal did Tissaphernes commit himself, effectively using an appeal to Artemis to win local loyalties. The successful defense of the west, then, required more than numerical superiority or strong fortifications: it meant knowing the geography, understanding local politics, and gathering accurate intelligence.

Had Cyrus not come to western Anatolia, Tissaphernes might have done better against the Spartans when they came in the 390s. As it was, the prince's failed attempt on the Achaemenid throne deprived Tissaphernes of skilled mercenaries and left behind a divisive legacy that hampered local defenses in Mysia and Aeolis, and probably elsewhere. But, it was not all Cyrus' fault. Tissaphernes did not appreciate the strategic significance of Magnesia as a counter to Spartan forces based at Ephesus. In 396–395, he may have neglected to guard the Tmolus passes or to prepare an adequate signaling system. Most importantly, even allowing for the bias in Xenophon's presentation of events, Tissaphernes failed to obtain accurate intelligence about Agesilaus' plans. This failure contrasts strikingly with Tissaphernes' earlier success in anticipating Thrasyllus' attack on Ephesus.

The Persian leaders who came after Tissaphernes showed they had learned from his successes. In 391 the satrap Strouthas (or Strouses) managed to defeat a Spartan force in the Maeander plain by waiting until it dispersed for plunder and then launching a surprise cavalry attack. Amongst the dead was the Spartan commander Thibron (Xen. *Hell.* 4.8.17–19). By then Sparta's attention had shifted to retaining power in mainland Greece, and soon after came the King's Peace of 386. The Persians were able to re-assert undisputed control over their western frontier, and Anatolia for a time enjoyed relative calm. It would be fifty years before another generation of Persian commanders had to face a new external threat to the west, in the person of Alexander of Macedon.

Bibliography

Adcock, F. (1956) *Caesar as a Man of Letters*. Cambridge.

Ager, S.L. (1996) *Interstate Arbitration in the Greek World: 337–90 B.C.* Berkeley.

Ait Amara, O. (2007) *Recherche sur les Numides et les Maures face à la guerre, depuis les guerrires puniques jusqu'à l'époque de Juba Ier*, diss. Université Jean Moulin Lyon 3.

Alcock, S., Bodel, J., and Talbert, R. (eds) (2012) *Highways, Byways, and Road Systems in the Pre-Modern World*. Chichester.

Alcock, S., D'Altroy, T., Morrison, K., and Sinopoli, C. (eds) (2001) *Empires: Perspectives from Archaeology and History*. Cambridge.

Alexander, J. (2001) "Defending Marye's Heights," in Cowley (2001) 160–176.

Alexander, M.C. (1990) *Trials in the Late Roman Republic: 149 B.C. to 50 B.C.* Toronto.

Alfody, G., Dobson, B., and Eck, W. (eds) (2000) *Kaiser Heer und Geselschaft in de Romischen Kaiserzeit*. Stuttgart.

Amit, M. (1975) "The disintegration of the Athenian Empire in Asia Minor (412–405 B.C.E.)," *Scripta Classica Israelica* 2: 38–72.

Anders, A. (2011) *Roman Light Infantry and The Art of Combat: The Nature and Experience of Skirmishing and Non-Pitched Battle in Roman Warfare 264 BC–AD 235*, diss. Cardiff University.

Ando, C. (2000) *Imperial Ideology and Provincial Loyalty in the Roman Empire*. Berkeley.

Armstrong, J. (2013) "Claiming victory: Triumphs in early Rome" in Spalinger and Armstrong (2013) 7–21.

——— (forthcoming) *War and Society in Early Rome: From Warlords to Generals*. Cambridge.

Arnheim, R. (1957) *Film as Art*. Berkeley.

Arnold, D., and Settgast, J. (1965) "Erster Vorbericht über die vom Deutschen Archäologischen Institut Kairo im Asasif unternommenen Arbeiten," *Mitteilungen des Deutschen Archäologischen Instituts, Abteilung Kairo* 20: 47–61.

Arnold, W.T. (1914) *The Roman System of Provincial Administration to the Accession of Constantine the Great*. 3rd edn. Oxford.

Aslaksen, O. (2007) *Paths and Places: The Landscape Identity of Colophon, 1300–302 BC*. MA Thesis, University of Oslo.

Asprey, R.B. (1975) *War in the Shadows: The Guerrilla in History, Vol. 1*. New York.

Assmann, J. (1975) *Zeit und Ewigkeit im Alten Ägypten: ein Beitrag zur Geschichte der literarischen Komminikation*. Heidelberg.

——— (2011) *Steinzeit und Sternzeit. Altägyptische Zeitkonzepte*. Munich.

Auberbach, E. (1953) *Mimesis: The Representation of Reality in Western Literature*. W. Trask (trans.) Princeton.

Austin, N.J.E. (1979) *Ammianus on Warfare. An Investigation into Ammianus' Military Knowledge*. Brussels.

Awerbruch, M. (1981) "Imperium: Zum Bedeutungswandel des Wortes im staatsrecht-lichen und politischen Bewußtsein der Römer," *Archiv für Begriffsgeschichte* 25: 162–184.

Badian, E. (1954) *"Lex Acilia Repetundarum," American Journal of Philology*, 75: 374–384.

——— (1958) *Foreign Clientelae (264–70 B.C.)*. Oxford.

——— (1989) "The *scribae* of the Roman Republic," *Klio* 71: 582–603.

——— (1993) "The Legend of the Legate who Lost his Luggage," *Historia* 42: 203–210.

Bahrani, Z. (2008) *Rituals of War: The Body and Violence in Mesopotamia*. New York.

Bakır, T., Sancisi-Weedenburg, H., Gürtekin, G., Briant, P., and Henkelman, W. (eds) (2001) *Achaemenid Anatolia*. Leiden.

Balcer, J. (1984) *Sparda by the Bitter Sea: Imperial Interaction in Western Anatolia*. Chico.

——— (1985) "Fifth century B.C. Ionia: A frontier redefined," *Revue des études anciennes* 87.1–2: 31–42.

Barnes, T.D. (1982) *The New Empire of Diocletian and Constantine*. Cambridge, Mass.

Baumgarten, J. (1992) "Urbanization in the Late Bronze Age," in Kempinski and Reich (1992) 143–150.

Bean, G. (1966) *Aegean Turkey*. New York.

Beard, M., North, J., and Price, S.R.F. (1998) *Religions of Rome*. Cambridge.

Bechtold, E., Gulyás, A., and Hasznos, A. (eds) (2011) *From Illahun to Djeme: Papers Presented in Honour of Ulrich Luft*. *BAR International Series 2311*. Oxford.

Beck, H. (2012) "Consular Power and the Roman Constitution" in Beck, Dupla, Jehne, and Pina Polo (2012) 77–96.

Beck, H., Dupla, A., Jehne, M., and Pina Polo, F. (eds) (2012) *Consuls and the Res Public: Hold High Office in the Roman Republic*. Cambridge.

Beck, H., and Serrati, J. (eds) (2016) *Money and Power in the Roman Republic*. Brussels.

Beden, H. (2005) "Une ville inconnue en Ionie," *Numismatica e antichità classiche* 34: 107–117.

Bekker-Nielsen, T., and Hannestad, L. (eds) (2001) *War as a Cultural and Social Force*. Copenhagen.

Bell, M. (1988) "Excavations at Morgantina, 1980–1985: Preliminary Report XII" *American Journal of Archaeology* 92: 313–342.

——— (2007a) "An archaeologist's perspective on the *lex Hieronica*," in Dubouloz and Pittia (2007) 187–204.

——— (2007b) "Apronius in the agora: Sicilian Civil Architecture and the *lex Hieronica*," in Prag (2007) 117–134.

Beloch, K. (1886) *Die Bevölkerung der griechisch-roemischen Welt*. Leipzig.

Benveniste, E. (1966a) "La nature des pronouns," in Benveniste (1966) 251–257.

——— (ed.) (1966b) *Problèmes de linguistique générale*. Paris.

Berlev, O. (1967) "The Egyptian Navy in the Middle Kingdom," *Palestinskij Sbornik* 80: 6–20.

BIBLIOGRAPHY

Bettini M. (2010) *The Ears of Hermes: Communication, Images, and Identity in the Classical World*. W.M. Short (trans.) Columbus.

Beylage, P. (2002) *Aufbau der königlichen Stelentexte vom Beginn der 18. Dynastie bis zur Amarnazeit. Teil 1: Transcription und Übersetzung der Texte*. Wiesbaden.

Bietak, M., and Sharz, M. (eds) (2002) *Krieg und Sieg. Narrative Wanddarstellungen von Altägypten bis ins Mittelalter*. Vienna.

Bingöl, O. (2007) *Magnesia on the Meander/Magnesia ad Maeandrum*. Istanbul.

Bintliff, J., and Sbonias, K. (eds) (1999) *Reconstructing Past Population Trends in Mediterranean Europe (3000 BC–AD 1800)*. Oxford.

Birks, P., Rodger, A., and Richardson, J.S. (1984) "Further Aspects of the *Tabula Contrebiensis*," *Journal of Roman Studies* 74: 45–73.

Bissa, E.M.A. (2009) *Governmental Intervention in Foreign Trade in Archaic and Classical Greece*. Leiden.

Blackhurst, A. (2004) "The House of Nubel: Rebels or Players?," in Merrils (2004) 59–75.

Blackman, D. (1969) "The Athenian Navy and Allied Naval Contributions in the Pentecontaetia," *Greek, Roman and Byzantine Studies* 10(3): 179–216.

Blamire, A. (2001) "Athenian Finance, 454–404 BC," *Hesperia* 70: 99–126.

Blaum, P.A. (1994) *The Days of the Warlords. A History of the Byzantine Empire, A.D. 969–991*. Lanham.

Bleckmann, B. (2011) "Roman Politics in the First Punic War" in Hoyos (2011) 167–183.

Bloch, K. (1938) "Warlordism: A Transitory Stage in Chinese Government." *The American Journal of Sociology* 43(5): 691–703.

Bonner, R. (1923) "The Commercial Policy of Imperial Athens," *Classical Philology* 18: 193–201.

Bonnet, C., and Valbelle, D. (eds) (1998) *Le Sinaï durant l'Antiquité et le Moyen-Âge: 4,000 ans d'historie pour un desert*. Paris.

Boot, M. (2013) *Invisible Armies: An epic history of guerrilla warfare from ancient times to the present*. New York.

Bose, P. (2004) *Alexander The Great's Art Of Strategy*. New Dehli.

Bouvier-Closse, K. (2003) "Les noms propres de chiens, chevaux et chats de l'Égypte ancienne: le rôle et le sens du nom personnel attribué à l' animal," *Anthropozoologica* 37: 11–37.

Bradeen, D. (1960) "The Popularity of the Athenian Empire," *Historia* 9: 257–269.

Bragg, E. (2010) "Roman Seaborne Raids During the Mid-Republic: Sideshow or Headline Feature?," *Greece & Rome* 57: 47–64.

Brand, C.E. (1968) *Roman Military Law*. Austin.

Brennan, P. (1996) "The Notitia Dignitatum," in Nicolet (1996) 147–178.

Brennan, T.C. (2000) *The Praetorship in the Roman Republic*. Oxford.

——— (2004) "Power and Process under the Roman 'Constitution'," in Flower (2004) 31–65.

Bresson, A. (2000) *La cité marchande*. Bordeaux.

——— (2008) *L'économie de la Grèce des cités*, 2 vols. Paris.

Briant, P. (2002) *From Cyrus to Alexander: A History of the Persian Empire*. Winona Lake.

——— (2012) "From the Indus to the Mediterranean: The administrative organization and logistics of the great roads of the Achaemenid Empire," in Alcock, Bodel, and Talbert (2012) 185–201.

Brice, L.L. (2011) "Disciplining Octavian: An Aspect of Roman Military Culture during the Triumviral Wars 44–30 BCE," in Lee (2011) 35–60.

——— (forthcoming) "Use of Insurgency and Terrorism in the Ancient Mediterranean," in Howe and Brice (forthcoming).

——— (ed.)(forthcoming) *Before, During and After Battle: Case Studies in the New Military History*. Ann Arbor.

Brice, L.L., and Roberts, J.T. (2011a) "Introduction," in Brice and Roberts (2011b) 1–10.

Brice, L.L., and Roberts, J.T. (eds) (2011b) *Recent Directions in the Military History of the Ancient World*. Regina, CA.

Britt, T., Castro, C., and Adler, A. (eds) (2006) *Military Life: The Psychology of Serving in Peace and Combat*. London.

Britt, T., and Dickinson, J. (2006) "Morale during Military Operations: A Positive Psychology Apporach" in Britt, Castro, and Adler (2006) 157–184.

Brocato, P., and Terrenato, N. (2012) *Nuove ricerche nell'area archeologica di S. Omobono a Roma*. Rome.

Brock, R. (2009) "Did the Athenian Empire Promote Democracy?," in Ma, Papazarkadas and Parker (2009) 149–166.

Brown, R. (2004) "'*Virtus Consili Expers*' An Interpretation of the Centurion's Contest in Caesar *de Bello Gallico* 5.44," *Hermes* 132: 292–308.

Broughton, T.R.S. (1951–1986) *The Magistrates of the Roman Republic*. New York.

Brückner, H., Müllenhoff, M., Vött, A., Gehrels, R., Herda, A., Knipping, M., and Gehrels, W. (2006) "From archipelago to floodplain—geographical and ecological changes in Miletus and its environs during the past six millennia (Western Anatolia, Turkey)," *Zeitschrift für Geomorphologie N.F.* (Suppl.) 142: 63–83.

Bruns-Özgan, C., Gassner, V., and Muss, U. (2011) "Kolophon: Neue Untersuchungen zur Topographie der Stadt," *Anatolia Antiqua* 19: 199–239.

Brunt, P. (1971) *Italian Manpower*. Oxford.

Buckler, J. (2003) *Aegean Greece in the Fourth Century BC*. Leiden.

Bühler, K. (1934) *Sprachtheorie. Die Darstellungsfunktion der Sprache*. Jena.

Bunse, R. (2002) "Die klassische Prätur und die Kollegialität (*par potestas*)," *Zeitschrift der Savigny-Stiftung für Rechtsgeschichte, Romanistische Abteilung* 119: 29–43.

Burgess, R.W. (1993) *The Chronicle of Hydatius and the Consularia Constantinopolitana. Two contemporary accounts of the final years of the Roman Empire*. Oxford.

Burrer, F., and Müller, H. (eds) (2008) *Kriegskosten und Kriegsfinanzierung in der Antike*. Darmstadt.

BIBLIOGRAPHY 287

Bury, J.B. (1923) *A History of the Later Roman Empire from the Death of Theodosius I to the Death of Justinian*, vol. 1. London.

Butler, S. (2002) *The Hand of Cicero*. London.

Callendar, V., Bares, L., and Bárta, M. (eds) (2011). *Times, Signs and Pyramids. Studies in Honour of Miroslav Verner on the Occasion of his Seventieth Birthday*. Prague.

Cameron, Al. (1970) *Claudian: Poetry and Propaganda at the Court of Honorius*. Oxford.

Cameron, Av. (1985) *Procopius and the sixth century*. Berkeley.

Campbell, B. (1984) *The Emperor and the Roman Army 31 B.C.–A.D. 235*. Oxford.

――― (1987) "Teach Yourself How to be a General," *The Journal of Roman Studies* 77: 13–29.

Carandini, A. (1997) *La nascita di Roma. Dèi, Lari, eroi e uomini al'alba di una civiltà*. Turin.

Carcopino, J. (1914) *La Loi de Hiéron et les Romains*. Paris.

Cargill, J. (1981) *The Second Athenian League: Empire or Free Alliance?* Berkeley.

Caritoux, L. (1996) *Le nom des attelages royaux au Nouvel Empire*. MA Thesis, University Paul Valéry: Montpellier.

――― (1998) "Les chevaux de pharaon," *Égypte: Afrique et Orient* 11: 21–26.

Carradice, I. (ed.) (1987) *Coinage and Administration in the Athenian and Persian Empires*. Ann Arbor.

Cartledge, P. (1987) *Agesilaos and the Crisis of Sparta*. London.

Cartledge, P., Cohen, E., and Foxhall, L. (eds) (2002) *Money, Labour and Land: Approaches to the Economies of Ancient Greece*. London.

Caspar, T.W. (2011) *Recovering the Ancient View of Founding: A Commentary on Cicero's De Legibus*. Plymouth.

Cawkwell, G. (2005) *The Greek Wars: The Failure of Persia*. Oxford.

Cavillier, G. (1998) "Some Notes about Thel," *Göttinger Miszellen* 166: 9–18.

――― (2001a) "The Ancient Military Road between Egypt and Palestine Reconsidered: A Reassessment," *Göttinger Miszellen* 185: 23–31.

――― (2001b) *Il faraone guerriero: I sovrani del Nuovo Regno alla conquista dell'Asia tra mito, strategia bellica e realtà archeologica*. Turin.

Centre Jean Bérard (1994) *Le ravitaillement en blé de Rome et des centres urbains des débuts de la République jusqu'au haut Empire: actes du Colloque international, Naples, 14–16 février 1991*. Naples/Rome.

Ch'en, J. (1968) "Defining Chinese Warlords and Their Factions," *Bulletin of the School of Oriental and African Studies, University of London* 31 (3): 563–600.

Chaplin, J.D. (2000) *Livy's Exemplary History*. Oxford.

Charles, M.B. (2007) *Vegetius in Context. Establishing the Date of the Epitoma Rei Militaris*. Stuttgart.

Clackson, J. (2007) *Indo-European Linguistics: An Introduction*. Cambridge.

Claes, W., de Meulenaere, H., and Hendrickx, S. (eds) (2009) *Elkab and Beyond. Studies in honour of Luc Limme* (*OLA 191*). Leuven.

Cline, E., and O'Connor, D. (eds) (2006). *Thutmose III: A New Biography*. Ann Arbor.

Clover, F.M. (1993a) "The Pseudo-Boniface and the Historia Augusta," in Clover (1993b) 79–89.

——— (1993b) *The Late Roman West and the Vandals*. Aldershot.

Coarelli, F. (1985) *Il Foro romano: Periodo repubblicano e augusteo*. Rome.

Cobban, J.M. (1935) *Senate and Provinces 78–49 B.C.: Some Aspects of the Foreign Policy and Provincial Relations of the Senate during the Closing Years of the Roman Republic*. Cambridge.

Colburn, H. (2013) "Connectivity and Communication in the Achaemenid Empire," *Journal of the Economic and Social History of the Orient* 56: 29–52.

Collier, M., and Snape, S. (eds) (2011). *Ramesside Studies in Honour of K.A. Kitchen*. Bolton.

Collingwood, R. (1946) *The Idea of History*. Oxford.

Collins, J.H. (1972) "Caesar as a Political Propagandist," *Aufstieg und Niedergang der Römischen Welt* 1: 922–966.

Connolly P. (1998) *Greece & Rome At War*. London.

Cooley, C. (1909) *Social Organization: A study of the Larger Mind*. New York.

Cornell, T. (1995) *The Beginnings of Rome: Italy and Rome from the Bronze Age to the Punic Wars (C. 1000–264 BC)*. London.

——— (1996) "Hannibal's Legacy: the effects of the Hannibalic war in Italy," in Cornell, Rankov, and Sabin (1996) 97–117.

——— (2005) "The Value of the Literary Tradition Concerning Archaic Rome," in Raaflaub (2005) 47–74.

Cornell, T., Rankov, B., and Sabin, P. (eds) (1996) *The Second Punic War: a reappraisal* (*BICS* supp. 67). London.

Costa, B., and Fernández, J.H. (eds) (2005) *Guerra y Ejército en el mundo fenicio-púnico. XIX Jornadas de Arqueología fenicio-púnica*. Eivissa.

Covino, R. (2011) "The Fifth Century, the Decemvirate, and the Quaestorship," *ASCS 32* [*2011*] *Selected Proceedings* at http://www.ascs.org.au/news/ascs32/Covino.pdf.

——— (2013) "*Stasis* in Roman Sicily," *Electryone* 1.1: 18–28.

Cowan, R. (2007) *Roman Battle Tactics 109 BC–AD 313*. Oxford.

Cowley, R. (ed.) (2001) *With my face to the enemy: perspectives on the Civil War*. New York.

Crawford, M. (1977) "Rome and the Greek World: Economic Relationships," *Economic History Review* 30: 42–52.

——— (1985) *Coinage and Money under the Roman Republic. Italy and the Mediterranean Economy*. London.

——— (ed.) (1996) *Roman Statutes*, *BICS* Supplement 64.1.

BIBLIOGRAPHY

Crisa, A. (2011) "Heroic Cults in Northern Sicily: Between Numismatics and Archaeology," in Holmes (2011) 114–122.

Crowley, J. (2012) *The Psychology of the Athenian Hoplite*. Cambridge.

Crown, A. (1974) "Tidings and instructions: how news travelled in the Ancient Near East," *Journal of the Economic and Social History of the Orient* 17: 244–271.

Crump, G.A. (1975) *Ammianus Marcellinus as a military historian*. Wiesbaden.

Cuniberti, G. (2009) "Annotazioni sui *Fragmenta Cairensia* delle *Elleniche di Ossirinco*," *Bulletin of the American Society of Papyrologists* 46: 69–74.

Curchin, L.A. (1991) *Roman Spain: Conquest and Assimilation*. London.

Czerny, E., Hein, I., Hunger, H., Melmen, D., and Schwab, A. (eds) (2006). *Timelines: Studies in Honour of Manfred Beitak*. Leuven.

Daly, G. (2002) *Cannae: The experience of battle in the Second Punic War*. London.

Dandamaev, M., and Lukonin, V. (1989) *The Social and Cultural Institutions of Ancient Iran*. Cambridge.

Darnell, J. (1991) "Supposed Depictions of Hittites in the Amarna Period," *Studien zur Altägyptischen Kultur* 18: 113–140.

——— (2011) "A Stela of Seti I from the Region of Kurkur Oasis," in Collier and Snape (2011) 127–141.

Davidson, J. (1991) "The Gaze in Polybius' Histories," *Journal of Roman Studies* 81: 10–24.

Davies, W. (2009) "The tomb of Ahmose Son-of-Ibana at Elkab: documenting the family and other observations," in Claes, de Meulenaere, and Hendrickx (2009) 139–175.

——— (1982) "The Origin of the Blue Crown," *Journal of Egyptian Archaeology* 68: 69–76.

Debord, P. (1999) *L'Asie mineure au IVe siècle (412–323 A.C.): pouvoirs et jeux politiques*. Bordeaux.

de Blois, L. (2000) "Army and Society in the Late Roman Republic: Professionalism and the Role of the Military Middle Cadre," in Alfody, Dobson, and Eck (2000) 11–30.

de Blois, L., Bons, J., Kessels, T., and Schenkeveld, D. (eds) (2005) *The Statesman in Plutarch's Works, Volume II: The Statesman in Plutarch's Greek and Roman Lives. Mnemosyne Suppl. 250/II*. Leiden.

de Blois, L., Erdkamp, P., Hekster, O., de Kleijn, G., and Mols, S. (eds) (2003) *The Representation and Perception of Roman Imperial Power: Proceedings of the Third Workshop of the International Network Impact of Empire (Roman Empire, c. 200 B.C.–A.D. 476)*. Leiden.

De Lepper, J.L.M. (1941) *De Rebus Gestis Bonifatii Comitis Africae et Magistri Militum*. Breda.

De Ligt, L. (2012) *Peasants, Citizens, and Soldiers. Studies in the Demographic History of Roman Italy 225 BC–AD 100*. Cambridge.

De Ligt, L., and Northwood, S. (eds) (2008) *People, Land, and Politics: Demographic Developments and the Transformation of Roman Italy 300 BC–AD 14*. Leiden.

Der Manuelian, P. (1987) *Studies in the Reign of Amenhotep II*. Hildeshein.

De Sanctis, G. (1907–1923) *Storia Dei Romani*. Rome.

Deussen, P.W. (1994) "The Granaries of Morgantina and the *Lex Hieronica*," in Centre Jean Bérard (1994): 231–235.

Diemke, W. (1934) *Die Entstehung hypotaktischer Saetze. Dargestellte an der Entwicklung des Relativsatzes in der Sprache der alten Aegypter*. PhD Dissertation, University of Vienna.

Diesner, H.J. (1963a) "Die Laufbahn des Comes Africae Bonifatius und seine Beziehungen zu Augustin," in Diesner (1963) 100–126.

——— (ed.) (1963b) *Kirche und Staat im spätrömischen Reich*. Berlin.

——— (1972) "Das Buccellariertum von Stilicho und Sarus bis auf Aëtius (454–455)," *Klio* 54: 321–350.

Doğer, E. (2001) "İlkçağ'da İzmir'in stratejik konumu," *Yüzyıl Eşiğinde İzmir Sempozyumu* 21: 16–27.

Doğer, E., and Gezgin, I. (1998) "Arkaik ve Klasik dönemde Smyrna'nın dış savunması üzerine gözlemler," in Ülker (1998) 5–30.

Dolger, F.J. (1930) *"Sacramentum Militiae," Antike und Christentum* 2: 268–280.

Domingez Monedero, A. J. (2005) "Los Mercenarios Baleáricos," in Costa and Fernández (eds) (2005) 163–180.

Drinkwater, J.F., and Elton H. (eds) (1992) *Fifth-Century Gaul: Crisis of Identity?* Cambridge.

Drogula, F. (2007) "Imperium, Potestas, and the Pomerium in the Roman Republic," *Historia* 56: 419–452.

——— (2011) "The *Lex Porcia* and the Development of Legal Restraints on Roman Governors," *Chiron* 41: 91–124.

Dubouloz, J., and Pittia, S. (eds) (2007) *La Sicile de Cicéron: Lectures des Verrines*. Franche-Comté.

Dupuy, R.E. (1939) "The nature of guerilla warfare," *Pacific Affairs* 12: 138–148.

Dusinberre, E. (2003) *Aspects of Empire in Achaemenid Sardis*. Cambridge.

——— (2013) *Empire, Authority, and Autonomy in Achaemenid Anatolia*. Cambridge.

Dwyer, P.G., and Ryan, L. (2012) "The massacre and history," in Dwyer and Ryan (eds) xi–xxv.

Dwyer, P.G., and Ryan, L. (2012) *Theatres of Violence: Massacre, Mass Killing and Atrocity throughout History*. New York.

Eaton-Krauss, M. (2000) "Review: W. Stevenson Smith's The Art and Architecture of Ancient Egypt—3rd edn.," *Orientalistische Literaturzeitung* 95: 399.

Eckstein, A. M. (1987) *Senate and General: Individual Decision-Making and Roman Foreign Relations, 264–194 B.C.* Berkeley.

——— (1995) *Moral Vision in the* Histories *of Polybius*. Berkeley.

Eddy, S. (1973) "The Cold War between Athens and Persia, ca. 448–412 B.C.," *Classical Philology* 63.4: 241–258.

BIBLIOGRAPHY

Edel, E. (1953) "Die Stelen Amenophis' II. aus Karnak und Memphis," *Zeitschrift des Deutschen Palästina-Vereins* 69: 97–176.

Edgerton, W., and Wilson, J. (1936) *Historical Records of Ramses III*. Chicago.

Eichler, E. (1993) *Untersuchungen zum Expeditionswesen des ägyptischen Alten Reiches*. Wiesbaden.

Eisenstein, S. (1949) *Film Form: Essays in Film Theory*. J. Ledya (trans.). San Diego.

Elayi, J. (1992) 'La présence grecque dans les cités phéniciennes sous l'Empire perse achéménide,' *Revue des Etudes Grecques* 105: 305–327.

Elison, G., and Smith, B.L. (1987) *Warlords, Artists and Commoners: Japan in the sixteenth century*. Honolulu.

El-Maksoud, M.A. (1998a) "Tjarou, porte de l'Orient," in Bonnet and Valbelle (1986) 61–65.

———— (1998b) *Tell Heboua (1981–1991). Enquête archéologique sur la Deuxième Période Intermédiaire et le Nouvel Empire à l'extrémité orientale du Delta*. Paris.

Elton, H. (1996) *Warfare in Roman Europe: AD 350–425*. Oxford.

———— (2007) "Military Forces in the Later Roman Empire," in Sabin, van Wees, and Whitby (eds) (2007): 270–309.

Emery, W. (1965) *Egypt in Nubia*. London.

Endesfelder, E., Priese, K.-H., Reineke, W.-F., and Wenig, S. (eds) (1977). *Ägypten und Kusch*. Berlin.

Enmarch, R. (2011) "Of Spice and Mine: The Tale of the Shipwrecked Sailor and Middle Kingdom Expedition Inscriptions," in Hagen, Johnston, Monkhouse, Piquette, Tait, Worthington (2011) 97–121.

Epigraphic Survey (1985). *Reliefs and Inscriptions at Karnak IV: The Battle Reliefs of King Sety I*. Chicago.

Erdkamp, P. (1998) *Hunger and the Sword. Warfare and Food Supply in Roman Republican Wars (268–30 B.C.)*. Amsterdam.

———— (ed.) (2007) *A Companion to the Roman Army*. Oxford.

———— (2011) "Manpower and food supply in the First and Second Punic Wars," in Hoyos (ed.) (2011) 58–76.

Eshmawy, A. (2007) "Names of Horses in Ancient Egypt," in Goyon and Cardin (2007) 665–676.

Eyre, C. (ed.) (1998) *Proceedings of the Seventh International Congress of Egyptologists*. Leuven.

———— (1999) "Is Egyptian Historical Literature 'Historical' or 'Literary'," in Loprieno (1996) 415–433.

Ezzamel, M. (2012) *Accounting and Order*. New York.

Fagan, G., and Trundle, M. (eds) (2010) *New Perspectives on Ancient Warfare*. Leiden.

Faucher, T., Marcellesi, M.-C., and Picard, O. (eds) (2011) *Nomisma: la circulation monetaire dans le monde grec antique*. Paris.

Feldherr, A. (1998) *Spectacle and Society in Livy's History*. Princeton.

———— (ed.) (2009) *The Cambridge Companion to the Roman Historians*. Cambridge.

Ferrari, G. (ed.) (2007) *The Cambridge Companion to Plato's Republic*. Cambridge.

Ferrary, J.-L. (1983) "Sulla legislatzione *de reptetundis*," *Labeo* 29: 70–77.

Ferrill, A. (1988) *The Fall of the Roman Empire: The Military Explanation*. London

Fields, N. (1994) "Apollo: God of War, Protector of Mercenaries," in Sheedy (1994) 95–113.

———— (2001) "Et in Arcadia Ego," *Ancient History Bulletin* 15: 102–138.

Figueira, T. (1981) *Aegina: Society and Politics*. Salem.

———— (1998) *The Power of Money: Coinage and Politics in the Athenian Empire*. Philadelphia.

Finley, M. (1968). *A History of Sicily: Ancient Sicily to the Arab Conquest*. London.

———— (1978) "The Fifth-Century Athenian Empire: A Balance-Sheet" in Garnsey and Whittaker (1978) 103–126.

Fischer, H. (1977) "More Ancient Egyptian Names of Dogs and Other Animals," *Metropolitan Museum Journal* 12: 177–178.

Fisher, N., and van Wees, H. (eds) (1998) *Archaic Greece: New Approaches and New Evidence*. London.

Flament, C. (2011) "Faut—il suivre les chouettes? Reflexions sur la monnaie comme indicateur d'echange a partir du athennien d'epoque classique," in Faucher, Marcellesi, and Picard (2011) 39–51.

Flower, H.I. (ed.) (2004) *The Cambridge Companion to the Roman Republic*. Cambridge.

Fornara C.W. (1983) *Archaic Times to the End of the Peloponnesian War*. Cambridge.

Forsythe, G. (1988) "The Political Background of the *Lex Calpurnia* of 149 B.C.," *Ancient World* 17: 109–119.

———— (2005) *A Critical History of Early Rome: From Prehistory to the First Punic War*. Berkeley.

Foster, E., and Lateiner, D. (eds) (2012) *Thucydides and Herodotus*. Oxford.

François, P., Moret, P., and Péré-Noguès, S. (eds) (2006) *L'Hellénisation en Méditerranée Occidentale au temps des guerres puniques. Actes du Colloque International de Toulouse, 31 mars–2 avril 2005*. Toulouse.

Frank, T. (1933) *An Economic Survey of Ancient Rome*. Baltimore.

———— (ed.) (1959) *An Economic Survey of Ancient Rome, Vol. III*. Baltimore.

Frazel, T.D. (2009) *The Rhetoric of Cicero's 〉〉In Verrem〈〈*. Göttingen.

Frederiksen, R. (2011) *Greek City Walls of the Archaic Period*. Oxford.

Freeman, P., and Kennedy, D. (eds) (1986) *The Defence of the Roman and Byzantine East: Proceedings of a Colloquium held at the University of Sheffield in April 1986, BAR International Series 297*. Oxford.

Fronda, M. P. (2010) *Between Rome and Carthage: Southern Italy during the Second Punic War*. Cambridge.

BIBLIOGRAPHY

Gabba, E. (1949) "Le origini dell'esercito professionale in Roma: i proletari e la riforma di Mario," *Athenaum n.s.* 27: 173–209.

——— (1976) *Republican Rome, the Army and the Allies.* Oxford.

Gaballa, G.A. (1976) *Narrative in Egyptian Art.* Mainz am Rhein.

Gabrielsen, V. (1994) *Financing the Athenian Fleet.* Baltimore.

——— (2001) "Naval Warfare: Its Economic and Social Impact in Greek Cities," in Bekker-Nielsen and Hannestad (2001) 72–98.

——— (2008) "Die Kosten der athenischen Flotte in klassischer Zeit," in Burrer and Muller (2008) 46–73.

Gardiner, A. (1951) *The Theory of Speech and Language.* 2nd ed. Oxford.

——— (1957) *Egyptian Grammar.* 3rd edn. London.

Gardiner, A., and Peet, T. (1920) "The Ancient Military Road between Egypt and Palestine," *Journal of Egyptian Archaeology* 6: 99–116.

Gargola, D.J. (1995) *Lands, Laws, & Gods: Magistrates & Ceremonies in the Regulation of Public Lands in Republican Rome.* Chapel Hill.

Garnsey, P. (1988) *Famine and Food Supply in the Graeco-Roman World: Responses to Risk and Crisis.* Cambridge.

Garnsey, P., and Whittaker, C. (eds) (1978) *Imperialism in the Ancient World.* Cambridge.

Gat, A. (2006) *War in Human Civilization.* Oxford.

Gelzer, M. (1912) *Die Nobilität der römischen Republik.* Leiden.

Genette, G. (1980) *Narrative Discourse: An Essay in Method.* J. Lewin (trans.) Ithaca.

Gezgin, İ. (2001) "Defensive systems in Aiolis and Ionia regions in the Achaemenid period," in Bakır, Sancisi-Weedenburg, Gürtekin, Briant and Henkelman (2001) 181–188.

Gil Egea, M.E. (1999) "Warrior retinues in Late Antiquity: the case of Pelagia," *Studies in Latin literature and Roman history* 11: 493–503.

Gilliver, C. (1999) *The Roman Art of War.* London.

——— (2007) "Battle," in Sabin, Van Wees, and Whitby (2007) 122–157.

Gilroy, T. (2002) "Outlandish Outlanders: Foreigners and Caricature in Egyptian Art," *Göttinger Miszellen* 191: 35–52.

Giovannini, A. (1983) *Consulare Imperium.* Basel.

Giustozzi, A. (2005) "The Debate on Warlordism: The Importance of Military Legitimacy," *Crisis States Discussion Papers* 13.

——— (2012) *Empires of Mud: Wars and Warlords in Afghanistan.* New York.

Gladigow, B. (1972) "Die sakralen Funktionen der Liktoren. Zum Problem von institutioneller Macht und sakraler Präsentation," *Aufstieg und Niedergang der Römischen Welt* 1.2: 295–314.

Goedicke, H. (1996) "The Thutmosis I Inscription Near Tomâs," *Journal of Near Eastern Studies* 55: 161–176.

——— (2000) *The Battle of Megiddo.* Baltimore.

Goldsworthy, A. (1996) *The Roman Army at War, 100 BC–AD 200*. Oxford.

———— (2000) *The Punic Wars*. London.

———— (2003) *The Complete Roman Army*. London.

Gonen, R. (1984) "Urban Canaan in the Late Bronze Period," *Bulletin of the American Schools of Oriental Research* 253: 61–73.

———— (1987) "Megiddo in the Late Bronze Age—Another Reassessment," *Levant* 10: 83–100.

González Romanillos, J.A. (2003) "Antecedentes de la *Quaestio de repetundis*," *Iura* 54: 136–156.

Gorman, V. (2001) *Miletos, the Ornament of Ionia: A History of the City to 400 B.C.E.* Ann Arbor.

Goyon, J.-C., and Cardin, C. (eds) *Proceedings of the Ninth International Congress of Egyptologists, Grenoble, 6–12 September 2004*. Leuven.

Graham, A. (1992) "Thucydides 7.13.2 and the Crews of Athenian Triremes," *Transactions of the American Philological Association* 122: 257–270.

———— (1998) "Thucydides 7.13.2 and the Crews of Athenian Triremes: An Addendum," *Transactions of the American Philological Association* 128: 89–114.

Grapow, H. (1949) *Studien zu den Annalen Thutmosis des Dritten und zu ihnen verwandten historischen Berichten des Neuen Reiches*. Berlin.

———— (1936) *Sprachliche und schriftliche Formung ägyptischer Texte*. Glückstadt.

Gray, V. (1979) "Two different approaches to the battle of Sardis in 395 B.C.," *California Studies in Classical Antiquity* 12: 183–200.

Greaves, A. (2002) *Miletus: A History*. London.

Griffith, G. (1935) *The Mercenaries of the Hellenistic World*. Cambridge.

Grimal, N. (1986) *Les termes de la propagande royale égyptienne de la XIXe dynastie à la conquête d'Alexandre*. Paris.

Grünewald, T. (2004) *Bandits in the Roman Empire: Myth and Reality*. London.

Haas, C. (1985) "Athenian Naval Power before Themistocles," *Historia* 34: 29–46.

Habachi, L. (1972) *The Second Stela of Kamose and his Struggle against the Hyksos Ruler and his Capital*. Glückstadt.

Hagen, F., Johnson, W., Monkhouse, W., Piquette, K, Tait, J., and Worthington, W. (eds) (2011). *Narratives of Egypt and the Ancient Near East: Literary and Linguistic Approaches*. Leuven.

Halsall, G. (2007) *Barbarian Migrations and the Roman West, 376–568*. Cambridge.

Hands, A.R. (1965) "The Political Background of the *lex Acilia de Repetundis*," *Latomus* 24: 225–237.

Hansen, M. (2006a) *Studies in the Population of Aigina, Athens and Eretria. Historisk-filosofiske Meddelelser 94*. Copenhagen.

———— (2006b) *The Shotgun Method: the Demography of the Ancient Greek City-State Culture*. Columbia.

BIBLIOGRAPHY

Hansen, M., and Nielsen, T. (2004) *An Inventory of Archaic and Classical Poleis*. Oxford.

Hansen, O. (1991) "The Province, not the Island of Sicily (*Note de lecture 354*)," *Latomus* 60: 184.

Hanson, V. (1989) *The Western Way of War: Infantry Battle in Classical Greece*. Berkeley.

——— (1995) *The Other Greeks: The Family Farm and the Agrarian Roots of Western Civilisation*. New York.

——— (1998) *Warfare and Agriculture in Ancient Greece*. Berkeley.

——— (1999) "The Status of Ancient Military History: Traditional Work, Recent Research and On-Going Controversies," *Journal of Military History* 63: 379–414.

Hardwick, T. (2003) "The Iconography of the Blue Crown in the New Kingdom," *Journal of Egyptian Archaeology* 89: 117–141.

Haring, B.J. (1997) *Divine Households: Administrative and Economic aspects of the New Kingdom Royal Memorial Temples in Western Thebes*. Leiden.

Harries, J. (1994) *Sidonius Apollinaris and the Fall of Rome, AD 407–485*. Oxford.

Harris, W.V. (1976) "The Development of the Quaestorship, 267–81 B.C.," *Classical Quarterly* 26: 92–106.

——— (1979) *War and Imperialism in Republican Rome 327–70 B.C.* Oxford.

Hartley, B., and Wacher, J. (eds) (1983) *Rome and Her Northern Provinces*. Gloucester.

Harvey, P. (1986) "The new harvests reappear: The impact of war on agriculture," *Athenaeum* 74: 205–218.

Hasel, M. (2006) *Military Practice and Polemic: Israel's Laws of Warfare in Near Eastern Perspective*. Berrien Springs.

Hayes, H., Feder, F., and Morenz, L. (eds) (2014). *Interpretations of Sinuhe: Inspired by Two Passages (Proceedings of a Workshop Held at Leiden University, 27–29 November 2009)*. Leuven.

Hayes, W.C. (1953) *The Sceptre of Egypt—A Background for the Study of the Egyptian Antiquities in The Metropolitan Museum of Art. Part I: From the Earliest Times to the End of the Middle Kingdom*. New York.

Heather, P.J. (2005) *The Fall of the Roman Empire: A New History of Rome and the Barbarians*. Oxford.

Heinz, S.C. (2001) *Die Feldzugsdarstellungen des Neuen Reiches: eine Bildanalyse*. Vienna.

Helck, W. (1974) *Altägyptische Aktenkunde des 3. und 2. Jahrtausends v. Chr.* Munich.

Herbst, J. (1997) "Responding to State Failure in Africa," *International Security* 21 (3): 120–144.

Helgeland, J. (1978a) "Christians and the Roman Army from Marcus Aurelius to Constantine," *Aufstieg und Niedergang der Römischen Welt* II.23.1: 724–834.

——— (1978b) "Roman Army Religion," *Aufstieg und Niedergang der Römischen Welt* II.16.2: 1470–1505.

Herrmann, P. (1968) *Der römische Kaisereid: Untersuchungen zu seiner Herkunft und Entwicklung.* Gottingen.

Heuss, A. (1944) "Zur Entwicklung des Imperiums der römichen Oberbeamten," *Zeitschrift der Savigny-Stiftung für Rechtsgeschichte* 64: 57–133.

Hin, S. (2008) "Counting Romans," in de Ligt and Northwood (2008) 187–238.

——— (2013) *The Demography of Roman Italy. Population Dynamics in an Ancient Conquest Society 201 BCE–14 CE.* Cambridge.

Hindmoor, A. (2006) *Rational Choice.* New York.

Hobbes, T. (1651/1968) *Leviathan.* London.

Hoffmann, D. (1969) *Das Spätrömische Bewegungsheer und die Notitia Dignitatum,* 2 vols., *Epigraphische Studien 1.* Dusseldorf.

Hölkeskamp, K.-J. (1987) *Die Entstehung der Nobilität: Studien zur sozialen und politischen Geschichte der Römischen Republik im 4. Jhdt. V. Chr.* Stuttgart.

——— (2010) *Reconstructing the Roman Republic: An Ancient Political Culture and Modern Research.* Princeton.

Holloway, R.R. (2009) "Praetor Maximus and Consul," in Marangio and Laudizi (2009) 71–75.

Holmes, N. (ed.) (2011) *Proceedings of the XIV International Numismatic Congress, Glasgow 31 August–4 September 2009.* Glasgow.

Hopwood, K. (ed.) (1998) *Organized Crime in Antiquity.* London.

Hornblower, S. (1991-1996-2008) *A Commentary on Thucydides, Volumes I–III.* Oxford.

——— (2002) *The Greek World 479–323 BC,* 3rd edn. London.

Hornung, E. (1982) "Pharao Ludens," *Eranos Jahrbuch* 51: 479–516.

Howe, T., and Brice, L.L., (eds) (forthcoming). *Brill Companion to Insurgency and Terrorism in the Ancient Mediterranean.* Leiden.

Howgego, C. (1995) *Ancient History from Coins.* London.

Hoyos, D. (1973) "*Lex Provinciae* and Governor's Edict," *Antichthon* 7: 47–53.

——— (2003) *Hannibal's Dynasty: Power and Politics in the Western Mediterranean, 247–183 BC.* London.

——— (ed.) (2011) *A Companion to the Punic Wars.* Chichester.

Hsieh, J. (2012) "Discussions on the Daybook Style and the Formulae of Malediction and Benediction Stemming from Five Middle Kingdom Rock-Cut Stelae from Gebel el-Girgawi," *Zeitschrift für Ägyptische Sprache und Altertumskunde* 139: 116–135.

Humphreys, S. (1970) "Economy and Society in Classical Athens," *Annali della Scuola Normale Superiore di Pisa* 39: 1–26.

Hunt, D.W.S. (1947) "Feudal survivals in Ionia," *Journal of Hellenic Studies* 67: 68–76.

Hunt, P. (1998) *Slaves, Warfare and Ideology in the Greek Historians.* Cambridge.

Hyland, J. (2007) "Thucydides' portrait of Tissaphernes re-examined," in Tuplin (2007) 1–25.

Isaac, B. (1995) "Hierarchy and Command Structure in the Roman Army," in La Bohec (1995) 23–31.

BIBLIOGRAPHY

Jackson, A. (1969) "The Original Purpose of the Delian League," *Historia* 18: 12–16.

Jacobs, B., and Rollinger, R. (eds) (2010) *Der Achämenidenhof/The Achaemenid Court.* Wiesbaden.

Janos, C. (1963) "Unconventional Warfare: Framework and Analysis," *World Politics* 15: 636–646.

Jenkins, G. (1972) *Ancient Greek Coins.* London.

Johnson, J. (1980) "NIMS in Middle Egyptian," *Serapis* 6: 69–73.

Johnson, W.R. (1992) *An Asiatic Battle Scene of Tutankhamun from Thebes: A Late Amarna Antecedent of the Ramesside Battle-Narrative Tradition.* Unpublished Ph.D. Dissertation, Chicago.

Jones, A. (1987) *The Art of War in the Western World.* New York.

Jones, A.H.M. (1964) *The Later Roman Empire, 284–602: A Social, Economic, and Administrative Survey.* Oxford.

Jones, A.H.M., Martindale, J., and Morris, J. (eds) (1971) *Prosopography of the Later Roman Empire (AD 260–395)*, vol. 1. Cambridge.

Jordan, B. (2000) "The Crews of Athenian Triremes," *L'Antiquite Classique* 69: 81–101.

Junge, F. (1989) *"Emphasis" and Sentential Meaning in Middle Egyptian.* Harrassowitz.

Kagan, D. (1974) *The Archidamian War.* Ithaca.

——— (1987) *The Fall of the Athenian Empire.* Ithaca.

Kagan, J. (2006) "The Beginnings of Coinage at Abdera," in van Alfen (2006) 49–59.

Kagan, K. (2006) *The Eye of Command.* Ann Arbor.

Kahrstedt, U. (1913) *Geschichte der Karthager von 218–146.* Berlin.

Kaldellis, A. (2004) *Procopius of Caesarea: Tyranny, History, and Philosophy at the end of Antiquity.* Philadelphia.

Kallet, L. (1993) *Money, Expense, and Naval Power in Thucydides' History 1–5.24.* Berkeley.

——— (1999) "The Diseased Body Politic, Athenian Public Finance, and the Massacre at Mykalessos (Thucydides 7.27–29)," *The American Journal of Philology* 120: 223–244.

——— (2001) *Money and the Corrosion of Power in Thucydides: the Sicilian Expedition and its Aftermath.* Berkeley.

——— (2009) "Democracy, Empire and Epigraphy in the Twentieth Century," in Ma, Papazarkadas, and Parker (2009) 43–66

——— (2013) "The Origins of the Athenian Economic *Archê*," *Journal of Hellenic Studies* 133: 43–60.

Kawashima, R. (2011) "The Syntax of Narrative Forms," in Hagen, Johnston, Monkhouse, Piquette, Tait, Worthington (2011) 341–369.

Kay, P. (2014) *Rome's Economic Revolution.* Oxford.

Keaveney, A. (2005) "Sulla the warlord and other mythical beasts," in de Blois, Bons, Kessels, and Schenkeveld (2005) 297–302.

Keegan, J. (1994) *A History of Warfare.* Baltimore.

Keeley, L. (1997) *War Before Civilization: The Myth of the Peaceful Savage.* Oxford.

Keil, J. (1908) "Zur Topographie der ionischen Küste südlich von Ephesus," *Jahreshefte des Osterreichischen Archaologischen Instituts* 11: 135–202.

Kemp, B. (1986) "Large Middle Kingdom Granary Buildings," *Zeitschrift für Ägyptische Sprache und Altertumskunde* 113: 120–136.

———— (1989) *Egypt: Anatomy of a Civilization.* London.

Kemp, W. (1996) "Narrative," in Nelson and Shiff (1996) 58–69.

Kempinski, A. (1992) "Middle and Late Bronze Age Fortifications," in Kempinski and Reich, (1992) 127–142.

Kempinski, A., and Reich, R. (eds) (1992) *Architecture of Ancient Israel: From the Prehistoric to the Persian Periods.* Jerusalem.

Keppie, L. (1984) *The Making of the Roman Army.* London.

Kerschner, M., Kowalleck, I., and Steskal, M. (2008) *Archäologische Forschungen zur Siedlungsgeschichte von Ephesos in geometrischer, archaischer und klassischer Zeit: Grabungsbefunde und Keramikfunde aus dem Bereich von Koressos.* Wien.

Khatchadourian, L. (2012) "The Achaemenid provinces in archaeological perspective," in Potts (2012) 963–983.

Kim, H. (2001) "Archaic Coinage as Evidence for the Use of Money," in Meadows and Shipton (2001) 7–21.

———— (2002) "Small Change and the Moneyed Economy," in Cartledge, Cohen, and Foxhall (2002) 44–51.

King, A. (2006) "The Word of Command: Communication and Cohesion in the Military," *Armed Forces & Society* 32: 493–512.

Kirsten, E. (1964) "Aigina," in *Der Kleine Pauly.* Stuttgart.

Kitchen, K.A. (1964) "Some New Light on the Asiatic Wars of Ramesses II," *Journal of Egyptian Archaeology* 50: 47–70.

———— (1977) "Historical Observations on Ramesside Nubia," in Endesfelder, Priese, Reineke, and Wenig (1977) 213–225.

———— (1998) "Ramesside Egypt's Delta Defence Routes—the SE Sector," *Studi di Egittologia e di Antichità Puniche* 18: 33–37.

———— (1999) *Ramesside Inscriptions: Historical and Biographical: Notes and Comments. Translations and Annotations II.* Oxford.

Kitchen, K.A., and Gaballa, G. (1970) "Ramesside Varia II," *Zeitschrift für Ägyptische Sprache und Altertumskunde* 96: 14–28.

Klug, A. (2002) *Königliche Stelen in der Zeit von Ahmose bis Amenophis III.* Turnhout.

Koehler, U. (1869) *Urkunden und Untersuchungen zur Geschichte des delisch-attischen Bundes.* Berlin.

Kolb, F. (1977) "Zur Statussymbolik im antiken Rom," *Chiron* 7: 239–259.

Kolditz, T. (2006) "Research in In Extremis Settings: Expanding the Critique of Why They Fight," *Armed Forces & Society* 32: 655–658.

BIBLIOGRAPHY

Konuk, K. (2011) "Des chouettes en Asia Mineure quelques pistes de reflexion," in Faucher, Marcellesi and Picard (2011) 53–66.

Kornemann, E. (1930) *Doppelprinzipat und Reichsteilung im imperium Romanum*, Leipzig.

Kornhardt, H. (1936) *Exemplum: eine bedeutungsgeschichtliche Studie*. Göttingen.

Kraay, C. (1956) *Archaic Coins of Athens: Classification and Chronology*. London.

——— (1964) "Hoards, Small Change and the Origin of Coinage," *Journal of Hellenic Studies* 84: 76–91.

——— (1976) *Archaic and Classical Greek Coins*. London.

Kraus, C. (2010) "'Divide and Conquer' Caesar *de Bello Gallico* 7," in Kraus, Marincola, and Pelling (2010) 41–59.

Kraus, C.S., Marincola, J., and Pelling, C. (eds) (2010) *Ancient Historiography and its Contexts. Studies in Honour of A.J. Woodman*. Oxford.

Krentz, P. (1989a) "Athenian politics and strategy after Kyzikos," *Classical Journal* 84.3: 206–215.

——— (1989b) *Xenophon, Hellenika I–II.3.10*. Warminster.

——— (1995) *Xenophon, Hellenika II.3.11–IV.2.8*. Warminster.

——— (2007) "War" in Sabin, van Wees, and Whitby (2007) 147–185.

Kroll, J. (2009) "What about Coinage," in Ma, Papazarkadas, and Parker (2009) 195–209.

——— (2011) "Minting for Export: Athens, Aegina and Others," in Faucher, Marcellesi, and Picard (2011) 27–38.

Kruchten, J. (1999) "From Middle Egyptian to Late Egyptian," *Lingua Aegyptia* 6: 1–97.

Kuhrt, A. (2007) *The Persian Empire: A Corpus of Sources from the Achaemenid Period*. London.

Kulikowski, M. (2000) "The Notitia Dignitatum as an Historical Source," *Historia* 49: 358–377.

——— (2007) *Rome's Gothic Wars: From the Third Century to Alaric*. Cambridge.

Kutger J. P. (1960) "Irregular Warfare in Transition," *Military Affairs* 24: 113–123.

Lacqueur, W. (1998) *Guerrilla Warfare: A Historical and Critical Study*. New Brunswick.

Lawrence, A. (1965) "Ancient Egyptian Fortifications," *Journal of Egyptian Archaeology* 51: 69–94.

Lazenby, J. F. (1978) *Hannibal's War*. Warminster.

——— (1996) *The First Punic War*. London.

——— (2004) *The Peloponnesian War: A Military Study*. London.

Lazzeroni, R. (1994) "Rileggendo Benveniste: le relazioni di persona nel verbo," *Rivista di Linguistica* 6: 267–274.

Le Bohec, Y. (1994) *The Imperial Roman Army*. London.

——— (ed.) (1995) *La hierarchie (Rangordnung) de l'armee romaine sous le Haut-Empire*. Paris.

——— (2006) *L'armée romaine sous le Bas- Empire*. Paris.

Lee, A.D. (1996) 'Morale and the Roman Experience of Battle' in Lloyd (1996) 119–218.

——— (2007) *War In Late Antiquity: A Social History*. Oxford.

Lee, W. (ed.) (2011), *Warfare and Culture in World History*. New York.

Lendon, J. (1999) "The Rhetoric of Combat: Greek Military Theory and Roman Culture in Julius Caesar's Battle Descriptions," *Classical Antiquity* 18: 273–329.

——— (2004) "The Roman Army Now," *The Classical Journal* 99: 441–449.

——— (2005) *Soldiers and Ghosts A History of Battle in Classical Antiquity*. London.

Le Rider, G. (1989) "A propos d'un passage des Poroi de Xenophon: la question du change et les monnaies incuses d'Italie du sud," in Le Rider, Jenkins, Waggoner and Westermark (1989) 159–170.

Le Rider, G., Jenkins, K., Waggoner, N., and Westermark, U. (eds) (1989) *Kraay-Merkholm Essays: Numismatic Studies in Memory of C.M. Hraay and O. Merkholm*. Louvain.

Lewis, D. (1977) *Sparta and Persia*. Leiden.

Lichtheim, M. (1975) *Ancient Egyptian Literature Volume I: The Old and Middle Kingdoms*. Berkeley.

Liddel, P. (2009) "European Colonialist Perspectives on Athenian Power: Before and After the Epigraphic Explosion," in Ma, Papazarkadas, and Parker (2009) 13–42.

Liebeschüetz, J.H.W.G. (1986) "Generals, Federates and Buccellarii in Roman Armies around AD 400," in Freeman and Kennedy (1986) 463–474.

——— (1992) "Alaric's Goths: Nation or Army?," in: Drinkwater, J.F., and Elton, H. (eds) (1992) 75–83.

——— (1996) "The Romans Demilitarized: The evidence of Procopius," *Scripta Classica Israelica* 15: 230–239.

——— (2007) "Warlords and Landlords," in Erdkamp (2007) 479–494.

Lintott, A. (1968) *Violence in Republican Rome*. Oxford.

——— (1977) "Cicero on Praetors who Failed to Abide by their Edicts," *Classical Quarterly* 27: 184–186.

——— (1981a) "The *Leges de Repetundis* and Associate Measures under the Republic," *Zeitschrift der Savigny-Stiftung für Rechtsgeschichte* 98: 162–212.

——— (1981b) "What was the 'Imperium Romanum'?," *Greece & Rome* 28.1: 53–67.

——— (1992) *Judicial Reform and Land Reform in the Roman Republic*. Cambridge.

——— (1993) *Imperium Romanum: Politics and Administration*. London.

——— (1999) *The Constitution of the Roman Republic*. Oxford.

Little, P.D. (2003) *Somalia: Economy without State*. Bloomington.

Lloyd, A. (ed.) (1996) *Battle in Antiquity*. London.

Lo Cascio, E. (1994) "The Size of the Roman Population: Beloch and the Meaning of the Augustan Census Figures," *Journal of Roman Studies* 84: 22–40.

——— (1999) "The Population of Roman Italy in Town and Country," in Bintliff and Sbonias (1999) 161–171.

BIBLIOGRAPHY 301

———— (2001) "Recruitment and the Size of the Roman Population from the Third to the First Century BCE," in Scheidel (2001) 111–137.

Loomis, W. (1998) *Wages, Welfare Costs and Inflation in Classical Athens.* Ann Arbor.

Loprieno, L. (2003) "Temps des dieux et temps des homes en ancienne Égypte," in Pirenne-Delforge and Tunca (2003) 123–141.

Loprieno, A. (1988) "Verbal Forms and Verbal Sentences in Old and Middle Egyptian," *Göttinger Miszellen* 102: 59–72.

———— (1995) *Ancient Egyptian: A Linguistic Introduction.* Cambridge.

———— (ed.) (1996). *Ancient Egyptian Literature: History and Forms.* Leiden.

Lorton, D. (1974) *The Juridical Terminology of International Relations in Egyptian Texts through Dyn. XVIII.* Baltimore.

Ludwig, P. W. (2007) "Eros in the Republic" in Ferrari (2007) 202–231.

Luft, U. (1998) "Review: K.J. Seyfried: Beiträge zu den Expeditionen des Mittleren Reiches in die Ostwüste," *Orientalistische Literaturzeitung* 83: 274–282.

Lutkenhaus, W. (1998) *Constantius III: Studien zu seiner Tätigkeit und Stellung im Westreich 411–421.* Bonn.

Ma, J. (2009) "Afterword: Whither the Athenian Empire?" in Ma, Papazarkadas, and Parker (2009) 223–231.

Ma, J., Papazarkadas, N., and Parker, R. (eds) (2009) *Interpreting the Athenian Empire.* London.

MacCoun, R.J., and Hix, W.M. (2010) "Cohesion and performance," in Rosteker (2010) 137–158.

MacGeorge, P. (2002) *Late Roman Warlords.* Oxford.

MacMullen, R. (1984) "The Legion as Society" *Historia* 33: 440–456.

Maganzani, L. (2007) "L'Editto Provinciale alla luce delle Verrine: Profili Strutturali, Criteri Applicativi," in Dubouloz and Pittia (2007) 127–146.

Magdelain, A. (1968) *Recherches sur l'imperium: La loi curiate et les auspices d'investiture.* Paris.

———— (1979) "Le Suffrage Universel a Rome Au Ve Siecle Av J.C.," *Comptes rendus de l'Académie des inscriptions et belles-lettres*: 698–713.

Marangio, C., and Laudizi, G. (eds) (2009) *Palaia Philia: Studi di Topografia antica in honore di Giovanni Uggeri.* Milan.

Marshall, A.J. (1967) "Verres and Judicial Corruption," *Classical Quarterly* 17.2: 408–413.

———— (1969) "Romans under Chian Law," *Greek, Roman and Byzantine Studies* 10: 255–271.

———— (1984) "Symbols and Showmanship in Roman Public Life: The *fasces*," *Phoenix* 38: 120–141.

Marshall, S.L.A. (1947) *Men Against Fire: The Problem of Battle Command in Future War.* New York.

302   BIBLIOGRAPHY

Martin, G.T. (1989) *The Memphite Tomb of Horemheb Commander-in-Chief of Tut'ankha-mūn: The Reliefs, Inscriptions, and Commentary*. London.

Martindale, J. (ed.) (1971–1992) *Prosopography of the Later Roman Empire*. Cambridge

Masi Doria, C. (2000) *Spretum imperium: prassi constituzionale e momenti di crisi nei rapport tra magistrate nella media e tarda repubblica*. Naples.

Mastelloni, M.A. (2004) "Agatirno: l'eroe, il centro e la moneta. Le monete dal territorio," in Spigo (ed.) (2004) 23–32, 45–68.

Mathisen, R.W. (1999) "Sigisvult the Patrician, Maximinus the Arian, and political strategems in the Western Roman Empire c. 425–440," *Early Medieval Europe* 8(2): 173–196.

Matthews, J.F. (1975) *Western Aristocracies and Imperial Court (A.D. 364–425)*. Oxford.

———— (1989) *The Roman Empire of Ammianus*. Baltimore.

Mattingly, H.B. (1970) "The extortion law of the *tabula bembina*," *Journal of Roman Studies* 60: 154–168.

———— (1979) "The character of the *Lex Acilia Glabrionis*," *Hermes* 107: 478–488.

———— (1983) "*Acerbissima Lex Servilia*," *Hermes* 111: 300–310.

———— (1968) "Athenian Finance in the Peloponnesian War," *Bulletin de Correspondence Hellenique* 92: 450–485.

Mattingly, H. (1996) *The Athenian Empire Restored*. Ann Arbor.

Mattingly, H., and Sydenham, E.A. (1926) *The Roman Imperial Coinage*, Volume II. London.

Mazzarino, S. (1942) *Stilicone: la crisi imperiale dopo Teodosio*. Rome.

McCall, J. (2002) *The Cavalry of the Roman Republic*. London.

McCord, E.A. (1993) *The Power of the Gun: The Emergence of Modern Chinese Warlordism*. Berkeley.

McCoy, W.J. (1977) "Thrasyllus," *American Journal of Philology* 98.3: 264–289.

McGregor, M. (1987) *The Athenians and their Empire*. Vancouver.

McKechnie, P., and Kern, S. (1998) *Hellenica Oxyrhynchia*. Oxford.

Mead, M. (1940) "War is only an invention—not a biological necessity" *Asia Map* 40: 402–405.

Meadows, A., and Shipton, K. (eds) (2001) *Money and its Uses in the Ancient Greek World*. Oxford.

Meiggs, R. (1972) *The Athenian Empire*. Oxford.

Meiggs, R., and Lewis, D. (1969) *A Selection of Greek Historical Inscriptions*. Oxford.

———— (1989) *A Selection of Greek Historical Inscriptions*. 2nd edn. Oxford.

Mellink, M. (1976) "Archaeology in Asia Minor," *American Journal of Archaeology* 80.3: 261–289.

Meritt, B. (1925) "Tribute Assessments of the Athenian Empire from 454 to 440 BC," *American Journal of Archaeology* 29: 247–271.

Meritt, B., Wade-Gery, H., and McGregor, M. (1939–1953) *The Athenian Tribute Lists*. Cambridge, MA.

BIBLIOGRAPHY 303

Merrils, A. (ed.) (2004) *Vandals, Romans and Berbers: New Perspectives on Late Antique North Africa.* London.

Michaélidès, D. (1970) *Sacramentum Chez Tertullien.* Paris.

Miles, R. (2011) "Hannibal and propaganda," in Hoyos (ed.) (2011) 260–279.

Miller, M. (2011) "Town and country in the satrapies of western Anatolia: the archaeology of empire," in Summerer, Ivantchik, and Kienlin (2011) 319–344.

Millett, P. (1993) "Warfare, Economy and Democracy in Classical Athens," in Rich and Shipley (1993a) 177–196.

Minetti, A. (2003) "Efficiency of equine express postal systems," *Nature* 426.18: 785–786.

Mitchell, R.E. (2005) "The Definition of *patres* and *plebs*: An End to the Struggle of the Orders," in Raaflaub (2005a) 128–167.

Moles, J.L. (1993) "Livy's Preface," *Proceedings of the Cambridge Philological Society* 39: 141–168.

Mommsen, T. (1887–1888) *Römisches Staatsrecht.* Leipzig.

Morgan, L. (1997) "*Levi quidem de re ...*' Julius Caesar as Tyrant and Pedant," *Journal of Roman Studies* 87: 23–40.

Morket, R. (2001) "Egypt and Nubia: The Egyptian Empire in Nubia in the Late Bronze Age (c. 1550–1070 BCE)," in Alcock, D'Altroy, Morrison, and Sinopoli (2001) 227–251.

Morris, E. (2005) *The Architecture of Imperialism: Military Bases and the Evolution of Foreign Policy in Egypt's New Kingdom.* Leiden.

Morris, I. (2004) "Economic Growth in Ancient Greece," *Journal of Institutional and Theoretical Economics* 160: 709–742.

——— (2009) "The Greater Athenian State," in Morris and Scheidel (2009) 99–177.

Morris, I., and Powell, B. (eds) (2010) *The Greeks: History, Culture, and Society.* 2nd edn. Upper Saddle River.

Morris, I., and Scheidel, W. (eds) (2009) *The Dynamics of Ancient Empires: State Power from Assyria to Byzantium.* New York.

Morris, S., and Papadopoulos, J. (2005) "Greek towers and slaves: an archaeology of exploitation," *American Journal of Archaeology* 109.2: 155–225.

Mossakowski, W. (1993) "The *crimen repetundarum*: the Analysis of the Juridical Sources of Roman Republic [*sic*]," *Eos* 81: 213–221.

Muhlberger, S. (1990) *The Fifth-Century Chroniclers: Prosper, Hydatius, and the Gallic Chronicler of 452.* Leeds.

Müller, M. (2011) "Die administrativen Texte der Berliner Lederhandschrift," in Bechtold, Gulyás, and Kasznos (2011) 173–181.

Munson, R. (2012) "Persians in Thucydides," in Foster and Lateiner (2012) 241–280.

Murphy, R. (1957) "Intergroup Hostility and Social Cohesion," *American Anthropologist* 59: 1018–1035.

Murray, O., and Price, S. (eds) (1990) *The Greek City: From Homer to Alexander.* Oxford.

Ñaco del Hoyo, T. (2011) "Roman economy, finance, and politics in the Second Punic War," in Hoyos (ed.) (2011) 376–392.

Ñaco del Hoyo, T., and López-Sánchez, F. (eds) (forthcoming) *Warlords, War and Interstate Relations in the Ancient Mediterranean 404 BC–AD 14*. Leiden: Brill.

Na'aman, N. (2006) "Did Ramesses II Wage Campaign against the Land of Moab?" *Göttinger Miszellen* 209: 63–69.

Naville, E. (1907) *The XIth Dynasty Temple Deir el-Bahari*. London.

Nelson, R., and Shiff, R. (eds) (1996) *Critical Terms for Art History*. Chicago.

Newberry, P. (1893) *Beni Hasan I*. London.

——— (1894) *Beni Hasan II*. London.

Nicasie, M.J. (1998) *Twilight of Empire: The Roman Army from the Reign of Diocletian until the Battle of Adrianople*, Amsterdam.

Nicolet, C. (1966) *L'Ordre Équestre a l'époque républicaine (312–343 av. J.-C.)*. Paris.

——— (1976) Tributum: *recherches sur la fiscalité directe sous la république romaine*. Bonn.

——— (1980) *The World of the Citizen in Republican Rome*. Berkeley.

——— (ed.) (1996) *Les littératures techniques dans l'Antiquité romaine, Entretiens de la Fondation Hardt 42*. Geneva.

Nielsen, T., and Roy, J. (eds) (1999) *Defining Ancient Arkadia: Acts of the Copenhagen Polis Centre vol. 6*. Copenhagen.

Nippel, W. (1995) *Public Order in Ancient Rome*. Cambridge.

Nisbet, R. (1968) *Tradition and Revolt*. New Brunswick.

Nixon, L., and Price, S. (1990) "The Size and Resources of Greek Cities," in Murray and Price (1990) 137–170.

Nolan, D. (2014) *The Role of Battle Narrative in the Bellum Gallicum*. PhD Thesis, University of Tasmania.

Nordling, J.G. (1991) *Indirect Discourse and Rhetorical Strategies in Caesar's Bellum Gallicum and Bellum Civile*. PhD Thesis, University of Wisconsin.

Noth, M. (1943) "Die Annalen Thutmose III. als Geschichtequelle," *Zeitschrift des Deutschen Palästina-Vereins* 66: 156–174.

O'Connor, D. (2006) "Thutmose III: An Enigmatic Pharaoh," in Cline and O'Connor (2006) 1–38.

O' Flynn, J.M. (1983) *Generalissimos of the Western Roman Empire*. Edmonton.

Oakley, S. (1985) "Single Combat in the Roman Republic," *Classical Quarterly* 35: 392–410.

——— (1997) *A Commentary on Livy Books VI–X. Vol. I: Book VI*. Oxford.

——— (1998) *A Commentary on Livy Books VI–X. Vol. II: Books VII–VIII*. Oxford.

Ober, J. (1989) *Mass and Elite in Democratic Athens: Rhetoric, Ideology, and the Power of the People*. Princeton.

——— (2008) *Democracy and Knowledge: Innovation and Learning in Classical Athens*. Princeton.

BIBLIOGRAPHY

Ogilvie, R.M. (1965) *A Commentary on Livy Books 1–5*. Oxford.

Oliver, L., Harman, J., Hoover, E., Hayes, S., and Pandhi, N. (1999) "A Quantitative Integration of the Military Cohesion Literature," *Military Psychology* 11: 103–127.

Olson, M. (1965) *The Logic of Collective Action: Public Goods and the Theory of Groups*. Cambridge, Mass.

Oost, S. (1956) "The date of the *lex Iulia de Repetundis*," *American Journal of Philology* 77: 19–29.

——— (1966) "The revolt of Heraclian," *Classical Philology* 61: 236–242.

——— (1968) *Galla Placidia Augusta*. Chicago.

Oredsson, D. (2000) *Moats in Ancient Palestine*. Stockholm.

Oren, E. (1987) "The 'Ways of Horus' in North Sinai," in Rainey (1987) 69–119.

——— (ed) (1997). *The Hyksos: New Historical and Archaeological Perspectives*. Philadelphia.

——— (2006) "The Establishment of Egyptian Imperial Administration on the 'Ways of Horus': An Archaeological Perspective from North Sinai," in Czerny, Hein, Hunger, Melman, and Schwab (2006) 279–292.

Oren, E., and Shereshevski, J. (1989) "Military Architecture along the 'Ways of Horus'— Egyptian Reliefs and Archaeological Evidence," *Eretz Israel* 20: 8–22.

Osborne, R. (1996) *Greece in the Making, 1200–479 BC*. London.

Osborne, R., and Rhodes, P.J. (2003) *Greek Historical Inscriptions 404–323 BC*. Oxford.

Ostrom, E. (1990) *Governing the Commons: The Evolution of Institutions for Collective Action*. Cambridge.

Owen, T., and Strong, B. (eds) (2004) *The Vocation Lectures*. Cambridge.

Palmer, R. (1970) *The Archaic Community of the Romans*. Cambridge.

——— (1997) *Rome and Carthage at Peace (Historia Einzelschriften* 113). Stuttgart.

Palombi, D. (1996) "Hercules Victor, *aedes et signum*," *Lexicon Topographicum Urbis Romae* 3: 23–25.

Papazarkadas, N. (2009) "Epigraphy and the Athenian Empire: Reshuffling the Chronological Cards," in Ma, Papazarkadas, and Parker (2009) 67–88.

Parke, W. (1933) *Greek Mercenary Soldiers*. Oxford

Parrish, D. (ed.) (2001) *Urbanism in Western Asia Minor*. Portsmouth.

Peet, T. (1925) "The Legend of the Capture of Joppa and the Story of the Foredoomed Prince: Being a Translation of the Verso of Papyrus Harris 500," *Journal of Egyptian Archaeology* 11: 225–229.

Petit, T. (1990) *Satrapes et satrapies dans l'empire achéménide de Cyrus le Grand à Xerxès Ier*. Paris.

Phang, S.E. (2008) *Roman Military Service—Ideologies of Discipline in the Late Republic and Early Principate*. Cambridge.

Picard, O. (2011) "La circulation monetaire dans le monde Grec: Le cas de Thasos," in Faucher, Marcellesi, and Picard (2011) 79–109.

Pina Polo, F. (2011) *The Consul at Rome: The Civil Functions of the Consuls in the Roman Republic*. Cambridge.

Pirenne-Delforge, A., and Tunca, Ö. (eds) (2003). *Représentations du temps des religions: actes du colloque organisé par le Centre d'Histoire des Religions de l'Université de Liège*. Geneva.

Popper, K.R. (1945) *The Open Society and its Enemies*. Vol. 1: *The Spell of Plato*. London.

Potts, D.T. (ed.) (2012) *A Companion to the Archaeology of the Ancient Near East: Volume I*. Oxford.

Poucet, J. (1985) *Les Origines de Rome: Tradition et Histoire*. Brussels.

Powell, J., and Paterson, J. (eds) (2004) *Cicero the Advocate*. Oxford.

Prag, J.R.W. (ed.) (2007) *Sicilia Nutrix Plebis Romanae: Rhetoric, Law & Taxation in Cicero's Verrines*. London.

Price, B.J. (1975) *Paradeigma and Exemplum in Ancient Rhetorical Theory*. Berkeley.

Pritchard, D. (2010a) "The Symbiosis of Democracy and War: The Case of Ancient Athens," in Pritchard (2010b) 1–73.

——— (ed.) (2010b) *War, Democracy and Culture in Classical Athens*. Cambridge.

——— (2012) "Costing Festivals and War: Spending Priorities of the Athenian Democracy," *Historia* 61(1): 18–65.

Pritchard, R.T. (1970a) "G. Verres and the Sicilian Farmers," *Historia* 19: 224–238.

——— (1970b) "Cicero and the *lex Hieronica*," *Historia* 20: 352–368.

——— (1975) "*Perpaucae Siciliae Civitates*: Notes on *Verr*. II.iii.6.13," *Historia* 24: 33–47.

Pritchett, W. (1969) "The Transfer of the Delian Treasury," *Historia* 18: 17–21.

——— (1971) *The Greek State at War* I. Berkeley.

——— (1974) *The Greek State at War* II. Berkeley.

——— (1991) *The Greek State at War* V. Berkeley.

Probert, P. (2006) "Clause Boundaries in Old Hittite Relative Sentences," *Transactions of the Philological Society* 104: 17–83.

Pulvel, J. (ed.) (1970). *Myth and Law among the Indo-Europeans: Studies in Comparative Mythology*. Berkeley.

Quesada Sanz F. (2006) "Not so different: individual fighting techniques and small unit tactics of Roman and Iberian armies," in François, Moret, and Péré-Noguès (eds) (2006) 245–263.

——— (2011a) "Military Developments in the 'Late Iberian' culture (c. 237–c. 195 BC): Mediterranean Influences in the Far West via the Carthaginian Military," in Sekunda and Noguera (eds) (2011) 207–257.

——— (2011b) "*Guerrilleros* in Hispania? The myth of Iberian guerrillas against Rome," *Ancient Warfare* 5: 46–52.

Raaflaub, K. (1999) "Archaic and Classical Greece," in Raaflaub and Rosenstein (1999) 129–162.

BIBLIOGRAPHY 307

———— (ed.) (2005a) *Social Struggles in Archaic Rome: New Perspectives on the Conflict of the Orders*. Expanded and updated edn. Ed. Oxford.

———— (2005b) 'From Protection and Defense to Offense and Participation: Stages in the Conflict of the Orders', in Raaflaub (2005a) 185–222.

———— (2009) "Learning from the enemy: Athenian and Persian 'instruments of empire'," in Ma, Papazarkadas, and Parker (2009) 89–124.

———— (2011) "Riding on Homer's Chariot: The Search for a Historical 'Epic Society'," *Antichthon* 15: 1–34.

Raaflaub, K., and Rosenstein, N. (eds) (1999) *War and Society in the Ancient and Medieval Worlds*. Cambridge.

Rainey, A. (ed.) (1987) *Egypt, Israel, Sinai*. Tel Aviv.

Ramage, E.S. (2002) "The Populus Romanus, Imperium and Caesar's Presence in the *de Bello Gallico*," *Athenaeum* 90: 125–146.

Rambaud, M. (1953) *Cicéron et l'historie romaine*. Paris.

———— (1966) *L'Art de la Déformation Historique dans les Commentaires de César*. Paris.

Rasmussen, D. (1963) *Caesars Commentarii: Stil und Stilwandel am Beispiel der Direckten Rede*. Göttingen.

Rathbone, D. "The Census Qualifications of the *Assidui* and the *Prima Classis*," in Sancisi-Weerdenburg, Van Der Speck, and Teitler (1993) 121–152.

Rawlings, L. (1996) "Celts, Spaniards and Samnites: Warriors in a Soldiers' War," in Cornell, Rankov, and Sabin (eds) (1996) 81–95.

———— (1998) "Condottieri and Clansmen: Early Italian Raiding, Warfare and the State," in Hopwood (1998) 97–127.

———— (2005) "Hannibal and Hercules," in Rawlings and Bowden (eds) (2005) 153–184.

———— (2007a) *The Ancient Greeks at War*. Manchester.

———— (2007b) "Army and battle during the conquest of Italy (350–264 BC)," in Erdkamp (ed.) (2007) 45–62.

———— (2007c) "Hannibal the cannibal? Polybius on Barcid atrocities," *Cardiff Historical Papers* 9: 1–30.

———— (2010) "The Carthaginian Navy: Questions and Assumptions," in Fagan and Trundle (eds) (2010) 253–287.

———— (2011) "The War in Italy," in Hoyos (ed.) (2011) 299–319.

Rawlings, L., and Bowden, H. (eds) (2005) *Herakles and Hercules: Exploring a Graeco-Roman Divinity*. Swansea.

Rawson, E. (1971) "The Literary Sources for the Pre-Marian Army," *Publications of the British School at Rome* 39: 13–31.

Redford, D. (1986) *Pharaonic King-Lists Annals and Day Books: A Contribution to the Study of the Egyptian Sense of History*. Missassauga.

———— (1997) "Textual Sources for the Hyksos Period," in Oren, E. (1997) 1–44.

———— (2003) *The Wars in Syria and Palestine of Thutmose III.* Leiden.

Redford, D., and Grayson, A.K. (1974) *Papyrus and Tablet.* Engelwood Cliffs.

Rhodes, P. (1985) *The Athenian Empire.* Oxford.

———— (1993) *A Commentary on the Aristotelian Athenaion Politeia.* Oxford.

———— (2006) *A History of the Classical Greek World.* London.

———— (2008) "After the Three-Bar Sigma Controversy: The History of Athenian Imperialism Reassessed," *Classical Quarterly* 51: 500–506.

Rhodes, P., and Osborne, R. (2003) *Greek Historical Inscriptions, 404–323 BC.* Oxford.

Rich, J. (1983) "The supposed Roman manpower shortage of the later second century BC," *Historia* 32: 287–331.

———— (2007) "Warfare and the Army in Early Rome" in Erdkamp (2007) 7–23.

Rich, J., and Shipley, G. (eds.) (1993a) *War and Society in the Greek World.* London.

———— (1993b) *War and Society in the Roman World.* London.

Richardson, J.S. (1983) "The *Tabula Contrebiensis*: Roman Law in Spain in the early first century B.C.," *Journal of Roman Studies* 73: 33–41.

———— (1986) *Hispaniae: Spain and the Development of Roman Imperialism 218–82 B.C.* Cambridge.

———— (1987) "The Purpose of the *lex Calpurnia de repetundis*," *Journal of Roman Studies* 77: 1–12.

———— (1991) *"Imperium Romanum*: Empire and the Language of Power," *Journal of Roman Studies* 81: 1–12.

———— (2003) *"Imperium Romanum* between Republic and Empire," in de Blois, Erdkamp, Hekster, de Kleijn, and Mols (2003). Leiden.

Rickman, G. (1980) *The Corn Supply of Ancient Rome.* Oxford.

Riggsby, A.M. (1999) "Iulius Victor on Cicero's Defenses *De Repetundis*," *Rheinisches Museum für Philologie* 142: 427–429.

———— (2006) *Caesar in Gaul and Rome.* Austin.

Rihll, T. (1993) "War, Settlement and Slavery in Early Greece," in Rich and Shipley (1993a) 92–105.

Roller, M. (2004) "Exemplarity in Roman Culture: The Cases of Horatius Cocles and Cloelia," *Classical Philology* 99: 1–56.

———— (2009) "The Exemplary Past in Roman Historiography and Culture" in Feldherr (2009) 214–230.

Roosevelt, C. (2009) *The Archaeology of Lydia from Gyges to Alexander.* Cambridge.

Roselaar, S. (2008) *Public land in the Roman Republic: a social and economic history of the* ager publicus. PhD Thesis, Leiden University.

Rosenstein, N. (2004) *Rome at War. Farms, Families, and Death in the Middle Republic.* Chapel Hill.

———— (2007) "Military Command, Political Power and the Republican Elite," in Erdkamp (2007) 132–147.

BIBLIOGRAPHY 309

——— (2008) "Aristocrats and Agriculture in the Middle and Late Republic," *Journal of Roman Studies* 98: 1–26.

——— (2016) *"Bellum se ipsum alet?* Financing Mid-Republican Imperialism," in Beck and Seratti (2016).

Rostker, B. (ed.) (2010) *Sexual orientation and U.S. military policy: An update of RAND's 1993 study*. Santa Monica.

Roth, J. (1999) *The Logistics of the Roman Army at War (264 B.C.–A.D. 235)*. Leiden.

Roy, J. (1999) "The Economies of Arkadia," in Nielsen and Roy (1999) 320–381.

Ruschenbusch, E. (1983) "Tribut und Bürgerzahl im ersten athenischen Seebund," *Zeitschrift für Papyrologie und Epigraphik* 53: 125–143.

Russell, F. (1999) *Information Gathering in Classical Greece*. Ann Arbor.

Ruzicka, S. (1985) "Cyrus and Tissaphernes, 407–401 B.C.," *Classical Journal* 80.3: 204–211.

——— (2012) *Trouble in the West: Egypt and the Persian Empire, 525–332 BCE*. Oxford.

Sabin P. (2000) "The Face of Roman Battle," *Journal of Roman Studies* 90: 1–17.

Sabin, P., Van Wees, H., and Whitby, M (eds) (2007) *The Cambridge History of Greek and Roman Warfare Vol. II Rome from the Late Republic to the Late Empire*. Cambridge.

Sadek, A. (1980) *The Amethyst Mining Inscriptions of Wadi el-Hudi*. Warminster.

de Ste Croix, G. (1981) *The Class Struggle in the Ancient Greek World*. London.

Saller, R. (1994) *Patriarchy, Property, and Death in the Roman Family*. Cambridge.

Samons, L. (2000) *Empire of the Owl: Athenian Imperial Finance*. Stuttgart.

——— (1993) "Athenian Finance and the Treasury of Athena," *Historia* 42: 129–138.

Samons, L., and Fornara, C. (1991) *Athens from Cleisthenes to Pericles*. Berkeley.

Sancisi-Weerdenburg, H., Van Der Spek, R., and Teitler, H. (eds) (1993) De agricultura. *In memoriam Pieter Willem de Neeve (1945–1990)*. Amsterdam.

Sandberg, K. (2008) "The so-called division of the Roman Empire in AD 395. Notes on a Persistent Theme in Modern Historiography," *Arctos* 42: 199–213.

Sargent, R. (1927) "The Use of Slaves by the Athenians in Warfare," *Classical Philology* 22(3): 264–279.

Sarkesian, S. (ed.) (1980) *Combat Effectiveness: Cohesion, Stress, and the Volunteer Army*. London.

Sealey, R. (1976) *A history of the Greek City States, ca. 700–338 BC*. Berkeley.

Scherrer, P. (2001) "The historical topography of Ephesos," in Parrish (2001) 57–87.

Scheidel, W. (ed.) (2001) *Debating Roman Demography*. Leiden.

Schmitt, O. (1994) "Die Buccellarii: eine Studie zum militärischen Gefolgschaftswesen in der Spätantike," *Tyche* 9: 147–174.

Schönenberger, H. (1911) *Beispiele aus der Geschichte, ein rhetorisches Kunstmittel in Ciceros Reden*. Ausberg.

Schulman, A. (1964) "Siege Warfare in Pharaonic Egypt," *Natural History* 73: 12–21.

———— (1982) "The Battle Scenes of the Middle Kingdom," *Journal of the Society of the Study of Egyptian Antiquities* 12: 165–183.

Schulz, R. (2002) "Der Sturm auf die Festung: Gedanken zu einigen Aspekten des Kampfbildes im Alten Ägypten vor dem Neuen Reich," in Bietak and Schwarz (2002) 19–41.

Schwartz, A. (2010) *Reinstating the Hoplite: Arms, Armour and Phalanx Fighting in Archaic and Classical Greece*. Stuttgart.

Scramuzza, V.M. (1959) "Roman Sicily," in Frank (1959) 225–377.

Scullard, H.H. (1981) *Festivals and Ceremonies of the Roman Republic*. London.

Sealey, R. (1976) *A history of the Greek City States, ca. 700–338 BC*. Berkeley.

Seibert, J. (1993) *Hannibal*. Darmstadt.

Sekunda, N. (1985) "Achaemenid colonization in Lydia," *Revue des etudes anciennes* 87.1–2: 7–30.

Sekunda, N., and Noguera, A. (eds) (2011) *Hellenistic Warfare I. Proceedings Torun Conference, 2003*. Valencia.

Seltman, C. (1924/1955) *Greek Coins: A History of Metallic Currency Down to the Fall of the Hellenistic Kingdoms*. London.

Senk, H. (1957) "Zur Darstellung der Sturmleiter in der Belagerungsszene des Kaemhesit," *Annales du Service des Antiquités de l'Égypte* 54: 207–211.

Serrati, J. (2000) "Garrisons and Grain: Sicily between the Punic Wars," in Smith and Serrati (2000) 120–133.

Seyfried, K. (1981) *Beiträge zu den Expeditionen des Mittleren Reiches in die Ost-Wüste*. Hildesheim.

Shatzman, I. (1972) "The Roman General's Authority over Booty," *Historia* 21: 177–205.

Shaw, I. (1996) "Battle in Ancient Egypt: The Triumph of Horus or the Cutting Edge of the Temple Economy?" in Lloyd (1996) 239–269.

Sheedy, K. (ed.) (1994) *Archaeology in the Peloponnese*. Oxford.

Sheridan, J.E. (1966) *Chinese Warlord: The Career of Feng Yu-hsiang*. Stanford.

Sherk, R.K. (1984) *Rome and the Greek East to the Death of Augustus*. Cambridge.

Sherwin-White, A.N. (1982) "The *lex repetundarum* and the Political Ideas of Gaius Gracchus," *Journal of Roman Studies* 72: 18–31.

Shils, E., and Janowitz, M. (1948) "Cohesion and Disintegration in the Wehrmacht in World War II," *Public Opinion Quarterly* 12: 280–315.

Siebold, G. (2006) "Military Group Cohesion," in Britt, Castro, and Adler (2006) 185–201.

Simms, K.S. (1987) *From Kings to Warlords. The Changing Political Structure of Gaelic Ireland in the Later Middle Ages*. Woodbridge.

Simón, F.M. (2012) "The *Feriae Latinae* as Religious Legitimation of the Consuls' *imperium*," in Beck, Duplá, Jehne, and Pina Polo (2012): 116–132.

Simpson, W.K. (1956) "The Single-Dated Monuments of Sesostris I: An Aspect of the

BIBLIOGRAPHY

Institution of Coregency in the Twelfth Dynasty," *Journal of Near Eastern Studies* 15: 214–219.

Sjöqvist, E. (1960) "Excavations at Morgantina (Serra Orlando), 1959 Preliminary Report IV," *The American Journal of Archaeology* 64: 125–138.

Smith, C. (2006) *The Roman Clan: The Gens from Ancient Ideology to Modern Anthropology*. Cambridge.

Smith, C., and Serrati, J. (eds) (2000) *Sicily from Aeneas to Augustus: New Approaches in History and Archaeology*. Edinburgh.

Smith, H.S. (1976) *The Fortress of Buhen: The Inscriptions*. Oxford.

Smith, K.U. (1965) *Behaviour Organization and Work*. Madison.

Smith, R.E. (1958) *Service in the Post-Marian Roman Army*. Manchester.

Smith, S.T. (1991) "Askut and the Role of the Second Cataract Forts," *Journal of the American Research Center in Egypt* 28: 107–132.

——— (1995) *Askut in Nubia: The Economics and Ideology of Egyptian Imperialism in the Second Millennium B.C.* London.

Smith, W.S. (1965) *Interconnections in the Ancient near East: A Study of the Relationships between the Arts of Egypt, the Aegean, and Western Asia*. New Haven.

Smyth, P.A (1984) *Warlords and Holy Men. Scotland AD 80–1000*. Edinburgh.

Sokolicek, A. (2009) *Diateichismata: Zu dem Phänomen innerer Befestigungsmauern im griechischen Städtebau*. Wien.

Snape, S. (1997) "Ramesses II's Forgotten Frontier," *Egyptian Archaeology* 11: 23–24.

——— (1998) "Walls, Wells and Wandering Merchants: Egyptian Control of Marmarica in the Late Bronze Age," in Eyre (1998) 1081–1084.

——— (2003) "New Perspectives on Distant Horizons: Aspects of Egyptian Imperial Administration in Marmarica in the Late Bronze Age," *Libyan Studies* 34: 1–8.

Southern, P., and Dixon, K.R. (1996) *The Late Roman Army*. New Haven.

De Souza, P. (1998) "Towards Thalassocracy: Archaic Greek Naval Developments," in Fisher and van Wees (1998) 271–294.

Spalinger, A. (1977/78) "A Canaanite Ritual found in Egyptian Reliefs," *Journal of the Society of the Study of Egyptian Antiquities* 8: 47–60.

——— (1982) *Aspects of the Military Documents of the Ancient Egyptians*. New Haven.

——— (1995) "The Calendrical Importance of the Tombos Stela," *Studien zur Altägyptischen Kultur* 22: 271–281.

——— (1997) "Drama in History: Exemplars from Mid Dynasty XVIII," *Studien zur Altägyptischen Kultur* 24: 269–300.

——— (1998) "Orientations on Sinuhe," *Studien zur Altägyptischen Kultur* 25: 311–339.

——— (2002) *The Transformation of an Ancient Egyptian Narrative: P. Sallier III and the Battle of Kadesh*. Wiesabaden.

——— (2011) "Königsnovelle and performance," in Callender, Bareš, and Bárta (2011) 351–374.

——— (2012) *Icons of Power: A Strategy of Reinterpretation*. Prague.

Spalinger, A., and Armstrong, J. (eds) (2013) *Rituals of Triumph in the Mediterranean World*. Leiden.

Speidel, M.P. (1984) *Roman Army Studies Volume 1*. Amsterdam.

Spence I. G. (1990) "Perikles and the defence of Attika during the Peloponnesian War," *Journal of Hellenic Studies* 110: 91–109.

Spencer, P. (1997) *Amara West 1*. London.

Spigo, U. (ed.) (2004) *Archeologia a Capo d'Orlando. Studi per l'Antiquarium*. Milazzo.

Stäcker, J. (2003) *Princeps und Miles—Studien zum Bindungs- und Nahverhältnis von Kaiser und Soldat im 1. Und 2. Jahrhundert n. Chr*. Hildesheim.

Stasse, B. (2005) "La loi curiate des magistrats," *Revue Internationale des droits de l'Antiquité* 52: 375–400.

Stauder, A. (2013) "Linguistic Dissonance in Sinuhe," in Hays, Feder, and Morenz (eds) (2013) 173–188.

Steel, C. (2004) "Being Economical with the Truth: What Really Happened at Lampsacus?," in Powell and Paterson (2004) 233–251.

——— (2007) "The Rhetoric of the *de Frumento*," in Prag (2007): 37–48.

Steever, S., Walker, C., and Mufwene, S. (eds) (1976). *Papers from the Parasession on Diachronic Syntax, April 22, 1976*. Chicago.

Stein, E. (1947) *Histoire du bas-empire vol. 1, de l'état romain à l'état byzantin, 284–476*. (Palanque trans.). Amsterdam.

Stevenson, G.H. (1939) *Roman Provincial Administration Till the Age of the Antonines*. New York.

Stevenson Smith, W. (1965) *The Interconnections in the Ancient Near East. A Study of the Relationships between the Arts of Egypt, the Aegean, and Western Asia*. New Haven.

Stewart, R. (1998) *Public Office in Early Rome: Ritual Procedure & Political Practice*. Ann Arbor.

Stickler, T. (2002) *Aëtius: Gestaltungsspichraume eines Heermeisters im ausgehenden Weströmischen Reich*. Munchen.

Stoll, O. (2007) "The Religions of the Armies," in Erdkamp (2007) 451–477

Stouffer, S., Suchman, E., DeVinney, L., Star, S., and Williams, R. (1949) *The American Soldier: Adjustment During Army Life* (*Studies in Social Psychology in World War 11, Vol. 1*). Oxford.

Strassler, R. (ed.) (2009) *The Landmark Xenophon's Hellenika*. New York.

Strauss, L. (1952) "On Collingwood's Philosophy of History," *The Review of Metaphysics* 5: 559–586.

Summerer, L., Ivantchik, A., and Kienlin, A. (eds) (2011) *Kelainai-Apameia Kibotos: Développement urbain dans le context anatolien*. Bourdeaux.

Tait, J. (2001) "The Sinews of Demotic Narrative," in Hagen, Johnston, Monkhouse, Piquette, Tait, Worthington (2011) 397–411.

BIBLIOGRAPHY 313

Tandy, D. (1997) *Warriors Into Traders: The Power of the Market in Ancient Greece.* Berkeley.

Taylor, L.R. (1960) *The Voting Districts of the Roman Republic: The Thirty-Five Urban and Rural Tribes.* Rome.

Teeter, E., and Larson, J.A. (eds) (1999) *Gold of Praise: Studies on Ancient Egypt in Honor of Edward F. Wente.* Chicago.

Terrenato, N. (2001) "The Auditorium site in Rome and the origins of the villa," *Journal of Roman Archaeology* 14: 5–32.

Thomas, S. (2000) "Tell Abqa'in: A Fortified Settlement in the Western Delta: Preliminary Report of the 1997 Season," *Mitteilungen des Deutschen Archäologischen Instituts, Abteilung Kairo* 56: 371–376.

Thonemann, P. (2009) "Lycia, Athens and Amorges," in Ma, Papazarkadas, and Parker (2009) 167–194.

Thomas, D. (2009) "Chronological problems in the continuation (1.1.1–2.3.10) of Xenophon's *Hellenika*," in Strassler (2009) 331–339.

Tilly, C. (1990) *Coercion, Capital, and European States, AD 990–1990.* Oxford.

Torelli, M. (1999) *Tota Italia: Essays in the Cultural Formation of Roman Italy.* Oxford.

Toynbee, A. (1965) *Hannibal's Legacy: the Hannibalic War's effects on Roman life,* 2 vols. London.

Traill, D. (2001) "Boxers and Generals at Mount Eryx," *American Journal of Philology* 122: 405–414.

Trigger, B. (1982) "The Reasons for the Construction of the Second Cataract Forts," *Journal of the Society of the Study of Egyptian Antiquities* 12: 1–6.

Trundle, M. (2004) *Greek Mercenaries from the Late Archaic Age to Alexander.* London.

——— (2010) "Coinage and the Transformation of Greek Warfare," in Fagan and Trundle (2010) 227–253.

——— (forthcoming) "The Logistics of Greek Warfare: Food, Pay, Plunder and Money," in Brice (forthcoming).

Tuplin, C. (1987a) "Xenophon and the garrisons of the Achaemenid empire," *Archäologische Mitteilungen aus Iran* 20: 167–247.

——— (1987b) "The administration of the Achaemenid empire," in Carradice (1987) 109–158.

——— (ed.) (2007) *Persian Responses: Political and Cultural Interaction with(in) the Achaemenid Empire.* Swansea.

United States Department of Defense (2007) *Irregular Warfare (IW) Joint Operating Concept (JOC),* version 1.0.

Ülker, N. (ed.) 11. Uluslararası İzmir Sempozyumu. İzmir.

Van Alfen, P. (ed.) (2006) *Agoranomia: Studies in Money and Exchange, Presented to John H. Kroll.* New York.

Vandersleyen, C. (1971) *Les guerres d'Ahmosis, fondateur de la XVIIIe Dynastie.* Brussels.

Vego, M. (2000) *Operational Warfare*. Newport.

Venturini, C. (1997) "*Quaestiones perpetuae constitutae*: per una riconsiderazione della *lex Calpurnia repetundarum*," *Iura* 48: 1–76.

Vernus, P. (1978) "L'instance de la narration dans les phases anciennes de l'Égyptien," *Discussions in Egyptology* 9: 100–102.

——— (1990a) *Future as Issue. Tense, Mood and Aspect in Middle Egyptian: Studies in Syntax and Semantics*. New Haven.

——— (1990b) "Les espaces de l'écrit dans l'Égypte pharaonique," *Bulletin de la Société française d'Égyptologie* 119: 35–53.

——— (2009) "Reception linguistique et idéologique d'une nouvelle technologie: le cheval dans la civilization pharaonique," in Wissa (2009) 1–46.

Vinci, A. (2007) "'Like Worms in the Entrails of a Natural Man': A Conceptual Analysis of Warlords," *Review of African Political Economy* 34 (112): 313–331.

Vogel, C. (2004) *Ägyptische Festungen und Garnisonen bis zum Ende des Mittleren Reiches*. Hildesheim.

Von der Way, T. (1984) *Die Textüberlieferung Ramses' II. zur Qadeß-Schlacht: Analyse und Struktur*. Hildesheim.

Von Petrikovits, H. (1983) "Sacramentum," in Hartley and Wacher (ed.) (1983) 178–201.

Von Premerstein, A. (1937) *Von Werden und Wesen des Prinzipats*. Munich.

Waldron, A. (1991) "The Warlord: Twentieth-Century Chinese Understanding of Violence, Militarism, and Imperialism," *American Historical Review* 96 (4): 1073–1100.

Waldron, A. (2003) *From War to Nationalism: China's Turning Point, 1924–1925*. Cambridge.

Walbank, F.W. (1957–1989) *A Historical Commentary on Polybius*. 3 vols. Oxford.

Walbank, M. (1978) *Athenian Proxenies of the Fifth Century BC*. Toronto.

Walsh, P. (1963) *Livy, His Historical Aims and Methods*. Cambridge.

Ward, G. (2012) *Centurions: The Practice of Roman Officership*. Chapel Hill.

Watanabe, C. (2004) "The 'Continuous Style' in the Narrative Schemes of Assurbanipal's Reliefs," *Iraq* 66: 103–114.

Waters, M. (2010) "Applied royal directive: Pissouthnes and Samos," in Jacobs and Rollinger (2010) 817–828.

Watkins, C. (1970) "Languages of Gods and Languages of Men: Remarks on Some Indo-European Metalinguistic Traditions," in Puhvel (1970) 1–18.

——— (1976) "Towards Proto-Indo-European Syntax: problems and pseudo-problems," in Steever, Walker, and Mufwene (1976) 305–326.

Watson, G.R. (1969) *The Roman Soldier*. London.

Weber, M. (2004) "Politics as Vocation," in Owen and Strong (2004) 32–94.

Webster, G. (1998) *The Roman Imperial Army* 3rd Ed. Oklahoma.

Van Wees, H. (2004) *Greek Warfare, Myths and Realities*. London.

BIBLIOGRAPHY

——— (2013) *Ships, Silver, Taxes and Tribute: A Fiscal History of Archaic Athens*. London.

Welch, K. (1998) "Caesar and His Officers in the Gallic War Commentaries," in Welch and Powell (1998) 85–110.

Welch, K., and Powell, A. (1998) *Julius Caesar as Artful Reporter: The War Commentaries as Political Instruments*. London.

Welles, C., Fink, R., and Gilliam, J. (1959) *The Parchments and Papyri. The Excavations of Dura-Europos. Final report 5.1*. New Haven.

Westbrook, R. (1999) "Vitae Necisque Potestas," *Historia* 48: 203–223.

Westbrook, S. (1980) "The Potential for Military Disintegration," in Sarkesian (1980) 244–278.

Westlake, H.D. (1979) "Ionians in the Ionian War," *Classical Quartely n.s.* 29.1: 9–44.

——— (1981) "Decline and fall of Tissaphernes," *Historia* 30.3: 257–279.

——— (1986) "Spartan intervention in Asia, 400–397 B.C.," *Historia* 35.4: 405–426.

Wheeler, E. (2001) "Firepower," *Roman Military Studies* 5: 169–184.

Wheeler E. L. (1988) *Stratagem and the Vocabulary of Military Trickery, Mnemosyne Supplement* 108. Leiden.

——— (1998) "Battles and Frontiers," *Journal of Roman Archaeology* 11: 644–651.

Whittakker, C.R. (1993) "Landlords and Warlords in the later Roman Empire," in Rich and Shipley (1993b) 277–302.

Wijnendaele, J.W.P. (2011) "Apocalypse, Transformation or Much ado 'bout nothing? Western scholarship and the Fall of Rome (1776–2008)," *Iris* 24: 42–52.

——— (2015) *'The Last of the Romans'. Bonifatius: Warlord and* comes Africae. London.

Williams, B. (1999) "Serra East and the Mission of Middle Kingdom Fortresses in Nubia," in Teeter and Larson (1999) 435–453.

Williams, S., and Friell, G. (1999) *The Rome that did not Fall: the Survival of the East in the Fifth Century A.D.* Berkeley.

Wilson, R.J.A. (2000) "Ciceronian Sicily: An Archaeological Perspective," in Smith and Serrati (2000) 134–160.

Winlock, H.E. (1945) *The Slain Soldiers of Neb-hetep-Re Mentu-hotpe*. New York.

Wiseman, T.P. (1998) "The Publication of *de Bello Gallico*," in Welch and Powell (1998) 1–9.

Wissa, M. (ed.) (2009). *The Knowledge Economy and Technological Capabilities: Egypt, the Near East and the Mediterranean 2nd millennium B.C.–1st millennium A.D. Proceedings of a Conference Held at the Maison de la Chimie, Paris, France 9–10 December 2005*. Barcelona.

——— (ed.) (2014) *The Knowledge Economy and Technological Capabilities: Egypt, the Near East and the Mediterranean 2nd millennium B.C.–1st millennium A.D., Aula Orientalis Supplementa*. Barcelona.

Wolf, W. (1961) "Review: W. Stevenson Smith's The Art and Architecture of Ancient Egypt—1st edn.," *Orientalistische Literaturzeitung* 56: 18.

Wong, L. (2003) *Why They Fight: Combat Motivation in the Iraq War*. Carlisle.

Wylie, G. (1992) "Agesilaus and the Battle of Sardis," *Klio* 74: 118–130.

Yadin, Y. (1963) *The Art of Warfare in Biblical Lands*. London.

Zába, Z. (1974) *The Rock Inscriptions of Lower Nubia: Czechoslovak Concession*. Prague.

Zecchini, G. (1983) *Aezio: l'ultima difesa dell'Occidente romano*. Rome.

Index

Individuals

Aëtius 7, 186, 192–193, 196, 198–202
Agesilaus 278–280
Ahmose 13–15, 241
Ahmose Pen-Nechbet 255
Ahmose Son of Ibana 14, 241, 255, 257
Alaric 190–191, 193–194, 201–202
Amenemhet I 25–26, 246
Ammianus Marcellinus 128–133, 135, 138–142
Amunhotep I 14, 19
Amunhotep II 17, 19
Arbogastes 133
Aristotle 65, 68
Aristagoras 68
Artaxerxes II 8, 264, 274–275, 277, 280
Augustus (Octavian) 123–124, 126, 131, 138, 195, 201

Belisarius 192, 197
Bonifatius 7, 186, 191, 196–203

Castinus 197–199, 201
Conon 79
Constantine 128–129, 132–133, 138, 193
Constantius II 128, 132, 134–135, 138–140, 193–194
Constantius III 197, 201
Cyrus, the Great 262
Cyrus, the Younger 8, 72, 264, 274–275, 277, 279–280

Dercylidas 278

G. Duilius 82
G. Flaminius 37, 153
G. Julius Caesar 2, 85
G. Valerius Flaccus 155–156
G. Verres 146, 152–153, 157–159, 161–162
Galla Placidia 198–200
Gn. Manlius Vulso 82, 159

Hamilcar 205, 214, 216–217, 219, 222, 225, 228, 230–232
Hannibal 80–81, 96, 204–212, 214–226, 228–231

Hanno 209–210, 226–227, 231
Honorius 191, 194, 198, 201–202

Ioannes 198–199, 202

Julian 128–131, 135, 139–140, 142

Lars Porsenna 107
Livy (T. Livius Patavinus) 81–82, 84, 95, 105, 108–110, 112–113, 115–116, 121, 145, 147, 149–153, 155, 158–159, 161, 204–232
L. Aurunculeius Cotta 35, 47–48
L. Caecilius Metellus 145, 158–159, 161
L. Domitius 161
L. Mummius 156
L. Vorenus 34, 42, 46–50, 58, 60

Magnus Maximus 129, 192
M. Aquillus 161
M. Atilius Regulus 215, 227–229, 232
M. Minucius Rufus 215, 220, 222
M. Petronius 58–59
M. Tullius Cicero 6, 29, 35–37, 41, 50, 145, 148–149, 152, 154–157, 159–162, 217
M. Valerius Laevinus 158
Mago 218
Maharbal 218, 227, 230
Massinissa 207–208, 213–214, 218, 221, 224, 231–232
Maximus Thrax 124, 128–130, 134, 138
Miltiades 68, 169
Muttines 218, 222, 227

Nebridius 140

Pericles 69–71, 77
Pharnabazus 263, 266, 278–279
Pissuthnes 263, 265–266
Plato 65, 101
Polybius 53, 81, 83, 85, 92, 94, 121, 126, 149, 205, 212, 215–217, 219–220, 223, 225–229, 232–233
Procopius (usurper) 129, 131–132, 135, 141–142

P. Rupilius   153, 157, 159–160
P. Sextius Baculus   38–46, 60

Q. Fabius Maximus   112, 210, 217, 220
Q. Fabius Pictor   83
Q. Titurius Sabinus   35, 47

Ramesses II   20, 24–25, 28–29, 32, 242, 244–
      245, 249, 251–252, 254–255, 257, 259
Ramesses III   32, 242, 244–245, 250–251, 253,
      259

Sesostris I   26–27, 241, 246, 253
Seti I   20–22
Severus Alexander   128, 134
Stages (hyparch)   266, 272–273, 280
Stilicho   192–194, 201

Themistocles   68

Theodosius I   120, 128–129, 132–133, 135, 193–
      195
Theodosius II   191, 198
Thibron   275–278, 280
Thrasyllus   269–274, 276, 278, 280
Thucydides   65–67, 70–72, 77, 168–169, 171,
      178–179, 266, 271
Thutmose I   14–15
Thutmose II   15
Thutmose III   15–17, 19, 27, 30, 32, 240, 242,
      252–253, 255, 257, 260
Ti. Sempronius Gracchus   204–205, 209, 225–
      226
T. Manlius Torquatus   35, 145
T. Pomponius Veientanus   209, 211, 231
T. Pullo   34, 42, 46–50, 58, 60

Valentinian III   128, 191, 194–196, 198–199, 202
Vegetius   121, 128–132, 139, 195–196

Places

Abdera   75
Adys   215, 228–229, 231
Aegina   66, 73, 75–76
Agathyrna   208–211, 230–231
Amorges   263, 266, 268
Antandros   268
Andros   68, 268
Ardea   108
Aricia   108
Armant   19
Apulia   204–205, 211, 214
Agrigentum   158, 208, 214, 221, 227
Arpi   205
Aswan-Philae   15
Athens   4, 6, 65–70, 72–74, 76–77, 79, 111,
      164, 169–181, 263, 265, 269, 274
Attica   66
Avaris   14, 241, 257

Beit el Wali   24, 29, 245
Beni Hasan   241, 243, 246–249, 256, 258
Beth Shan   20–21
Black Sea   66, 69
Bolae   108–109
Bruttium   204, 206, 208–209, 213, 223,
      231

Callicula   215, 218
Campania   204, 210, 212, 219, 223, 226, 228,
      230–231
Capua   211–212, 214, 223, 226
Caria   262–264, 266, 269, 274–275, 278
Carystos   68
Chersonesus   66
Chios   69, 77, 262
Corcyra   66, 77
Corinth   66, 160

Djedku Canal   14

El Girgawi   25
Ephesus   265–266, 269–273, 275–276, 278–
      280
Eryx   205, 216, 219, 228, 232

Gaul   35–36, 193–194, 199, 201, 204
Gaza   16
Gebel Barkal   19, 30, 252–253, 255
Gergovia   35, 53–54, 57–58, 61
Gerunium   214–215, 218, 220, 222, 225, 227

Hatnub   27
Heircte   205, 216, 219, 232

INDEX 319

Hellespont   66, 71, 262–263, 269
Herdonia   211, 218, 228

Iasos   266, 268, 270

Kadesh   18, 20–23, 27, 249, 252, 257, 259
Karnak   17–22, 249, 251–252
Knidos   268

Lebanon   21, 127
Latium   108
Lesbos   69, 77, 262
Levant   73
Libya   21, 237, 239
Lucania   204, 209–210, 213–214
Luceria   205
Luxor   24, 251, 257, 259
Lydia   69, 262–263, 272

Magnesia-on-Maeander   264, 269, 271, 273,
    275–276, 278–280
Medinet Habu   32, 242, 244, 251, 259
Megiddo   15–17, 19, 240, 253–255, 257, 260–
    261
Memphis   17–19
Miletus   170, 265, 268–271, 274
Mitylene   73, 79

Naharain   14
Nubia   14–15, 21, 24, 29, 237–238, 244, 246–
    247, 255
Ny   18

Pa-Canaan   20–21
Paros   68, 70
Perunefer   19
Phocaea   72
Phrygia   262–263
Picenum   223, 230
Potidaea   72, 174
Pygela   266, 269–273, 280

Rhegium   209, 230

Samos   5, 66, 68, 70, 72, 171, 262, 266, 269–271
Salamis   68, 76
Samnium   223
Sardis   262–264, 266–267, 271, 279–280
Shamash-Edom   17
Sharuhen   14, 241
Sicily   66, 72, 77, 81–82, 85, 146, 148, 152–153,
    156–161, 205, 207–208, 214, 223–225, 227,
    230–231
Sparta   68, 70, 81, 172–173, 266
Syria   14, 25, 73
Syracuse   78, 146, 208, 210

Teos   264, 266, 268, 273
Thasos   69, 75
Ticinus   213–214
Tralles   264, 279

Wadi el Hudi   26, 253

Yenoam   21, 254, 257

Other

Achaemenid(s)   8, 262–265, 272–273, 275,
    278, 280
Aegates islands, battle of   231
Aegospotami, battle of   79
Anabasis, Xenophon   275
Arcadia, Arcadians   78
Archidamian War   67, 69–70
Argyrologoi   70
assiduus/assidui   4, 82–84, 86–97
auxiliaries   38, 59, 129, 193–194, 199, 201, 248,
    250–251, 256

Bellum Civile, Caesar   37, 52–53
Bellum Gallicum, Caesar   4, 34, 36–38, 51,
    53–54, 60–61
booty/spoils   55, 57–58, 65, 80–82, 93, 96, 118,
    207, 209, 253–254, 272, 276
Bruttian(s)   206, 209, 230–231
Buccellarii   7, 186, 197, 199, 202

Cayster, river   264–266, 273, 276, 279
census   82–84, 88–96, 224
chrêmata   67–71
clan(s)   109, 111, 114–115, 188–190

320   INDEX

comes Africae   7, 186, 196, 199, 201

dapanai   71
Dapur, battle of   25, 244, 250–251, 257, 259
De Legibus, Cicero   162
dekatê—tenth   71
dies imperii   125, 129
eikostê—twentieth   71, 76
eisphora—contributions   70

equites   83, 88–89, 116
Eurymedon River, battle of   169–171

Germani   41, 48, 51–52
Goths   193, 202

hastati   195, 216
Helvetii   39, 51
Hittite(s)   21, 24, 32, 254, 259
Huns   193, 200, 202
Hyksos   238, 241, 254

Imperium   6, 111–114, 116, 118, 145–153, 155–157, 162–163
indemnity   80–81
Inty, tomb of   241, 243, 248, 254, 256
Inyotef, tomb of   241, 243, 246, 249, 258
ius iurandum   120, 123–124

Lapis Satricanus   115
latrocinia/latrones   206–209, 211, 223, 230–231
Law of Nikophon   175
Leuktra, battle of   173
leves   212–213, 216–217
lex Hieronica   153, 157–163
lex Porcia   147–148, 153, 163
lex provinciae   147, 153–154, 157
lex Rupilia   153, 157, 159–163

Maeander, river   264–266, 269–271, 273, 275–276, 278–280
Magister utriusque militia   194, 197
maniple(s)   44, 59, 211, 216
metics   79

misthophoria—Wage-earning   72
misthos—Pay, Wage   69–70, 179

Nicias, Peace of   71
Notitia Dignitatum   193, 201
Numidian(s)   206–208, 213–214, 218–219, 221–222, 224, 226–229, 232

Orontes, river   17, 19

Panathenaea   69
paralus   79
patronage   188–191, 203
Peace of Kallias   172–173
Peisistratids   66
Peloponnesian War   70, 72, 78, 171–172, 255, 263, 272
peltast(s)   215, 269, 271, 276
phalanx   52, 102–103, 115, 215
phoros—tribute   66, 68, 74, 79
pomerium   108, 111, 118, 145, 148, 150
prolepsis   29

Sabis, river   39, 43–44, 46
Samnite(s)   208, 211, 223
satrap(s)   171, 262, 264, 266, 269, 273–274, 280
senatus consultum   155, 229
Shasu   21
signa militaria   135
slaves   67, 77–79, 84, 95–96, 154, 159, 161, 209
stipendium   4, 81, 85, 90, 92, 96
Sybota, battle of   77

Tabula Contrebiensis   155–156, 163
thalamioi   79
thranitai   79
Triumph (Roman)   81, 118, 148
trophê—food   68

velites   213, 227
ver sacrum   230

zygioi   79

Printed in the United States
By Bookmasters